GENETIC ENGINEERING OF OSMOREGULATION

Impact on Plant Productivity for Food, Chemicals, and Energy

BASIC LIFE SCIENCES

Alexander Hollaender, General Editor

Associated Universities, Inc.
Washington, D.C.

Volume 1 • GENE EXPRESSION AND ITS REGULATION
Edited by F. T. Kenney, B. A. Hamkalo, G. Favelukes, and J. T. August

Volume 2 • GENES, ENZYMES, AND POPULATIONS
Edited by A. M. Srb

Volume 3 • CONTROL OF TRANSCRIPTION
Edited by B. B. Biswas, R. K. Mandal, A. Stevens, and W. E. Cohn

Volume 4 • PHYSIOLOGY AND GENETICS OF REPRODUCTION (Parts A and B)
Edited by E. M. Coutinho and F. Fuchs

Volume 5 • MOLECULAR MECHANISMS FOR REPAIR OF DNA (Parts A and B)
Edited by P. C. Hanawalt and R. B. Setlow

Volume 6 • ENZYME INDUCTION
Edited by D. V. Parke

Volume 7 • NUTRITION AND AGRICULTURAL DEVELOPMENT
Edited by N. Scrimshaw and M. Béhar

Volume 8 • GENETIC DIVERSITY IN PLANTS
Edited by Amir Muhammed, Rustem Aksel, and R. C. von Borstel

Volume 9 • GENETIC ENGINEERING FOR NITROGEN FIXATION
Edited by Alexander Hollaender, R. H. Burris, P. R. Day, R. W. F. Hardy, D. R. Helinski, M. R. Lamborg, L. Owens, and R. C. Valentine

Volume 10 • LIMITATIONS AND POTENTIALS FOR BIOLOGICAL NITROGEN FIXATION IN THE TROPICS
Edited by Johanna Döbereiner, Robert H. Burris, Alexander Hollaender, Avilio A. Franco, Carlos A. Neyra, and David Barry Scott

Volume 11 • PHOTOSYNTHETIC CARBON ASSIMILATION
Edited by Harold W. Siegelman and Geoffrey Hind

Volume 12 • GENETIC MOSAICS AND CHIMERAS IN MAMMALS
Edited by Liane B. Russell

Volume 13 • POLYPLOIDY: Biological Relevance
Edited by Walter H. Lewis

Volume 14 • GENETIC ENGINEERING OF OSMOREGULATION: Impact on Plant Productivity for Food, Chemicals, and Energy
Edited by D. W. Rains, R. C. Valentine, and Alexander Hollaender

---

GENETIC ENGINEERING OF OSMOREGULATION

Impact on Plant Productivity for Food, Chemicals, and Energy

Edited by

D. W. Rains
and
R. C. Valentine

University of California
Davis, California

and

Alexander Hollaender

Associated Universities, Inc.
Washington, D.C.

PLENUM PRESS · NEW YORK AND LONDON

Library of Congress Cataloging in Publication Data

Symposium on Genetic Engineering of Osmoregulation: Impact on Plant Productivity for Food, Chemicals, and Energy, Brookhaven National Laboratories, 1979.
Genetic engineering of osmoregulation.

(Basic life sciences; v. 14)
Sponsored by the College of Agricultural & Environmental Sciences, University of California, Davis.
Includes index.
1. Genetic engineering—Congresses. 2. Osmoregulation—Congresses. 3. Plants—Hardiness—Congresses. 4. Plant-breeding—Congresses. 5. Salt-tolerant crops—Congresses. I. Rains, D. W. 1937- II. Valentine, Raymond Carlyle, 1936- III. Hollaender, Alexander, 1898- IV. California. University, Davis, College of Agricultural and Environmental Sciences.
QH442.S95 1979 581.1'33 80-14972
ISBN 0-306-40454-0

ADVISORY COMMITTEE FOR SYMPOSIUM

J. Boyer
A. Hollaender
D. W. Rains
G. G. Still
A. Szalay
R. C. Valentine

Proceedings of the Symposium on Genetic Engineering of Osmoregulation: Impact on Plant Productivity for Food, Chemicals and Energy, held at Brookhaven National Laboratories, November 4-7, 1979, and supported under NSF Grant OPA 79-21432 and Department of Energy to Associated Universities, Inc. Any opinions, findings, conclusions, or recommendations herein are those of the speakers and do not necessarily reflect the views of NSF. Co-sponsored by the College of Agricultural & Environmental Sciences, University of California, Davis, California.

A Division of Plenum Publishing Corporation
227 West 17th Street, New York, N.Y. 10011

Printed in the United States of America

Special thanks to H. W. Siegelman,
G. J. Wagner, E. Shaw, R. B. Setlow,
Mrs. Helen Kondratuk and her associates.
We greatly appreciate the efforts of
Suzanne Epperly and the staff of the
Plant Growth Laboratory

PREFACE

The plant world represents a vast renewable resource for production of food, chemicals and energy. The utilization of this resource is frequently limited by moisture, temperature or salt stress. The emphasis of this volume is on the molecular basis of osmoregulation, adaptation to salt and water stress and applications for plant improvement. A unified concept of drought, salt, thermal and other forms of stress is proposed and discussed in the publication.

The volume developed from a symposium entitled "Genetic Engineering of Osmoregulation: Impact on Plant Productivity for Food, Chemicals and Energy," organized by D. W. Rains and R. C. Valentine in cooperation with Brookhaven National Laboratory and directed by D. W. Rains and A. Hollaender. The program was supported by a grant from the National Science Foundation, Division of Problem Focused Research, Problem Analysis Group, and the Department of Energy.

This symposium is one of several in the past and pending which deal with potential applications of genetic engineering in agriculture. Since the question was raised several times during the meeting it is perhaps a convenient time to attempt to define genetic engineering in the context of the meeting.

- Genetic engineering of osmoregulation is simply the application of the science of genetics toward osmotically tolerant microbes and plants.

- Recombinant DNA is regarded as just another tool along with conventional genetics to be utilized for improvement of microbes and plants.

The symposium brought together molecular geneticists, biochemists, plant physiologists, and plant geneticists to focus on environmental stress as a major barrier to crop productivity on primary and marginal lands. The information discussed at this symposium and in this volume is applicable to the rapidly developing biosalinity program recently formalized within the National Science Foundation.

We could not cover all of the subject matter applicable to this field but hope the research topics and the ideas stimulate our colleagues to approach the area of biological stress on a broad but unified and integrated front. With this approach the potential use of marginal lands and environments for food production will be greatly enhanced.

D. W. Rains
R. C. Valentine
A. Hollaender

CONTENTS

FOREWORD*

TO THE SYMPOSIUM ON GENETIC ENGINEERING

OF OSMOREGULATION

Making the connection between basic research and applying it is often a difficult and complicated problem. Many of us have considered how to proceed on this. Basic research is generally done in the laboratory and is highly specialized, whereas applied research as it relates to this symposium, requires a broad knowledge of the basic principles as well as the mechanisms of growth and synthesis of plant systems. Of course, we are all very conscious that considerably more knowledge will be necessary before we can broadly apply genetic engineering to plant sciences. However, we can attempt to apply the little knowledge we have acquired to observe whether, as a result of its application, new ideas are developed. This is the usual process of events, especially in the areas of science with are of greatest consequence in affecting the well being of mankind.

About a year and a half ago, I had an opportunity to attend a workshop organized by the National Science Foundation at Kiawah Island where the applied aspects of biosaline research were discussed. As a result of this small workshop, a pamphlet was published which later grew, with the addition of chapters not included at the Kiawah Island conference, into a book entitled, The Biosaline Concept: An Approach to the Utilization of Underexploited Resources, published this year. Then a number of us met to continue discussions with particular emphasis on genetic engineering of osmoregulation and the application of this concept to microorganisms and plants. After several discussions, Drs. Rains and Valentine, with the support of a committee, developed the program for this symposium. My contribution was made on the basis of my experience in the development and editorship of the biosaline research volume and, more

---

* This work was supported in part by U. S. Department of Energy Contract EY-76-C-02-0016 with Associated Universities, Inc., Brookhaven National Laboratories.

importantly, my deep interest in genetic engineering and the practical application of its basic aspects. We feel that the problems discussed in this volume are of immediate interest, although extensive progress in applying genetic engineering has not yet taken place. This symposium was created to cite the work that has been accomplished in this field and the future areas of considerable potential.

A discussion of the government sources of support from the National Science Foundation (NSF), Department of Energy (DOE) and the United States Department of Agriculture (USDA) was organized as a roundtable forum and proved to be most informative and beneficial, demonstrating the productivity of close cooperation among different government agencies.

We express special thanks to our colleagues at the Brookhaven National Laboratory, especially the administrative staff, who magnificantly organized this symposium. Because of this assistance, advantage was taken of many of the excellent facilities offered on the BNL site.

Combining the expertise of our west coast colleagues at Davis with whom we have been interchanging ideas for a number of years, and the generous support of the government agencies involved, the Symposium on Genetic Engineering of Osmoregulation successfully contributed to the promotion of research in this progressive area of science.

Alexander Hollaender
Associated Universities, Inc.
1717 Massachusetts Avenue, N.W.
Washington, D.C. 20036

BIOLOGICAL STRATEGIES FOR OSMOREGULATION

D. W. Rains and R. C. Valentine

Plant Growth Laboratory/Department of Agronomy
and Range Science
University of California, Davis
Davis, CA 95616

Green plants provide mankind with enormous quantities of food, fiber, oils, chemicals, and energy. Today there is great interest in increasing the productivity of the plant world, with plants, in essence, behaving as solar energy machines for capturing radiant energy and converting it into essential products. Unfortunately, one of the major barriers to productivity in many environments is the harmful effects of salt and water (drought) stress. This is a severe national as well as world-wide problem.

The ancient art of preservation or curing of foods (e.g., salt pork or fish) in high salt solutions and modern man's attempt to grow crops using seawater irrigation are both applications of a phenomenon called osmoregulation. In essence water is essential for all forms of life. When the solutes in water increase in concentration cellular dehydration commences ultimately causing death of the organism. Man and animals combat the problem of osmotic adjustment using their kidneys whereas most plant and microbial life produce enough solutes of their own in the cell to balance the dehydrating pull of the surrounding environment. This adaptation to salt or drought is called osmoregulation. Osmoregulation is not a term familiar to most of us although its effects have shaped civilizations past and present. For example, the irrigated soils of the Euphrates Valley of Biblical times became so laden with salts that crops withered and died as did the culture which they supported. Similar devastation of once rich lands is now occurring on the border lands of Mexico which are irrigated by the increasingly saline water of the Colorado River. The salinization of soils of the globe is increasing at an alarming rate as water quality for irrigation continues to decline.

The search for solutions to some of these challenging problems was highlighted at a recent international symposium dealing with how microbes and green plants adapt to salt and drought stress. Discussion was developed on how the simplest forms of life adapt to water stress and progressed to include the plant world and the major crop plants.

The ultimate goal of the research discussed at this symposium was to enhance the productivity of biological systems in stress environments. This was expected to be accomplished through a basic understanding of the physiological, biochemical and genetic bases for stress tolerance. Genetic improvement of plants should enhance the potential of supplying food, chemicals and biomass to our society. The symposium attempted to "bridge" between the microbial systems and plants with application of knowledge and techniques as follows:

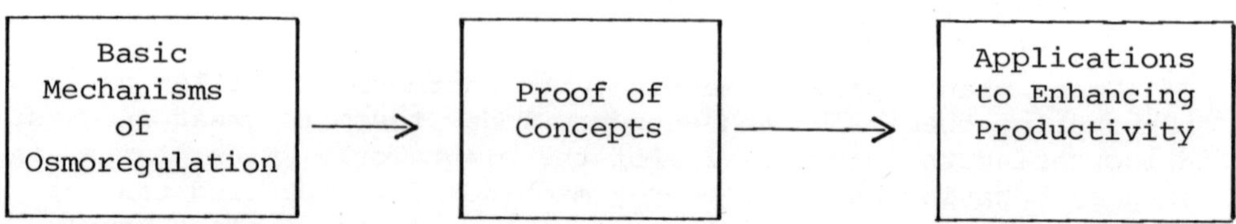

The microbial section was highlighted by the finding that the much studied lactose utilization (lac) operon and histidine (his) operon, among the best understood pieces of DNA, could serve as probes for monitoring the cellular changes in solutes which occur during adaptation to salt or drought stress. The unique mutant enzymes available for the lac or his operons, enzymes which bend and twist in and out of shape in response to a variety of solutes, were used to test the intracellular solute changes which take place during environmental challenges following salinity or water stress. These studies showed that marked changes occur in the cell during osmotic stress, changes that could be deciphered by following the activity patterns of the uniquely marked enzymes. In most dramatic cases enzymes that were completely inactive with cells in the normal environment became activated in the saline or water stress environment.

The use of the simple organisms has also led to the finding that L-proline may play a unique role as internal osmoregulatory protecting against the damages of water stress. This compound fulfills "Koch's Postulates" as the antistress agent in that L-proline which accumulates during osmotic stress increases resistance to stress when added to the growth medium. This has led to the construction of a salt (drought) tolerant plasmid which increases the stress tolerance of the cell which carries it. Although these studies represent "only the tip of the iceberg" they open the door to a fresh approach to osmoregulation, an approach using the powerful tools of molecular biology.

Biological systems exposed to environments with reduced water availability (salinity, drought, extreme temperatures) have adaptive responses which involve organic and inorganic osmoregulation. Osmotic adjustment by these organisms in response to stress involves regulation of the intracellular levels of various carbon compounds (sugars, organic acids, polyols), nitrogenous compounds (amino acids, tertiary nitrogenous substances) and inorganic ions.

Higher microbial forms such as yeast and algae utilize strategies similar to bacteria, employing so-called organic osmotica such as glycerol for maintaining water balance with these surroundings. Indeed, Israeli scientists spurred on by the scarcity of fresh water in their country are cultivating salt tolerant algae in ponds beside the Dead Sea in an attempt to extract from these algae their glycerol, a valuable chemical feedstock. These algae are composed of an amazing 80% of glycerol when grown on the highly saline water from the Dead Sea.

The metabolic energy expended in osmoregulation is expected to be greater for organisms surviving and growing in stress environments than the same organism in a low stress situation. The metabolic cost of ion transport and organic osmoregulation would be expected to influence productivity. The question of productivity in stress environments was suggested as a high priority research objective and estimates developed on the energy cost for organisms growing under stress conditions should be obtained. Model systems using microorganisms have been developed to determine the biological cost of nitrogen fixation. This approach should be extrapolated to determine the energy required for microorganisms and plants to tolerate stress. The technique of plant cell culture could be applied using similar approaches developed in microbial studies. These ideas are to be explored.

Among the plant physiologists reporting at the meeting, Dr. John Boyer, a USDA scientist at the University of Illinois, described the remarkable similarity between the mechanism of osmoregulation in young soybean seedling compared to microorganisms. During a round table discussion following the formal presentations, Dr. Boyer outlined two research priorities: a) the nature of the signal or trigger mechanism stimulating osmoregulation, and b) the biochemistry or enzymology of osmoregulation -- goals that strongly complement current work on microbial systems. Work on selection of salt tolerance plants using tissue culture where billions of individual cells may be screened raises the possibility for selection of new salt tolerant crops. The halophytic plants such as abound in salt marshes are being actively studied both from the standpoint of the mechanism of osmotic tolerance as well as being eyed as a potential source of biomass.

Microbes such as halobacterium, and plants such as mangrove which have become so specialized as to require high salt for growth were not emphasized at this meeting having been discussed in detail at a sister meeting two weeks previously sponsored by the National Science Foundation and Boyce Thompson Institute at Cornell University. Representatives from this meeting summarized recent advances in halophilic organisms. In projecting toward our major crop plants a number of workers pointed out that salt and drought tolerance were probably simpler to achieve in the short term whereas in the long term completely halophilic plants might be desirable.

The section on selection and breeding of salt resistant crop plants was highlighted by descriptions by scientists at UC Davis and the University of Arizona that crop plants may be bred to grow using seawater irrigation. This might lead to a greening of arid coastal regions of the world. In the case of barley, scientists have continued a practice started by the Sumarians of Biblical times where salt damaged soils supported cropping with barley but not wheat. The natural salt hardiness of barley is now being systematically explored by plant breeders. One of the more interesting stories to come from the meeting is that of the salt tolerant Galapagos tomato. During the collection of ancestral types of tomato Dr. Charles Rick, reknowned tomato geneticist at UC Davis, discovered an unusual plant growing along the cliffs of the Galapagos in a small pocket of sandy soil. The plant was growing above the tidal zone while receiving sufficient moisture from the continuous, highly saline sea mist. This plant, carrying salt tolerance traits, may be the forerunner for a new generation of drought or salt hardy plants. Indeed some of the early hybrids show high levels of tomato solids, a property of great interest to catsup manufacturers.

The genetic bases for stress tolerance in bacteria, algae and higher plants were discussed. Commonality in tolerance mechanisms and in genetic regulation of these mechanisms in these three organisms were evaluated. The potential of genetic engineering techniques in manipulating stress tolerance genes were discussed and the feasibility of applying genetic engineering techniques used in microbial systems to higher plants was proposed for plant cell culture systems. The possible application of novel genetic techniques to enhance the physiological and biochemical capability of plants exposed to stress was addressed at the symposium with a general consensus that in the area of osmoregulation in higher plants considerable progress has been made in describing the phenomenon and as a process it is reasonably well understood. It was suggested that a more basic understanding of the metabolic and genetic control of the process will be required before genetic manipulation of this characteristic is realized.

Energy was on the minds of several of the contributors particularly Dr. Robert Rabson of the DOE, one of the sponsors of the meeting, who pointed out the need to utilize marginal lands for energy crops. Drought is a perennial problem limiting biomass production on such lands. Representatives from the National Science Foundation, the primary sponsor of the meeting and a major sponsor of research on osmoregulation pointed out the interrelationships of developments of basic research on osmoregulation to such projects as stress tolerance in symbiotic nitrogen fixation to new technologies involving immobilized enzyme catalysts and cells operating at high temperatures and saline conditions.

RESPONSES OF PLANTS TO SALINE ENVIRONMENTS

Emanuel Epstein

Department of Land, Air and Water Resources
University of California, Davis
Davis, California 95616

Plants can only grow and reproduce when their cells are bathed by water and permeated by it. Algae and the most active organs of higher plants, the leaves and the roots, are 85-95 percent water by weight. Desiccation or freezing do not necessarily spell death of all plant cells but slow down to imperceptible rates the metabolic processes of growth and development. Water, then, is of the essence in the functioning of plants.

Water is abundant on Earth. The oceans occupy 71 percent of the surface of the globe, and much of the land area is kept supplied with water via the hydrologic cycle. Plant life exists wherever water reaches, and only utterly dry areas are barren. This abundance of water gives the impression that its acquisition must be relatively easy compared with the acquisition of the other essential materials, viz. carbon and the mineral nutrients. The concentration in the environment of these elements is low. Only 3 molecules out of 10,000 in air are carbon dioxide, and as for the mineral nutrients, the concentrations of most of them in the water from which plants have to acquire them are on the order of some parts per million.

But despite its relative abundance compared with that of the other nutrients, water is acquired by much of the plant kingdom only at great metabolic cost and through complicated structural and functional adaptations. There are two principal reasons for this. (1) The oceans are highly saline, with a concentration of sodium chloride of about 0.5 M and appreciable concentrations of other salts. The marine algae and other marine plant life must therefore maintain intracellular concentrations of solutes at least as high as that of seawater if they are to avoid osmotic desiccation. (2) On land, plants have to cope with the solid matrix of soil to acquire water through their roots, while at the same time presenting exposed surfaces, leaves, to the atmosphere and to light for the essential function of photosynthesis, with its integral feature of gas exchange. Where carbon dioxide can enter, water vapor can leave, and so it does, drawing moisture from the soil. If soil water is in short supply or difficult to extract from the soil, the plant may not make up its transpirational water losses and may suffer drought, wilting, and death. Plants have evolved numerous structural and functional adaptations to cope with these conflicting demands of essential and of potentially detrimental gas exchange.

Of all the plants on Earth, the terrestrial plants in arid and semi-arid regions confront the most severe problems of water economy. The reason for this is that these plants face the threat of both osmotic withdrawal of water and excessive transpirational loss of it. Soil salinity, common in these regions, gives rise to the former stress, and their hot, dry atmospheric conditions cause the latter.

Yet despite these precarious conditions, the arid and semi-arid regions are among the most promising ones to turn to in our quest to increase the production of food, fiber, chemicals, and biomass for energy. The relatively unleached soils of these regions are often inherently fertile, the growing season is long, temperature and light intensity are high, and the atmospheric humidity is low, reducing disease problems. All these features favor agricultural productivity if water and salinity problems can be solved, as is shown for example by California's Central Valley, one of the most productive agricultural areas of the world.

This paper presents a broad overview of the cardinal features of plant-salinity interactions. The emphasis on selective ion transport and on genotypic variation, and the examples given, reflect the particular interests of this laboratory.

SALINITY -- ITS NATURE AND EXTENT

To judge by some popular accounts of salinity one might gain the impression that salinity of soils and water is simply due to inappropriate schemes and management of irrigation. Irrigation is in many areas an important agent of salinization, but to keep things in perspective let us remember that salinity is one of the great phenomena of nature. The ocean is as salty as it is because it has for eons been the sink into which soluble salts from the continental masses have been leached. The extensive areas of land that represent uplifted ocean floors are therefore themselves saline. Rainwater and weathering slowly bring primary minerals into solution and add to the salt burden of soils, especially in relatively dry regions, and so does the formation of secondary minerals. Droplets of seawater whipped into the air evaporate and "cyclic salt" circles

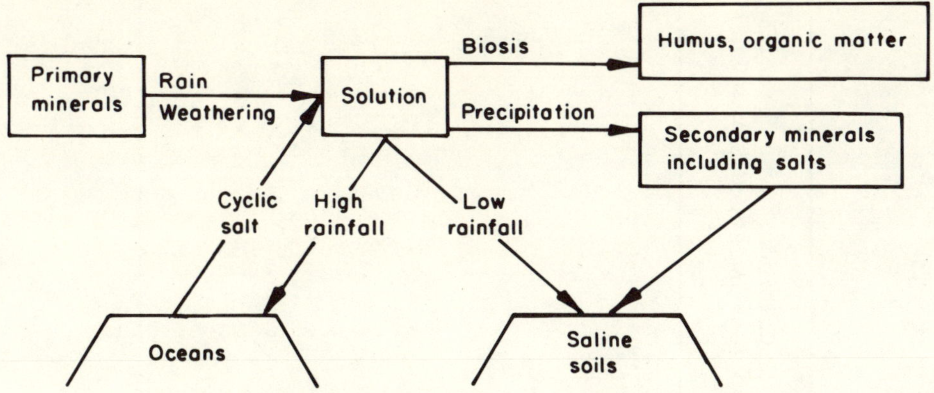

Fig. 1. Biogeochemical cycling of salt.

the globe, eventually to descend in rain and snow. So effective are these processes in distributing salt all over the world that no plant growing under the open sky has yet been found to suffer from chlorine deficiency, although chlorine is a micronutrient required by plants in greater amounts than, say, iron and manganese -- micronutrients which not infrequently are deficient in plants. Figure 1 is a highly simplified diagram of the main pathways of salt in nature.

The diagram omits the contribution made by man, mainly through irrigation agriculture in arid and semi-arid regions. All irrigation water carries some salt in solution, and water available in dry regions is often saline to a considerable extent. Table I gives the salinity of several water supplies, for purposes of comparison. It should be kept in mind that whether water of a given salinity is good, marginal, or poor depends on many factors which cannot be adequately included in such a table: the specific salts in solution, soil conditions, climate, the particular crop irrigated, cultural practices, and still others. It is nevertheless apparent from this table that even marginal irrigation water has a salinity of only about 10 percent that of seawater.

Of the land area of the Earth, which is 1.33×10^{10} hectares, Rodin et al. (1975) estimate that 46.5 percent is arid or semi-arid, or 6.2×10^{9} hectares. If we assume that 950×10^{6} hectares are salt affected (Massoud, 1974), then 15.4 percent of all arid and semi-arid land falls in this class. Irrigated land is estimated to occupy 230×10^{6} hectares (Wittwer, 1979), of which one-third is believed to be affected by salinity (Eckholm, 1975), or 76.7×10^{6} hectares. Thus by any measure salinity looms large in the chemical

Table I. Salinity of water supplies.*

Salinity measurement	Irrigation water, good quality	Irrigation water, marginal quality	Sacramento River (California)	Colorado River (Southwest, U.S.)	Pacific Ocean
Electrical conductivity (millimhos/cm)	0-2	2.5	0.2	1.3	46
Concentration of dissolved solids (parts per million)	0-1,000	1,000-3,000	300	850	35,000

* Based on various sources

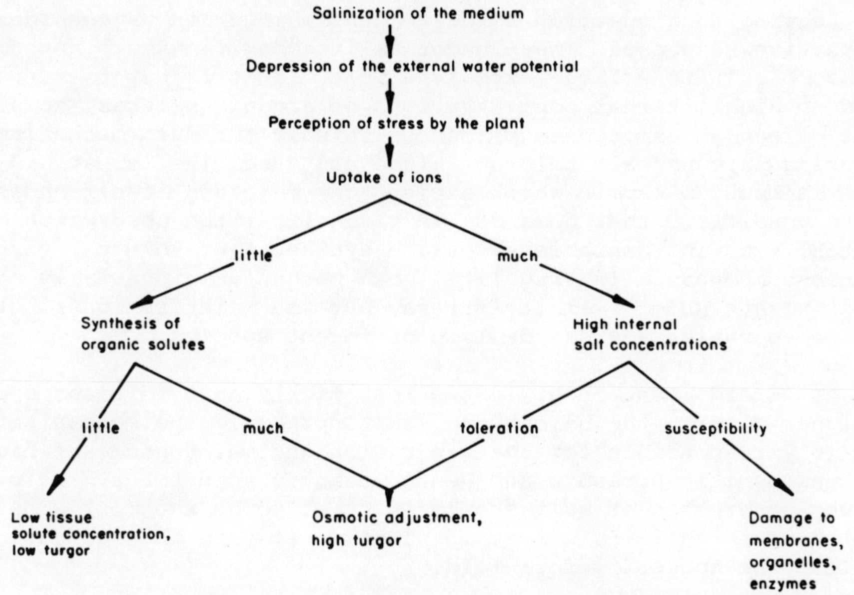

Fig. 2. Outline of the cardinal responses of plants to salinity.

economy of terrestrial plant life and the agricultural economy of mankind.

RESPONSES OF PLANTS TO SALINITY

It has been customary for agricultural scientists concerned with salinity to confine their research entirely or nearly so to crop plants, virtually none of which are salt tolerant to any high degree. It is, however, logical to give much emphasis to those plants adapted through the processes of evolution to an existence in saline soils and waters. Higher plants so endowed are called halophytes, or salt plants.

The main stress that these plants are adapted to cope with is the osmotic one of lowered external water potential. Figure 2 outlines in much simplified fashion the main types of responses that these and all other plants may make when exposed to this stress. High external concentrations of salt are tantamount to a high external osmotic pressure or a low external water potential. The plant, and most immediately its root system, faces the threat of osmotic withdrawal of water. Most halophytes respond by absorption of ions from the external medium, building up internal concentrations high enough to effect an osmotic adjustment and maintain turgor and growth. This implies that high internal concentrations of salt are tolerated, and therefore gives rise to the question of how this toleration is effected.

Leaving that question aside for the moment let us consider alternative responses. When under saline conditions a plant fails to absorb salt to a large extent it must, if it is not to succumb, build up high internal concentrations of organic solutes, so high as to effect an osmotic adjustment. This is the main mechanism operating in many salt tolerant algae and fungi (Hellebust, 1976). There is thus no way in which a plant can maintain itself under saline conditions that does not involve either the absorption of solutes, viz. inorganic ions, or the synthesis of solutes, viz. organic compounds. In actuality, both mechanisms invariably operate in plants under salinity stress, but the relative contribution each makes varies a great deal in different species.

As already mentioned, in the most highly salt tolerant species of higher plants, the halophytes, salt uptake is the mechanism that is mainly responsible for the generation and maintenance of high internal osmotic pressure and hence, osmotic adjustment. We might ask why absorption of salt should turn out to be the most widespread adaptation to salinity stress in plants that face this condition in nature. The answer is four-fold.

(1) Salt being the very agent in the substrate causing the stress, ions of salt are the most readily available solutes by which the plant can build up its internal osmotic pressure. No altogether novel or unique mechanisms need to be elaborated for ion uptake, such mechanisms and their operation being among the normal functions that roots perform.

(2) Long-distance transport of the ions to the shoot is via the metabolically inexpensive process of flow in the transpiration stream.

(3) To the extent that inorganic ions are used for osmotic adjustment, photosynthate is not needed for either the carbon skeletons of organic osmotica or the energy required to synthesize them.

(4) Any osmotic adjustment in the root by organic molecules instead of inorganic ions necessitates the transport of photosynthate from the shoot to the root. That transport is via the phloem, and costly in terms of the expenditure of metabolic energy.

It follows from the above argument that inorganic ions are superior to organic osmotica in terms of both their acquisition and their transport. Hence absorption and transport of inorganic ions seem to have emerged as the premier evolutionary adaptation of plants to saline conditions. The puzzle has been how such high internal concentrations of ions (several hundred millimoles/liter) can be reconciled with the unimpaired functioning of organelles and enzymes (Flowers et al., 1977). The answer to that question appears to be that a high concentration of inorganic ions is not the sole mechanism whereby a favorable osmotic gradient is established and maintained throughout the entire cell volume. It seems to apply predominantly to that very large fraction of mature plant

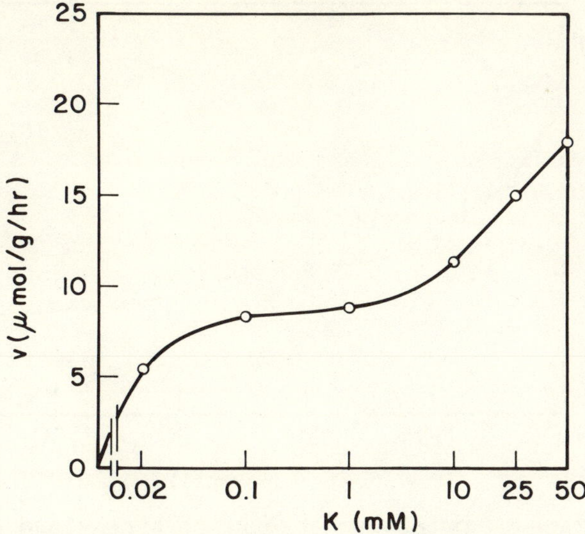

Fig. 3. The rate, v, of potassium absorption by excised barley roots as a function of the external potassium concentration, plotted logarithmically. After Epstein (1968).

cells (about 90%) occupied by the vacuole. The cytoplasm is then located between two compartments containing high concentrations of salt: the cell wall spaces into which the soil solution or the transpiration stream delivers its solutes, on the outside, and the vacuole within. Osmotic desiccation of the cytoplasm is avoided by synthesis of metabolically innocuous organic solutes (Hellebust, 1976; Flowers et al., 1977; Storey and Wyn Jones, 1979). This is metabolically expensive, as pointed out above, but applies to only a small fraction of the total cell volume. Thus, the "toleration" indicated in Figure 2 in all probability involves a component from the left side of that simplistic diagram: synthesis of organic solutes in the cytoplasm and maintenance of a high concentration of them in that compartment.

If this is a reasonably accurate view of halophytic adaptation to salinity it follows that the cell membranes delimiting the principal cell compartments play key roles. Specifically, the external membrane or plasmalemma must not only resist disorganization by the concentrated salt solution bathing it but in addition must have an extremely selective mechanism for absorption of potassium from a solution containing a preponderance of a potentially competing cation, sodium. This is so because many processes in the cell require potassium and are inhibited by sodium. There is evidence for highly selective potassium transport. Figure 3 shows the rate of

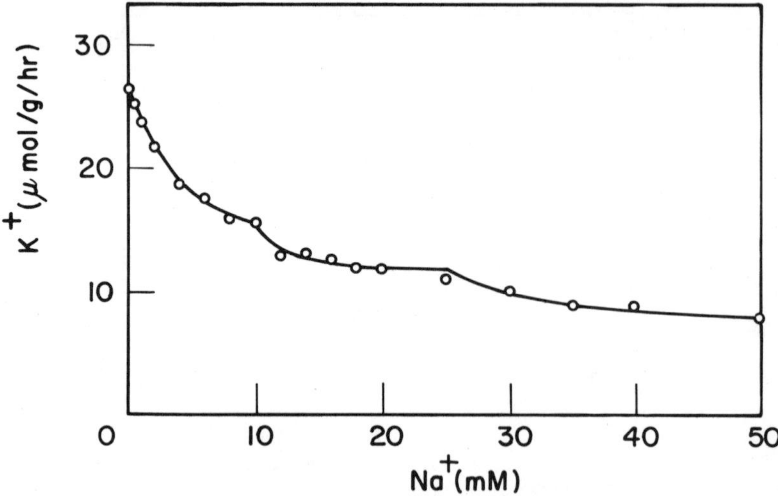

Fig. 4. The rate of potassium absorption by excised barley roots as a function of the external concentration of sodium. That of potassium was 50 mM. After Rains and Epstein (1967a).

absorption of potassium by barley roots as a function of the external potassium concentration. (Barley is fairly salt tolerant, though no halophyte.) The rate over the low range of concentrations, up to 1 mM potassium, shows saturation kinetics, but at higher concentrations the rate increases (Rains, 1972; Epstein, 1976). The mechanism operating at low concentrations turns out to have not only high affinity for potassium but also high selectivity for this element vis à vis sodium. Figure 4 shows the effect of increasing sodium concentrations on the rate of potassium absorption by barley roots, potassium being present in the experimental solution at 50 mM. At the highest sodium concentrations, the rate of potassium absorption remains fairly constant, and equal to the maximal rate attainable by the low-concentration mechanism (cf. Figure 3). It appears on this and other evidence that this low-concentration (high-affinity) mechanism is responsible for supplying the plant with adequate amounts of potassium even in the presence of high sodium concentrations, by virtue of its selective affinity for potassium.

In true halophytes, even more spectacular examples of preferential potassium absorption can be cited. Rains and Epstein (1967b) studied the absorption of potassium by leaf tissue of the mangrove, *Avicennia marina*, as a function of the external concentrations of potassium and sodium. The concentration of the former ion was

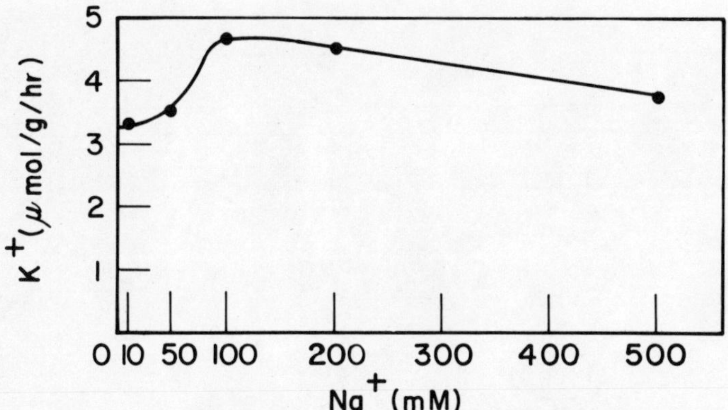

Fig. 5. The rate of potassium absorption by leaf tissue of the mangrove, Avicennia marina, as a function of the external concentration of sodium. That of potassium was 10 mM. After Rains and Epstein (1967b).

10 mM (close to its concentration in seawater) and that of the latter ranged up to 500 mM (the sodium concentration of seawater is 457 mM). Concentrations of sodium up to 100 mM not only failed to diminish the rate of potassium absorption but increased it, and even at 500 mM sodium the rate of potassium absorption remained above that of the zero sodium control.

This was a laboratory experiment with excised tissue. Storey and Wyn Jones (1979) grew the halophyte, Suaeda monoica, in the greenhouse in solution cultures salinized to various concentrations of sodium chloride. At 500 mM, sodium chloride increased the potassium concentration of the roots by a factor of 7. Other such instances have been noted.

Taken together, evidence from short-term experiments with excised tissues and long-term experiments with growing plants suggests that resistance to sodium salts of necessity involves potassium transport mechanisms that are highly indifferent to or even stimulated by sodium.

Calcium is essential for selective transport of alkali cations (Epstein, 1961; Läuchli and Epstein, 1970). If, as the above discussion suggests, selective transport of potassium in the presence of excess sodium concentrations is of the essence in salt toleration, calcium would be expected to play a role. LaHaye and Epstein (1969) grew the (salt sensitive) bean, Phaseolus vulgaris, in solution cultures salinized with sodium chloride at 50 mM, at various

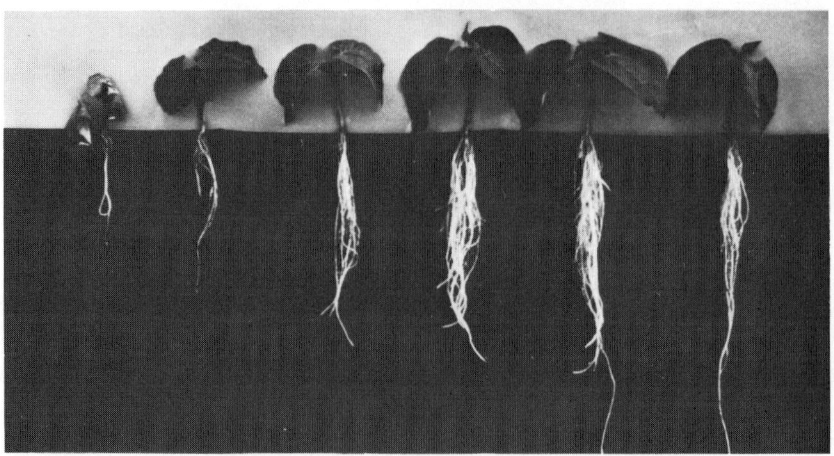

Fig. 6. Bean plants, Phaseolus vulgaris, grown in nutrient solutions at 50 mM sodium chloride. Calcium concentrations were (left to right) zero, 0.1, 0.3, 1.0, 3.0, and 10 mM. After LaHaye and Epstein (1969).

concentrations of calcium (Figure 6). At the higher calcium concentrations (3 and 10 mM) the plants grew as well as did those in the unsalinized control cultures. The presence of calcium kept the entry of sodium proportional to growth and at a level that the plants tolerated (Figure 7). At six weeks the plants looked healthy and produced beans in profusion (Figure 7 in LaHaye and Epstein, 1971). Other evidence of the crucial importance of calcium comes from experiments in which the performance of two species of wheatgrass in salinized solution cultures was compared: the relatively salt sensitive intermediate wheatgrass, Agropyron intermedium, and the more tolerant tall wheatgrass, A. elongatum (Elzam and Epstein, 1969). A. intermedium yielded dramatically less total mass at 5 mM salt than at 0.5 mM, and at the higher concentration contained much less calcium in the roots. A similar parallelism between yield decrement and lower calcium concentration in the roots also occurred in the tolerant A. elongatum, but at the much higher external salt concentration of 100 mM.

Like other observations the work just described suggests that more research into the role of calcium in salt toleration might be rewarding. The work deserves comment in another connection. Comparisons are frequently made between the performance under saline conditions of very different plants, such as a salt sensitive crop plant and a wild halophyte. Many differences of course are revealed in aspects of ion transport, water relations, photosynthesis, and a host of other features. But such contrasts are to be expected simply on the basis that the plants are taxonomically and phylogenetically not closely related. If, on the other hand, plants

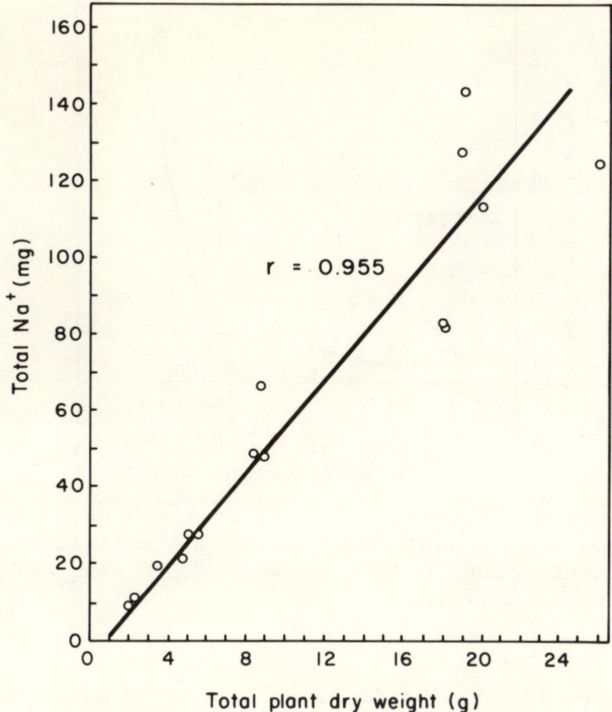

Fig. 7. Total sodium content of bean plants, *Phaseolus vulgaris*, as a function of the total dry weight of the plants. They were grown in nutrient solutions at 50 mM sodium chloride and 10 mM calcium sulfate. Plants were harvested after 1, 2, 3, 5, and 6 weeks. Each point represents a single plant. After LaHaye and Epstein (1971).

of closely related genotypes differing in salt tolerance are compared, such as species of one genus, or genotypes within a species, then chances are good that differential aspects of ion transport, water economy, organic acid metabolism and many more may bear a causal relationship to the salt tolerance and sensitivity of the plants, respectively.

A recent comparison of two species of the tomato is a case in point (Figure 8, after Rush and Epstein, 1976). The plants were grown in solution cultures which at weekly intervals were progressively salinized with a synthetic sea salt mix, each increment raising the salinity by 0.1 seawater. The exotic *Lycopersicon cheesmanii*, seed of which came from the coast of Isla Isabella in the Galapagos Islands, behaved like a halophyte, absorbing large amounts of sodium (up to nearly 6 percent of the dry weight of the leaf), and it tolerated that concentration. The commercial *L.*

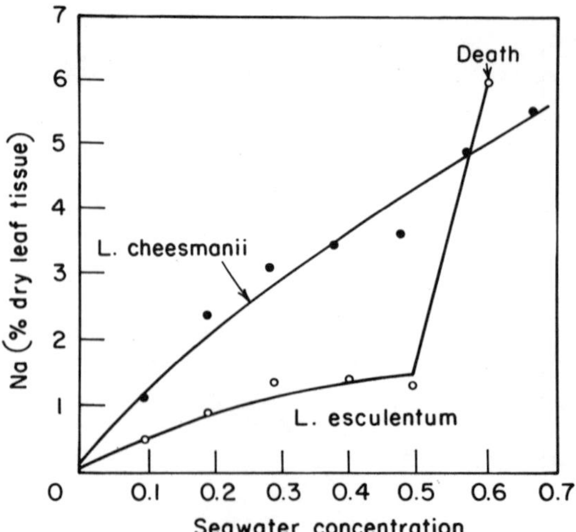

Fig. 8. Percentages of sodium in dry leaf tissue of two species of the tomato as a function of the salinization of the nutrient solution with seawater salt mix. After Rush and Epstein (1976).

esculentum under the same conditions tended to exclude sodium from the shoot, as many salt sensitive plants do, until at about 0.5 seawater salinity it succumbed, sodium breaking through into the shoot in large amounts. Even more useful for comparative investigations are genotypes within a species that respond differentially to salt, as discussed by J. D. Norlyn at this meeting.

It is even worthwhile to make deliberate selections within a cultivar for salt tolerance, on the one hand, and for sensitivity to salt, on the other. In this way, sharply contrasting genotypes can be obtained which in all other aspects of their genome are nearly identical. For investigations on mechanisms of salt tolerance and sensitivity in higher plants, such genotypes, "selected apart" in only that respect, can play a role similar to that which mutants of Escherichia coli, Neurospora crassa, and other organisms have attained in nutritional and metabolic investigations. The time is overdue for such comparative investigations of the salt relations of plants.

As for the vacuole, it is often assumed that inorganic ions in it are in equilibrium or nearly so with those in the cytoplasm, with neither large concentration nor electropotential gradients across the intervening membrane, the tonoplast. But if the view of adaptation to saline conditions outlined here is correct, the situation at the tonoplast of salt absorbing halophytes must be quite different (Dainty, 1979). The tonoplast must maintain vacuolar concentrations of sodium and often, chloride, of several hundred millimoles/liter, but cytoplasmic concentrations very much lower. And both it and the plasmalemma must confine within the cytoplasm the very high concentrations of the organic solute or solutes which serve as osmotica in that compartment. Regrettably, we do not know much about the permeability properties and transport capabilities of the tonoplast even in ordinary plants, let alone in halophytes.

CONCLUSION

The principal responses of halophytes and other plants that tolerate saline environments are of necessity those that lead to osmotic adjustment. Unless that premier adaptation is accomplished existence in saline substrates is impossible. The osmotic adjustment can be achieved by the buildup within the cells of high concentrations of either inorganic ions or organic solutes. In higher plants both invariably play some role, but the contribution made by each varies a great deal in different plants, and different cell compartments even in the same plant.

In the salt accumulating halophytes absorption and transport of inorganic ions predominate. They are largely sequestered within the vacuoles. In the much smaller volume occupied by cytoplasm, metabolically tolerable organic solutes are responsible for keeping the cytoplasm from becoming dehydrated through osmotic loss of water to the saline solution in the cell walls, on the one side, and the vacuole on the other.

In salt tolerant plants the principal cellular membranes, the outer cell membrane or plasmalemma and the one between the cytoplasm and the vacuole, the tonoplast, must possess extraordinary properties and capabilities. The plasmalemma must tolerate contact with highly saline water without damage or disintegration. It also must have at least one transport mechanism able to absorb the essential ion, potassium, in the face of a preponderance of the potentially competitive alkali cation, sodium. A spectacular degree of selectivity for potassium as against sodium is implied for that mechanism. Finally, the plasmalemma must minimize leakage and loss of the organic solutes which are present at high concentrations in the cytoplasm and maintain its osmotic adjustment.

The tonoplast must deliver to the vacuole inorganic ions and keep them there at high concentrations, implying maintenance of large concentration disequilibria between cytoplasm and vacuole. It also must minimize leakage of the cytoplasmic organic osmotica into the vacuole.

Comparative studies between closely related genotypes, especially genotypes within a species contrasting only in salt tolerance and sensitivity, are needed to identify the adaptations that render salt tolerant plants capable of life in the severe environments of saline soils and waters that are fatal to salt sensitive ones. Comparisons of the structure and function of the cellular membranes of such contrasting genotypes otherwise similar in their genomes might be especially rewarding.

ACKNOWLEDGMENTS

Work from this laboratory referred to in this paper was supported by the National Science Foundation and Office of Sea Grant, NOAA, U. S. Department of Commerce. I thank R. L. Bieleski for comments.

REFERENCES

Dainty, J., 1979, The ionic and water relations of plants which adjust to a fluctuating saline environment, in: "Ecological Processes in Coastal Environments," R. L. Jefferies and A. J. Davy, eds., Blackwell Scientific, Oxford.

Eckholm, E. P., 1975, Salting the earth, Environ., 17(7):9.

Elzam, O. E., and Epstein, E., 1969, Salt relations of two grass species differing in salt tolerance. I. Growth and salt content at differing salt concentrations, Agrochim., 13:187.

Epstein, E., 1961, The essential role of calcium in selective cation transport by plant cells, Plant Physiol., 36:437.

Epstein, E., 1968, Microorganisms and ion absorption by roots, Experientia, 24:616.

Epstein, E., 1976, Kinetics of ion transport and the carrier concept, Encyclopedia of Plant Physiology New Series, 2, B:70.

Flowers, T. J., Troke, P. F., and Yeo, A. R., 1977, The mechanism of salt tolerance in halophytes, Ann. Rev. Plant Physiol., 28:89.

Hellebust, J. A., 1976, Osmoregulation, Ann. Rev. Plant Physiol., 27:485.

LaHaye, P. A., and Epstein, E., 1969, Salt toleration by plants: enhancement with calcium, Science, 166:395.

LaHaye, P. A., and Epstein, E., 1971, Calcium and salt toleration by bean plants, Physiol. Plant., 25:213.

Läuchli, A., and Epstein, E., 1970, Transport of potassium and rubidium in plant roots: the significance of calcium, Plant Physiol., 45:639.

Massoud, F. I., 1974, Salinity and Alkalinity as Soil Degradation Hazards. FAO/UNEP Expert Consultation on Soil Degradation, FAO, Rome.

Rains, D. W., 1972, Salt transport by plants in relation to salinity, Ann. Rev. Plant Physiol., 23:367.

Rains, D. W., and Epstein, E., 1967a, Sodium absorption by barley roots: its mediation by mechanism 2 of alkali cation transport, Plant Physiol., 42:319.

Rains, D. W., and Epstein, E., 1967b, Preferential absorption of potassium by leaf tissue of the mangrove, Avicennia marina: an aspect of halophytic competence in coping with salt, Aust. J. Biol. Sci., 20:847.

Rodin, L. E., Bazilevich, N. I., and Rozov, N. N., 1975, Productivity of the world's main ecosystems, in: "Productivity of World Ecosystems," National Academy of Sciences, Washington, D.C.

Rush, D. W., and Epstein, E., 1976, Genotypic responses to salinity: differences between salt-sensitive and salt-tolerant genotypes of the tomato, Plant Physiol., 57:162.

Storey, R., and Wyn Jones, R. G., 1979, Response of Atriplex spongiosa and Suaeda monoica to salinity, Plant Physiol., 63:156.

Wittwer, S. H., 1979, Future technological advances in agriculture and their impact on the regulatory environment, BioSci., 29: 603.

AGROBACTERIUM TI PLASMIDS AS A TOOL FOR GENETIC ENGINEERING IN PLANTS

Mary-Dell Chilton

Department of Biology
Washington University
St. Louis, MO 63130

The technology of cloning poses issues and presents possibilities that were almost unimaginable only a few years ago. With the aid of restriction endonucleases and ligase, the biochemist can literally be a genetic architect. If a means can be devised for detecting its presence, a desired gene can be cloned in *Escherichia coli*, and plasmid DNA can be isolated that contains the gene in high concentration and purity. The gene can then be excised from the *E. coli* vector plasmid and inserted into other vectors for introduction into a eukaryotic cell. Monkey kidney tissue culture cells, for example, have been shown to produce mouse globulin after introduction of the mouse globulin gene attached to SV40 viral DNA (Hamer and Leder, 1979). In the view of some critics (Goodfield, 1977; Packard, 1977) experiments of this sort pose ethical question and raise the specter of manipulation of the human gene pool. However, the application of recombinant DNA technology to agricultural crop improvement has not been subjected to such criticism. Man has been manipulating the genomes of crop plants for centuries by more traditional methods. Crop improvement by direct manipulation of DNA appears to be a socially acceptable extension of plant breeding efforts. Thus the obstacles that beset the plant genetic engineer are chiefly technical rather than social ones.

VECTORS FOR PLANTS

With the goal of crop improvement in mind, one can list a number of attributes required of a suitable vector for introduction of cloned DNA fragments into the plant cell. Feasibility is of paramount importance: the vector must have a route of entry into the plant cell and must be stably maintained. It should be capable of systemic dispersal within the plant, in order to ensure that the introduced DNA will find its way to the desired plant organs. The vector must be compatible with healthy, vigorous plant phenotype. It would be desirable for the vector to be seed-transmissible. Finally, wide host range would be ideal, with particular emphasis on the principal food crops: cereals, maize and legumes.

The most promising vectors under investigation are agents of plant diseases, for example, the DNA-containing plant viruses and *Agrobacterium* Ti plasmids. The advantage of these systems is that the genetic engineer can exploit the pathogen's highly evolved and efficient route of entry into the plant cell. Other methods of introducing DNA into plants are inevitably based on the idea of transforming plant cells with DNA, an approach that has proven extremely troublesome in the past (Kleinhofs and Behki, 1977). The importance of plant cell transformation by DNA as a research objective is clear, but it seems equally clear that this class of approaches will be of later value to the genetic engineer and more limited applicability: a method of introducing DNA will have to be developed for each type or even strain of plant to be transformed. Some plants may prove refractory to all efforts at transformation.

AGROBACTERIUM TI PLASMID DNA IN PLANT CELLS

Among plant pathogens, *Agrobacterium tumefaciens* appears uniquely adapted to the role of vector for genetic engineering. This soil bacterium incites cancers called crown gall tumors in a wide variety of dicotyledonous plants. Oncogenic strains contain large plasmids called Ti (tumor-inducing) plasmids (Zaenen et al., 1974) that carry genetic information essential for tumor induction (Van Larebeke et al., 1974; Watson et al., 1975). Transformed (tumorous) plant tissue that is free from the inciting bacterium grows *in vitro* without exogenously supplied auxin and cytokinin, and produces novel amino acid derivatives called opines that are specified by the Ti plasmid (Petit et al., 1970; Bomhoff et al., 1976; Montoya et al., 1977). These altered properties comprise the tumorous phenotype, and are apparently ascribable to a genetic change in the plant cell brought about by the oncogenic bacterium. As indicated schematically in Figure 1, the transformed plant cells contain and maintain copies of a small part of the Ti plasmid called T-DNA (Chilton et al., 1977; Merlo et al., 1980; Yang et al., 1980; Van Montagu and Schell, 1979). The exact boundaries of T-DNA on the Ti plasmid can vary (Merlo et al., 1980), but in all cases studied thus far, T-DNA includes a highly conserved genetic region that is common to all broad host range Ti plasmids (Chilton et al., 1978; Depicker et al., 1978). This common DNA region has been shown to code for oncogenic functions: cointegration of RP4 with the Ti plasmid at a site in this region renders the Ti plasmid non-oncogenic (Depicker et al., 1978). It is clear that some parts of T-DNA are non-essential, for they are missing in certain tumor lines (Merlo et al., 1980).

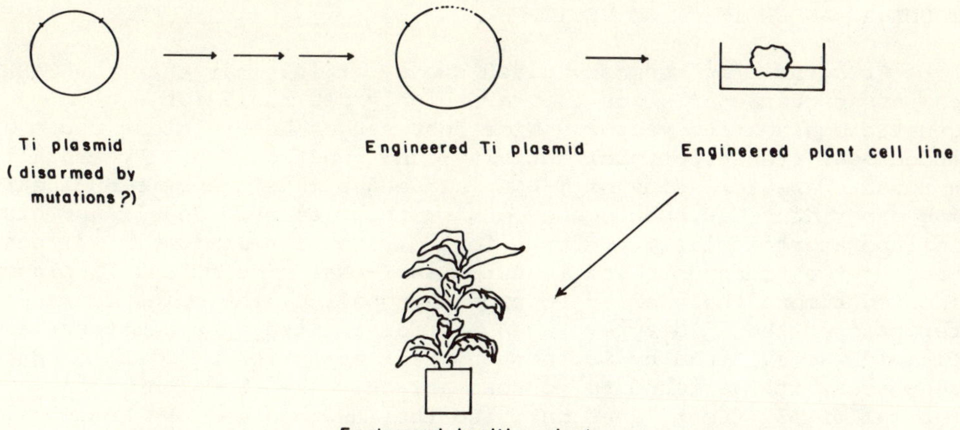

Fig. 1. Tumor line incited by Agrobacterium tumefaciens maintains copies of foreign DNA. Oncogenic Agrobacterium stains carry tumor-inducing (Ti) plasmids. A small specific region of the Ti plasmid, called T-DNA, is stably maintained by the transformed plant cells. The tumor tissue, grown axenically (free from bacteria) as a tissue culture line, maintains T-DNA copies for many years. Other parts of the Ti plasmid are not detected in the tumor cells; it is not known whether they are never incorporated into the plant cells, or whether they are incorporated but fail to transform the cells to the tumorous state.

Transcripts of T-DNA are present in tumor cells (Drummond et al., 1977; Lederboer, 1978; Gurley et al., 1979), and are polyadenylated (Lederboer, 1978) and found on polysomes (Chilton et al., 1979a). It seems highly probable, therefore, that T-DNA is functionally active in the turmor cell.

The location of T-DNA has been studied in detail in a T-37 teratoma called B-T37. Pure chloroplast and mitochondrial DNAs from this tumor line were found to be free from T-DNA (Chilton et al., 1979a,b) while nuclear DNA contains T-DNA (Chilton et al., 1979b). Studies of T-DNA reisolated from the plant tumor cell by molecular cloning in Charon lambda bacteriophage have shown that plant DNA is covalently joined to T-DNA (M. Thomashow, R. Nutter, and K. Postle, unpublished data; N. Yadav, K. Postle, and R. Saiki, unpublished data). Thus the available data support the view that T-DNA forms a part of the chromosomal DNA of the plant cell and is covalently inserted into the genome of the plant. Consistent with this view, T-DNA is stably maintained by the tumor line in tissue culture.

T-DNA AS AN ENGINEERING VECTOR

From the brief summary given above, it is clear that the T-DNA possesses characteristics that are highly desirable for a plant genetic engineering vector. Wide host range, an efficient route of infection, transcriptional activity, and genetic stability are all natural characteristics of T-DNA. Indeed a model genetic engineering experiment has been achieved with this vector. Schell and his collaborators isolated a mutant form of the Ti plasmid pTi T37 containing the transposon Tn7 inserted in T-DNA. The mutant Ti plasmid incited tumors that failed to produce nopaline, the opine characteristic of the wild type. When a tumor incited by the mutant Ti plasmid was examined by Southern hybridization for Ti plasmid DNA sequences, it was found to possess not only the expected T-DNA typical of T37 tumor lines but also the inserted Tn7 DNA sequences (Schell, 1979). Thus, extra DNA introduced into a non-vital part of T-DNA can be carried into the plant cell, inserted and maintained in the resulting tumor line.

While T-DNA clearly has great promise as a vector for insertion of DNA into crown gall tumors, the problem remains that we wish to obtain altered healthy plants. Two types of manipulation have been used to generate plants from cloned B-T37 tumor cells. The first such experiment, performed by Braun and his collaborators (Braun and Wood, 1976; Turgeon et al., 1976), produced normal appearing shoots by grafting the teratomatous shoots to decapitated tobacco plants (Figure 2). The grafted shoot retained most or all of the T-DNA of the parental tumor line (Yang et al., 1980; Chilton et al., 1979b; R. K. Saiki and M. D. Chilton, in preparation), and retained tumorous characteristics: explants grew as teratomas _in vitro_, and plant parts as well as explants were found to contain nopaline (Braun and Wood, 1976; Turgeon et al., 1976). In contrast, M. P. Gordon regenerated normal appearing plants from the same T37 tumor line by manipulation of cytokinin levels _in vitro_ (M. P. Gordon, private communication). The Gordon plants lack nearly all of the T-DNA of the parental line but retain a small segment (F.-M. Yang and M. P. Gordon, private communication) and are normal (non-tumorous when explanted; not nopaline-producing).

An unexplained and disturbing feature of the shoot regenerated by grafting is that its gametes do not transmit T-DNA to F1 progeny plants (Yang et al., 1980). Indeed, a plant derived from haploid anther tissue of the grafted shoot was also found to be free from T-DNA (Yang et al., 1980; R. K. Saiki and M. D. Chilton, in preparation). Thus the meiotic loss of T-DNA seems either actively to remove or to select against T-DNA. It will be extremely interesting to obtain comparable data for the progeny of Gordon's _in vitro_ regenerated plants.

The generalization that one can draw thus far is that T-DNA gene products make the plant cell tumorous, and that true recovery

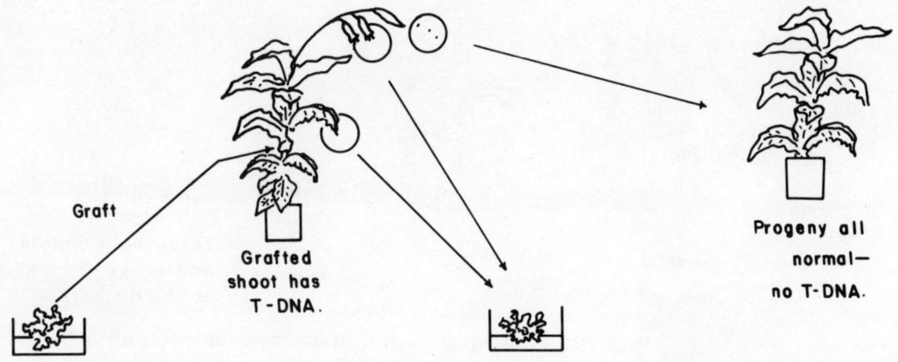

Fig. 2. Regeneration of a tobacco shoot from a crown gall teratoma by grafting. Braun and his collaborators (Braun and Wood, 1976; Turgeon et al., 1976) grafted cloned T37 Havana tobacco tumor tissue to a decapitated Turkish tobacco plant. More or less normal-looking Havana shoots were obtained, most of which were sterile. A rare fertile shoot gave rise to completely normal progeny that were free from T-DNA. However the shoot itself retained tumorous characteristics. When parts of the shoot (leaves, flower, petals, filaments) were planted on tissue culture medium, they grew as teratomas once again. These teratomas contained T-DNA, which must therefore be maintained silently by the normal-looking grafted shoot. One tissue of the grafted shoot was normal and free from T-DNA: a tissue line derived from haploid anther cells. This finding hints that meiosis may be the stage at which T-DNA is lost or excised. Alternatively, it is possible that only normal cells can participate in meiosis. Thus most grafted shoots are sterile, and only one rare grafted shoot had a deletion in T-DNA that allowed expression of fertility.

is achieved by loss of most or all of the T-DNA. The tumorous state seems not to be compatible with regeneration of a normal plant with roots. If we are to employ T-DNA as an engineering vector, it will be necessary to disarm it (either before or after insertion into the plant genome) by inactivation of the genes that code for tumor maintenance functions. A scheme for such an approach is outlined in Figure 3. Whether such disarmed T-DNA would be seed transmitted remains to be seen.

TARGET GENES

If the problem of a vector is solved, there remains the more difficult problem of choice and isolation of the DNA one wishes to introduce into the plant genome. What genetic trait must one

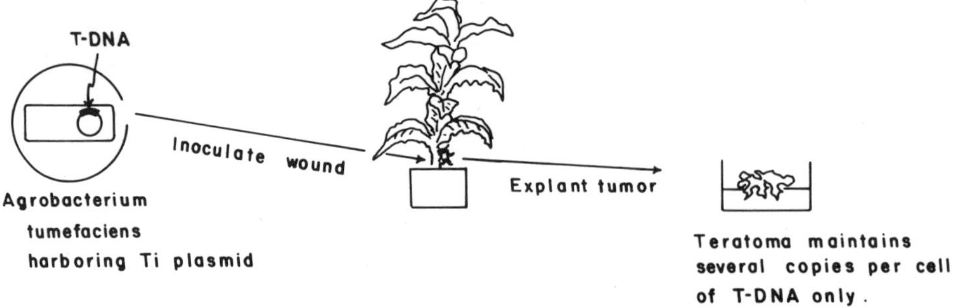

Fig. 3. Scheme for production of an "engineered" plant containing desired DNA introduced by a Ti plasmid vector. By suitable recombinant DNA manipulations, one can obtain Ti plasmid with a desired DNA fragment (symbolized by the dotted line) inserted into the T-DNA (the marked sector of the Ti plasmid). The Ti plasmid used could be "disarmed" by introduction of mutations into its T-DNA such that it does not impair the ability of the tumor cell to regenerate into a plant. When such T-DNA is incorporated into the plant genome, an engineered plant cell line can be obtained, and from the latter, an engineered plant. However it is not clear whether such a disarmed T-DNA would be lost at meiosis (Figure 2) or would be seed-transmissible.

modify in order to make the plant hardier (disease resistant, drought resistant, salt tolerant, cold tolerant, pest resistant) or more nourishing (higher lysine or sulfur amino acid content, higher protein content in leaves or seeds) or more productive (more and/or earlier seeds, leaves or fruits)? Can the desired trait be localized to a particular gene product and gene in the plant genome? Alternatively, is it possible to isolate a bacterial gene that can confer the desired trait?

It is clear that most of the traits one might wish to modify in crop plants are determined by the interplay of many genes rather than a single, simple gene product. It is unlikely that genetic engineering will succeed in gross alteration of such complex attributes. Even simple attributes conferred by a single plant gene or gene cluster will present great challenge: the prospect of trying to clone in E. coli the DNA responsible for most simple plant traits appears today to be formidable. However the lesson of recent history in the field of nucleic acids is clearly one of innovation, ingenuity and optimism. One can foresee a general approach to the cloning of simple plant genes that will exploit mobile DNA elements analogous to bacterial transposons. Transposons (Kleckner, 1977) have proven a powerful tool in bacterial genetics for physically marking genetic loci of interest. Analogous transposable elements apparently exist in eukaryotic cells (Nevers and Saedler, 1977; Potter et al., 1979; Strobel, 1979; Cameron et al., 1979). DNA of transposable elements can in principle be exploited in plant cells as DNA hybridization probes for isolation of any gene into which they have inserted. The cloning of simple plant genes is a goal that we can confidently expect to see attained. Traits in the plant that can be manipulated by such simple genes are the most promising targets for genetic engineering efforts.

The possibility of modifying plants by introduction of extremely foreign DNA such as bacterial or yeast DNA is more speculative. Although isolation of the cloned DNA fragments from these relatively small genomes would be simpler, it is not clear how their function could be regulated in the plant cell, or indeed whether they would function at all. Foreign enzymes, even if correctly produced in the plant cell, might exhibit little activity because the conditions of temperature, pH and salt are not suitable or their substrates are unavailable in sufficient concentration.

Introduction of synthetic DNA that would code for a nutritionally balanced protein is a seemingly simple approach. However this type of foreign DNA might prove detrimental to the plant if it were expressed at significant levels. It would be desirable to regulate its expression so that the product would be formed only in the desired plant organ.

The most attractive approach to crop improvement is that of modification of existing plant genes rather than introduction of totally foreign ones. One could isolate a normal seed storage protein gene by molecular cloning, analyze the coding DNA sequence, and alter *in vitro* the DNA coding for some limited portion of the protein. The resulting DNA, together with the flanking elements that regulate its expression, would probably be expressed in the desired way after reintroduction into the plant of origin.

PROSPECTS

The T-DNA of *Agrobacterium* Ti plasmids is but one of several agents under active investigation as potential vectors for introducing desirable DNA into higher plants. Success of at least some of the vectors from a technical point of view seems very probable. Whether the application of such vectors will prove of practical use for crop improvement is less certain. Limited applications for simple alterations of plant DNA can be foreseen already, and future progress in our understanding of plant gene regulation may extend the scope of this approach to crop improvement at the DNA level.

REFERENCES

Bomhoff, G., Klapwijk, P. M., Kester, H. C. M., Schilperoort, R. A., Hernalsteens, J. P., and Schell, J., 1976, Octopine and nopaline synthesis and breakdown genetically controlled by a plasmid of Agrobacterium tumefaciens, Molec. Gen. Genet., 145:177.

Braun, A. C. and Wood, H. N., 1976, Suppression of the neoplastic state with the acquisition of specialized functions in cells, tissues and organs of crown-gall teratomas of tobacco, Proc. Natl. Acad. Sci. USA, 73:496.

Cameron, J. R., Loh, E. Y., and Davis, R. W., 1979, Evidence for transposition of dispersed repetitive DNA families in yeast, Cell, 16:739.

Chilton, M.-D, Drummond, M. H., Merlo, D. J., Sciaky, D., Montoya, A. L., Gordon, M. P., and Nester, E. W., 1977, Stable incorporation of plasmid DNA into higher plant cells: the molecular basis of crown gall tumorigenesis, Cell, 11:263.

Chilton, M.-D., Drummond, M. H., Merlo, D. J., and Sciaky, D., 1978, Highly conserved DNA of Ti plasmids overlaps T-DNA maintained in plant tumors, Nature, 275:147.

Chilton, M.-D., McPherson, J., Saiki, R. K., Thomashow, M. F., Nutter, R. C., Gelvin, S. B., Montoya, A. L., Merlo, D. J., Yang, F. M., Garfinkel, D. J., Nester, E. W., and Gordon, M. P., 1979a, in: "Emergent Techniques for the Improvement of Crop Plants," I. Rubenstein, ed., Univ. of Minnesota Press, (in press).

Chilton, M.-D., Yang, F. M., Saiki, R. K., Postle, K., Montoya, A. L., Nester, E. W., Quetier, F., and Gordon, M. P., 1979b, in: "Genome Organization and Expression in Plants," C. J. Leaver, ed., Plenum Press, (in press).

Depicker, A., Van Montagu, M., and Schell, J., 1978, Homologous DNA sequences in different Ti plasmids are essential for oncogenicity, Nature, 275:150.

Drummond, M. H., Gordon, M. P., Nester, E. W., and Chilton, M.-D., 1977, Foreign DNA of bacterial plasmid origin is transcribed in crown gall tumors, Nature, 269:535.

Goodfield, J., 1977, Playing God: genetic engineering and the manipulation of life, Harper and Row, New York.

Gurley, W. B., Kemp, J. D., Albert, M. J., Sutton, D. W., and Callis, J., 1979, Transcription of Ti plasmid-derived sequences in three octopine-type crown gall tumor lines, Proc. Natl. Acad. Sci. USA, 76:2828.

Hamer, D. H. and Leder, P., 1979, Expression of the chromosomal mouse β^{maj}-globin gene cloned in SV40, Nature, 281:35.

Kleckner, N., 1977, Translocatable elements in procaryotes, Cell, 11:1123.

Kleinhofs, A. and Behki, R., 1977, Prospects for plant genome modification by nonconventional methods, Ann. Rev. Genet., 11: 79.

Lederboer, A., 1978, Ph.D. Thesis, University of Leiden.

Merlo, D. J., Nutter, R. C., Montoya, A. L., Garfinkel, D. J., Drummond, M. H., Chilton, M.-D., Gordon, M. P., and Nester, E. W., 1980, The boundaries and copy numbers of Ti plasmid T-DNA vary in crown gall tumors, Molec. Gen. Genet., (in press).

Montoya, A. L., Chilton, M.-D., Gordon, M. P., Sciaky, D., and Nester, E. W., 1977, Octopine and nopaline metabolism in Agrobacterium tumefaciens and crown gall tumor cells: Role of plasmid genes, J. Bacteriol., 129:101.

Nevers, P. and Saedler, H., 1977, Transposable genetic elements as agents of gene instability and chromosomal rearrangements, Nature, 268:109.

Packard, V., 1977, The people shapers, Little, Brown and Company.

Petit, A., Delhaye, S., Tempé, J., and Morel, G., 1970, Recherches sur les guanidines des tissues de crown gall. Mise en évidence d'une relation biochimique spécifique entre les souches d'Agrobacterium tumefaciens et les tumeurs qu'elles induisent, Physiol. Vég., 8:205.

Potter, S. S., Brorein, W. J., Dunsmuir, P., and Rubin, G. M., 1979, Transposition of elements of the 412, copia and 297 dispersed repeated gene families in Drosophila, Cell, 17:415.

Schell, J., 1979, in: "Nucleic Acids in Plants," T. Hall, et., Academic Press, (in press).

Strobel, E., Dunsmuir, P., and Rubin, G. M., 1979, Polymorphism in the chromosomal locations of elements of the 412, copia and 297 dispersed repeated gene families in Drosophila, Cell, 17:429.

Turgeon, R., Wood, H. N., and Braun, A. C., 1976, Studies on the recovery of crown gall tumor cells, Proc. Natl. Acad. Sci. USA, 73:3562.

Van Larebeke, N., Engler, G., Holsters, M., Van den Elsacker, S., Zaenen, I., Schilperoort, R. A., and Schell, J., 1974, Large plasmid in Agrobacterium tumefaciens essential for crown gall-inducing activity, Nature, 252:169.

Van Montagu, M. and Schell, J., 1979, The plasmids of Agrobacterium tumefaciens, in: "Plasmids of Bacteria of Medical Environmental and Commercial Importance, K. Timmis, ed., Amsterdam, Eslevier/ North Holland Medical Press.

Watson, B., Currier, T. C., Gordon, M. P., Chilton, M.-D., and Nester, E. W., 1975, Plasmid required for virulence of Agrobacterium tumefaciens, J. Bacteriol., 123:255.

Yang, F. M., Montoya, A. L., Merlo, D. J., Drummond, M. H., Chilton, M.-D., Nester, E. W., and Gordon, M. P., 1980, Foreign DNA sequences in crown gall teratomas and their fate during loss of the tumorous traits, Molec. Gen. Genet., (in press).

Zaenen, I., Van Larebeke, N., Teuchy, H., Van Montagu, M., and Schell, J., 1974, Supercoiled circular DNA in crown-gall inducing Agrobacterium strains, J. Mol. Biol., 86:109.

SECTION I

OSMOREGULATION IN PROKARYOTIC ORGANISMS

THE ROLE OF L-PROLINE IN RESPONSE TO OSMOTIC STRESS IN *SALMONELLA TYPHIMURIUM*: SELECTION OF MUTANTS WITH INCREASED OSMOTOLERANCE AS STRAINS WHICH OVER-PRODUCE L-PROLINE

Laszlo N. Csonka

Plant Growth Laboratory/Agronomy & Range Science Dept.
University of California, Davis
Davis, CA 95616

INTRODUCTION

When bacteria of a wide variety of species are stressed by high osmolarity in the growth medium, a common response they exhibit is that there is a pronounced elevation in the intracellular concentration of free L-proline, or L-glutamate, or, in a few instances, γ-amino butyric acid. In a number of cases in which the bacteria were grown in the absence of exogenously added amino acids, this accumulation was due to the enhanced net rate of synthesis of these metabolites (Tempest et al., 1970; Brown and Stanley, 1972; Makemson and Hastings, 1979), but in others, in which the cells were grown in a complex media it was not clear whether the increases in concentration were due to the stimulation of the synthesis or uptake of these compounds (Measures, 1975; Koujima et al., 1978). It was shown by Britten and McClure (1962) that in *E. coli* the accumulation of L-proline from the medium is stimulated in direct proportion to the external osmolarity. An explanation offered for the elevation of the concentration of these compounds is that it serves to balance the intracellular osmolarity against the osmolarity of the growth medium (Brown and Stanley, 1972; Measures, 1975).

There is some experimental evidence, though indirect, that the increase in the intracellular concentration of L-proline is indeed of physiological importance for the organisms to cope with osmotic stress. In 1955, Christian discovered that L-proline, added exogenously, greatly stimulated the growth rate of *Salmonella oriaenburg* in media of high osmolarity (Christian, 1955a,b). Since enteric bacteria are able to concentrate L-proline at least 100-fold over the extracellular levels (Anderson et al., 1979), and since the magnitude of the concentration gradient is increased in direct proportion to the osmolarity of the growth medium (Britten and McClure, 1962), it seems possible that L-proline could function as an osmotic balance, at least when it is provided in the growth medium.

In this paper we describe experiments with Salmonella typhimurium which further characterize the phenomenon of growth stimulation by L-proline in media of inhibitory osmotic strength, first reported by Christian. We also describe the isolation and partial characterization of mutants which produce abnormally high levels of L-proline, and which, as a consequence, prove to exhibit increased osmotolerance.

MATERIALS AND METHODS

Bacterial Media and Growth Conditions

The complex medium used was LB, containing in one liter: 10 g Bacto tryptone (Difco), 5 g Bacto yeast extract (Difco), 10 g NaCl, and 0.72 g NaOH. The minimal medium used was medium 63 (Cohen and Rickenberg, 1956) consisting of 0.1 M KH_2PO_4, 0.075 M KOH, 0.015 M $(NH_4)SO_4$, 1.6 x 10^{-4} M $MgSO_4 \cdot 7H_2O$, 3.9 x 10^{-6} M $FeSO_4 \cdot 7H_2O$ with the pH being 7.2. The osmolarity of this basal medium was increased by the addition of sucrose or inorganic salts, as indicated. In the case of the inorganic salts, it was necessary to readjust the pH to 7.2, which was accomplished by the addition of the respective cognate bases, in amounts that were always less than 10% of the salt concentration. D-glucose was the carbon source, at a concentration of 10 mM unless stated otherwise; L-amino acid supplements, when required were 0.1 mM. Kanamycin sulfate, used for one step in strain construction, was at a concentration of 75 µg/ml. Media were solidified by the addition of agar (Difco) in the amount of 20 g per liter. Cultures were grown with areation at 37 C.

Bacterial strains. The genotypes of the bacterial strains used are listed in Table I. All strains employed are derivatives of Salmonella typhimurium LT-2, except ProAB47, TR3290, and JL2468 which are derived from Salmonella typhimurium LT-7. The generalized transducing phage P22 HT 105/1 int201, obtained from H. Johnston, was used for some steps of strain construction, and transductions were carried out as described by Johnston and Roth (1979).

Determination of growth rates. Single colony isolates were inoculated into liquid LB cultures and grown to saturation, then subcultured at a dilution of 1:100 into minimal medium and grown to saturation, over night. Then 2.5 ml were subcultured into 20 ml of minimal medium, the cells allowed to double, and 1.0 ml aliquots inoculated into the final medium containing the desired concentration of sucrose or inorganic salts. The cultures were incubated

Table I. Bacterial strains.

Strain		Source or Method of Construction
LT-2	wild type	From B. Ames, via S. Küstü and J. Ingraham
ProAB47	del (proAB) 47	From R. Menzel
JL2468	del (proAB) 47 argI537 leuD798 ara-9 fol-1/F'128 $proA^+B^+$ $argF^+$ $lacI^q$::$lacZ^+Y^+$	From J. Ingraham
TR3290	del (proAB) 47 argI539 del (trp) 130 metE338 amtA1	From J. Ingraham
TT2601	putA842::Tn5	From R. Menzel
TC106	putA842::Tn5	LT-2 transduced to kanamycin resistance by P22 grown on TT2601
TC117	del (proAB) 47	LT-2 transduced to resistance to 300 μg/ml of 8-azaguanine (cf. Sanderson and Hartman, 1978)
TC81	del (proAB) 47 argI537 leuD798 ara-9 fol-1/F' 128 pro-81 $argF^+$ $lacI^q$:: $lacZ^+Y^+$	L-azetidine-2-carboxylate resistant derivative of JL2468. See text.
TC82	del (proAB) 47 argI537 leuD798 ara-9 fol-1/ F'128 pro-82 $argF^+$ $lacI^q$::$lacZ^+Y^+$	L-azetidine-2-carboxylate resistant derivative of JL2468. See text.
TC83	del (proAB) 47 argI537 leuD798 ara-9 fol-1/ F'128 pro-83 $argF^+$ $lacI^q$::$lacZ^+Y^+$	L-azetidine-L-carboxylate resistant derivative of JL2468. See text.
TC84	del (proAB) 47 argI537 leuD798 ara-9 fol-1/ F'128 pro-84 $argF^+$ $lacI^q$::$lacZ^+Y^+$	L-azetidine-2-carboxylate resistant derivative of JL2468. See text.
TC86	del (proAB) 47 argI537 leuD798 ara-9 $argF^+$	L-azetidine-2-carboxylate resistant derivative of JL2468. See text.

TABLE I. Bacterial strains (continued)

Strain		Source or Method of Construction
	F'128 pro-86 $argF^+$ $lacI^q$::$lacZ^+Y^+$	
TC88	del (proAB) 47 argI537 leuD798 ara-9 fol-1/ F'128 pro-88 $argF^+$ $lacI^q$::$lacZ^+Y^+$	L-azetidine-2-carboxylate resistant derivative of JL2468. See text.
TC126	del (proAB) 47/F'128 pro-88 $argF^+$$lacI^q$:: $lacZ^+Y^+$	Pro^+ Leu^+ progeny of mating TC88 and TC117
TC128	del (proAB) 47/F'128 $proA^+B^+$ $argF^+$ $lacI^q$:: $lacZ^+Y^+$	Pro^+ Leu^+ progeny of mating J12468 and TC117

Most genetic symbols are defined in the article by Sanderson and Hartman (1978). The ones important for this work are: del (proAB) 47 is a deletion that removes all of the proA and proB genes, whose products catalyze the second and first reaction of the L-proline biosynthetic pathway, respectively; leuD798 is a lesion in the L-leucine biosynthetic pathway conferring a requirement for the amino acid; the putA842::Tn5 is a mutation caused by the insertion of Tn5, a transposable genetic element conferring kanamycin resistance into the gene coding for a bifunctional enzyme which catabolizes L-proline to L-glutamate. The alleles pro-81, pro-82, pro-83, pro-84, pro-86, and pro-88 have been isolated in this work as putative regulatory mutants in the L-proline biosynthetic pathway.

at 37 C, and growth was monitored by measurement of the optical density at 420 nm, applying a correction factor to account for the fact that the absorbance is not proportional to cell density at high cell densities.

RESULTS

The Stimulation of Growth Rate by L-Proline in Media of High Osmolarity

As the osmolarity of the standard medium 63 is increased, the growth rate of Salmonella typhimurium is diminished, but this inhibition can be partly relieved by the addition of relatively low concentrations of L-proline. So, for instance, the growth rate of Salmonella typhimurium is reduced, from its value of 1.0 generation per hour in the normal medium, to 0.13 and 0.014 generation per hour in the presence of 0.7 and 0.9 M NaCl, respectively (cf. Figure 1). L-proline, which has little effect in the absence of the added NaCl, at a concentration of 0.5 mM increased the growth rate to 0.28 and 0.10 generation per hour in 0.7 and 0.9 M NaCl, respectively (cf. Figure 1). The effect is more general, because as can be seen from Figures 2 and 3, L-proline is stimulatory in the presence of $(NH_4)_2SO_4$, $K_{1.75}H_{1.25}PO_4$ (pH 7.2), and sucrose. The growth rates with 0.7 M $(NH_4)_2SO_4$ and 0.4 M $K_{1.75}H_{1.25}PO_4$ in the absence of L-proline were 0.10 and 0.44 generation per hour, whereas in the presence of 0.5 mM of the imino acid, the values were boosted to 0.24 and 0.59 generation per hour, respectively (Figure 2). As was also noted by Christian (1955b), the situation in the presence of sucrose is slightly more complicated because the growth rate is not constant throughout the logarithmic phase of growth: in 0.8 M sucrose without L-proline the growth rate initially was 0.075 generation per hour and after about twelve hours it increased to 0.175 generation per hour (Figure 3). If L-proline was added, the growth rate was 0.37 generation per hour throughout.

Thus, L-proline can alleviate, at least in part, the inhibitory effects of four different substances present at concentrations which are inhibitory. The effect is seen with sucrose, a neutral compound impermeable to Salmonella typhimurium (Kohno et al., 1978) and also with three salts, two of which (potassium phosphate and ammonium sulfate) supply essential ions, and thus are taken up by Salmonella typhimurium. So, because of the general nature of the phenomenon, we conclude that L-proline can counteract, at least partially, the inhibitory effects of high osmolarity, regardless of the cause. Of course it may be that each substance might also have other more specific inhibitory properties (e.g., sucrose increases the viscosity of the medium), which might be the reason why the relief of growth inhibition by L-proline is not complete (see below).

Characterization of the response to L-proline in media of high osmolarity. The effect of L-proline at several NaCl concentrations was examined, and the results are depicted in Figure 4. At any given NaCl concentration the growth rate in the presence of 0.5 mM L-proline was greater than at the same NaCl concentration in the absence of imino acid, the relative stimulation being more pronounced as the salt concentration (hence the inhibition) is increased. So at 0.65 M NaCl there is approximately a two-fold stimulation of the growth rate, while at 0.9 M the effect is about seven-fold. But, at these high NaCl concentrations the growth rate is severely inhibited even in the presence of L-proline, in comparison to that found in the absence of the extra NaCl. Results obtained using varying concentrations of $(NH_4)_2SO_4$, $K_{1.75}H_{1.24}PO_4$, or sucrose were qualitatively similar to those in Figure 4 in that at any solute concentration, growth in the presence of L-proline was faster than in its absence, with the relative stimulation being greater at higher solute concentrations (data not shown).

One of the interesting aspects of the phenomenon of growth rate stimulation by L-proline in media of high osmolarity is that in comparison to the osmolarity in the medium, very small concentrations of the imino acid are sufficient to bring about the effect. The dependence of growth rate on the L-proline concentration at a constant (0.65 M) concentration of NaCl is presented in Figure 5: approximately 1 mM of L-proline causes maximal stimulation (which is two-fold at this particular NaCl concentration), with approximately 0.1 mM being sufficient to bring about half maximal effect.

In order to answer the question whether L-proline was specific in its ability to alleviate the inhibitory effect of high osmolarity,

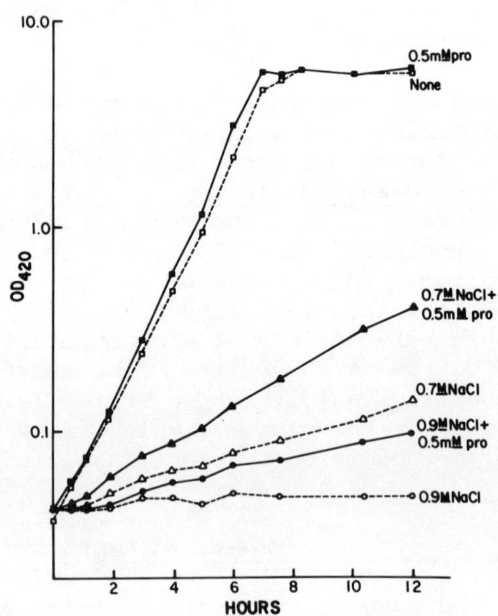

Fig. 1. The effect of L-proline on the growth of *Salmonella typhimurium* in the presence of NaCl. Optical density of the cultures is plotted as a function of incubation time. Broken lines and open symbols, growth without L-proline; solid lines and closed symbols, growth with 0.5 mM L-proline; squares, triangles and circles, growth without NaCl, with 0.7 M, and 0.9 M NaCl, respectively.

all the 19 common L-amino acids were tested in like manner. The results, summarized in Table II, show that in the presence of NaCl, growth was notably stimulated only by L-proline and all the other compounds had only slight stimulatory, and in some cases, inhibitory effects. After L-proline, the second most effective stimulation was caused by L-glutamine and by the combination of L-serine plus glycine, which was slight and did not prove to be reproducible. L-cysteine presented a curious finding in that in the presence of the high level of NaCl it was completely inhibitory. (The reason is not clear. L-cysteine has a slight inhibitory effect on *Salmonella typhimurium* in the standard minimal medium (Filutowicz et al., 1979; Csonka, unpublished observation) and it may be that this inhibition is enhanced by high osmolarity.) The effects of all the 19 common L-amino acids and L-proline were also tested in 0.7 *M* $(NH_4)_2SO_4$ and the results were similar: only L-proline caused a major stimulation, L-cysteine was totally inhibitory, and the others had only small stimulatory or inhibitory effects.

We have carried out one more experiment to establish whether the stimulation of growth in media of high osmolarity is unique to L-proline. There are mutants of *Salmonella typhimurium* lacking the

TABLE II. The effects of amino acids on the growth of *Salmonella typhimurium* in 0.65 *M* NaCl.[1]

Amino Acid	Generation Per Hour
None	0.20
L-proline	0.40
L-glutamate	0.24
L-serine + glycine	0.23
L-arginine	0.22
L-glutamate	0.22
Others	0.21-0.18
L-cysteine	No growth in 4 days

[1] All 19 common L-amino acids and L-proline were added at a final concentration of 0.5 *mM* to cultures of *Salmonella typhimurium* LT-2 grown in the presence of 0.65 *M* NaCl. The amino acids were added individually except for L-isoleucine, L-leucine, and L-valine which were added as a trio, as were the pairs of amino acids: L-asparagine plus L-aspartate, L-serine plus glycine, and L-histidine plus L-lysine.

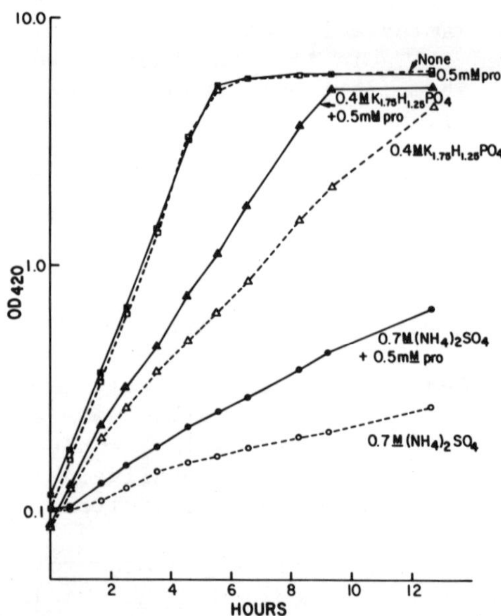

Fig. 2. The effect of L-proline on the growth of Salmonella typhimurium LT-2 in the presence of $(NH_4)_2SO_4$ and $K_{1.75}H_{1.25}PO_4$. Optical density is plotted as a function of incubation time. Broken lines and open symbols, growth without L-proline; solid lines and closed symbols, growth with 0.5 mM L-proline; squares, growth in normal medium 63 without extra osmoticum; triangles, growth with 0.4 M $K_{1.75}H_{1.25}PO_4$; circles, growth with 0.7 M $(NH_4)_2SO_4$.

bifunctional enzyme which oxidizes L-proline to L-glutamate in two steps, encoded by the putA gene (Ratzkin and Roth, 1978). We determined the growth rate of a $putA^-$ strain in high concentration of NaCl with varying concentrations of L-proline and found it to be as stimulatory to the mutant as to the wild type (Figure 6), indicating that it is not necessary to convert L-proline to L-glutamate to elicit the stimulatory response.

Selection of L-Proline Over-Producing Mutants

Since L-proline, supplied exogenously in low concentrations can overcome some of the inhibition due to high osmolarity, we reasoned that mutants which produce elevated levels of the compound might grow faster than the parental strain in media of high osmolarity. In general, bacterial mutants that over-produce a given metabolite can be obtained by selecting derivatives that can grow in the presence of a toxic analogue of the metabolite in question (Umbarger, 1978). L-proline over-producing strains of E. coli and

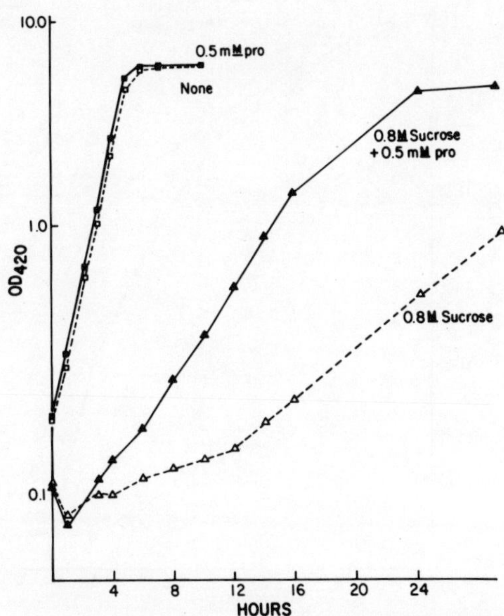

Fig. 3. The effect of L-proline on the growth of Salmonella typhimurium LT-2 in the presence of sucrose. Optical density of the cultures is plotted as a function of incubation time. Broken lines and open symbols, growth without L-proline; solid lines and closed symbols, growth with 0.5 mM L-proline; squares, growth without sucrose; triangles, growth with 0.8 M sucrose.

Salmonella typhimurium have been obtained by selection for growth in the presence of L-azetidine-2-carboxylate (Condamine, 1971; Ratzkin et al., 1978), a four carbon analogue of L-proline, which is poisonous becuase it is incorporated in place of L-proline into proteins (Grant et al., 1975).

A complication on this selection scheme arises from the fact that there is a second, and much more likely, mutation which also confers resistance to the analogue: the loss of the transport system taking up the analogue (which normally functions in the transport of L-proline (Ratzkin et al., 1978)). However, mutants of this kind can be distinguished from the class of L-proline over-producers on the basis of two characteristics. First, because they have suffered a defect in L-proline transport, they are unable to utilize that imino acid as the sole source of carbon or nitrogen. Second, the chromosomal location of a set of mutations leading to L-proline overproduction isolated previously was shown to be closely linked to the cluster of genes called proB proA which code for the first and second enzymes of L-proline biosynthesis, and which are at some distance removed from putP, the gene which codes for

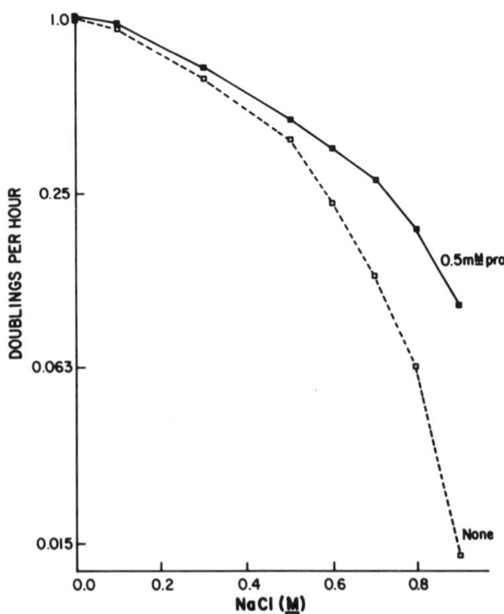

Fig. 4. The effect of L-proline on the growth rate of Salmonella typhimurium LT-2 in the presence of NaCl. Growth rate is plotted as a function of NaCl concentration. Broken line and open squares, growth without L-proline; solid line and closed squares, growth with 0.5 mM L-proline.

the L-proline (and L-azetidine-2-carboxylate) transport system (Sanderson and Hartman, 1978). We have made use of both of these characteristics in the isolation of the desired mutants.

In order to facilitate the isolation of these mutants, the parental strain employed was JL2468, and L-leucine auxotroph with a chromosomal deletion of the entire proA and proB genes. In addition, the strain harbored the extrachromosomal DNA element, F'128, which is a self replicating, transmissible plasmid carrying the wild type proAB genes derived from E. coli K12. The reason for using this strain was that, because the F' plasmid could be readily transferred by conjugation to other strains, it facilitated the mapping of mutations conferring L-azetidine-2-carboxylate resistance that might be localized in the neighborhood of the proAB genes.

The selection was carried out by plating approximately 10^8 cells of JL2468, on glucose minimal medium supplemented with L-leucine and 0.25, or 0.5, or 1.0 mM L-azetidine-2-carboxylate. Mutants resistant to the analogue appeared, regardless of the analogue concentration, at the order of magnitude of 1 per 10^5 cells plated. The colonies which grew up were replica plated to a lawn on 10^8 cells of TR3290 (del (proAB)) spread on plates containing

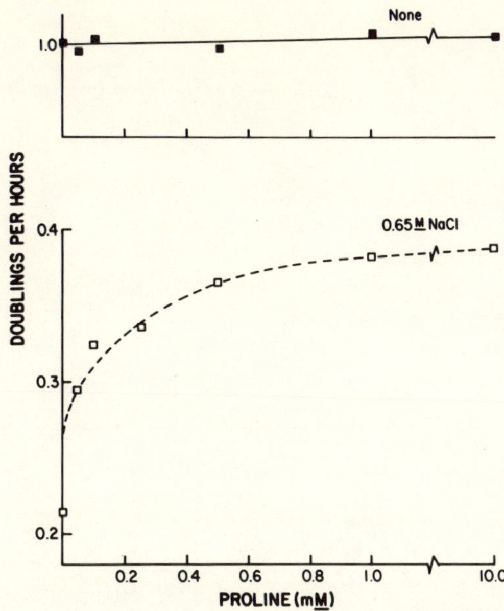

Fig. 5. The effect of L-proline on the growth rate of Salmonella typhimurium LT-2 in the presence of NaCl. Growth rate is plotted as a function of L-proline concentration at a constant concentration of NaCl. Solid line and closed squares, growth without NaCl; broken line and open squares, with 0.65 M NaCl.

glucose and L-tryptophan, L-methionine, and L-azetidine-2-carboxylate at the same concentration as was used in the first step of the selection. This medium was permissive neither to TR3290 (because of the presence of L-azetidine-2-carboxylate and the absence of L-proline) nor to the mutant derivatives of JL2468 (because of the absence of L-leucine). But, any derivatives of JL2468, which have acquired mutations located in the proAB region conferring analogue resistance could be recognized easily because they would be able to transfer the mutation to TR3290 via the F', and thus enable the progeny of the mating to grow under the replica image of the original colony. In the initial screening, a total of 119 mutants with the desired phenotype were spotted in this manner. They were recovered from the original plate on which they were selected, and tested again for the ability to transfer the L-azetidine-2-carboxylate resistant phenotype to ProAB47, another strain with a deletion running through the proAB genes. The phenotype bred true for only 34 of the strains: 7 of which were selected for resistance to 0.25 mM L-azetidine-2-carboxylate, 17 for resistance to 0.5 mM, and 10 for resistance to 1.0 mM. These 34 were tested for their ability to grow on L-proline as sole source of carbon, and all were able to do so, lending a further corroboration to the fact that the

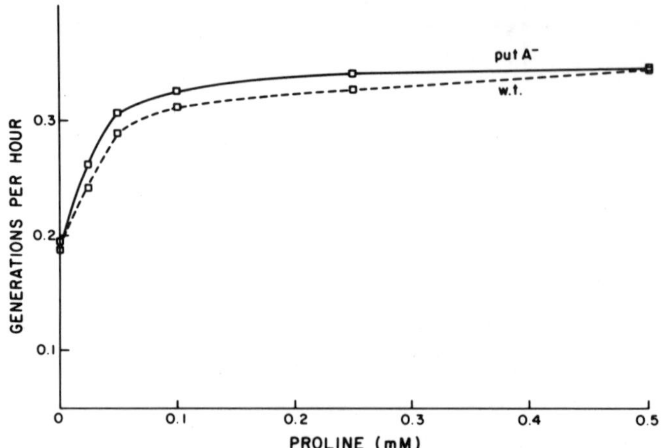

Fig. 6. The effect of L-proline on the growth rate of Salmonella typhimurium LT-2 and TC106 (putA$^-$) in the presence of NaCl. Growth rate is plotted as a function of L-proline concentration in the presence of 0.65 M NaCl. Broken line, strain LT-2; solid line, TC106.

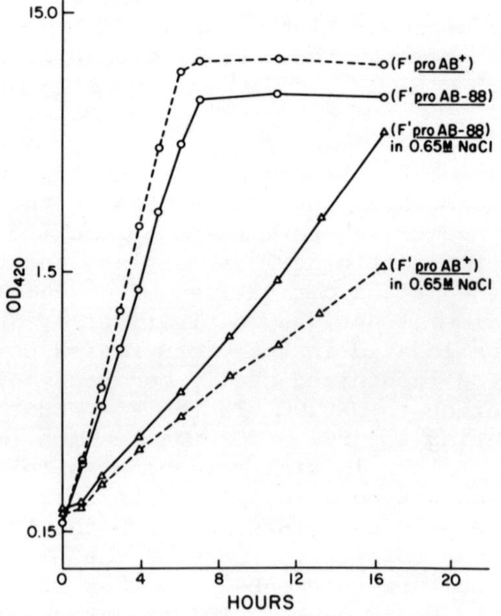

Fig. 7. The effect of the pro-88 allele on the growth of Salmonella typhimurium. Broken lines, growth of strain TC128 (del (proAB) 47/F'proA$^+$B$^+$); solid lines, TC126 (del (proAB) 47/F'pro-88); circles, in the absence of NaCl; triangels, in the presence of 0.65 M NaCl.

mutation to analogue resistance was not in the putP gene.

L-Proline Over-Production Results in Enhanced Osmotolerance

The 34 mutants thus culled were streaked to single colonies on glucose, L-leucine plates containing 0.65 M NaCl and their growth compared with that of JL2468. In this initial test, there were six strains which appeared to give larger colonies than JL2468: four, named TC81, TC82, TC83, and TC84, having been selected for resistance to 1.0 mM L-azetidine-2-carboxylate, and two, named TC86 and TC88, selected for resistance to 0.5 mM concentration of the analogue. Subsequent screening, as judged by colony sizes on plates containing 0.65 M NaCl indicated that TC81, TC82, TC83, and TC84 grew marginally faster than JL2468 and TC86 slight faster. Only TC88 gave rise to colonies that were markedly bigger than those of JL2468. The growth rates of these strains were checked in liquid media in the presence of 0.65 M NaCl, and they were in accord with the NaCl tolerance as estimated from growth on plates: the growth rate of TC88 being 0.41 generations per hour, of TC86 0.26, of TC81 0.24, and those of the others were the same as that of JL2468: 0.22. Thus, only one of the L-azetidine-2-carboxylate resistant mutants, namely TC88, is notably more NaCl-tolerant than the parental strain, while the others have only very slight (if any) increase in growth rate in the presence of NaCl. In order to show that TC88 indeed overproduced L-proline, we have grown it to saturation in glucose, L-leucine minimal medium and added aliquots of the sterilized cell supernatant, to cultures in glucose minimal medium of a proC⁻ mutant of Salmonella typhimurium. The latter mutant, being blocked in the last step of L-proline biosynthesis (Sanderson and Hartman, 1978) specifically requires L-proline. Judged by the cell yield of this mutant that were supported by aliquots of the culture medium of TC88, TC88 excreted L-proline, so that its final concentration in the cell supernatant was at least 0.2 mM, while aliquots of the culture medium of the parental strain JL2468, treated in like manner caused no detectable growth of the proC⁻ auxotroph.

The pro-88 allele, isolated in TC88 as the mutation conferring increased NaCl tolerance, was transferred by F' mediated conjugation from its original genetic background (derived from Salmonella typhimurium LT-7) into TC117 (a derivative of LT-2). The growth rate of the resultant strain, TC126, was compared with that of an isogenic strain TC128, which carried the parental F'128 with the wild type pro genes. As can be seen from the curves in Figure 7, the pro-88 mutation does indeed confer a major growth advantage in the presence of 0.65 M NaCl: the growth rate of TC126 was 0.31 generation per hour, as compared with the value of 0.18 generation per hour determined for TC128. Thus, this result shows that it is possible to isolate rather simple spontaneous mutations conferring increased NaCl tolerance, which map in the proAB region, and which can be transferred, with full phenotypic consequences between strains of Salmonella.

DISCUSSION

The Phenomenon of Growth Rate Stimulation by L-Proline in Media of High Osmolarity

As we have seen, whether the inhibition of the growth rate of *Salmonella typhimurium* is caused by high concentrations of sucrose, sodium chloride, ammonium sulfate, or potassium phosphate, L-proline can reverse, though not completely, the inhibition. Since this effect of L-proline is general, irrespective of the solute used, we propose that it counteracts, by an unknown mechanims the inhibition due to the reduced water activity in the growth medium. That L-proline cannot restore the growth rate fully to its uninhibited value might be because of a number of reasons such as the maximal levels to which L-proline is concentrated by the cells, or that each compound might have multiple inhibitory effects in addition to lowering the water activity.

This ability to stimulate growth in media of high osmolarity is specific to L-proline because none of the 19 common L-amino acids were capable of eliciting a response of the same magnitude, though other substances besides amino acids were not tested. The notion that stimulation is unique to L-proline is further buttressed by the fact that it is found in the *putA*⁻ mutant (which cannot convert L-proline to L-glutamate) implying that catabolism of the imino acid is not necessary for the growth stimulation.

The Isolation of Mutations Conferring Increased NaCl Tolerance

In order to obtain an independent confirmation that elevation of the intracellular levels of L-proline indeed confers increased osmotolerance, we have set out to obtain mutants which produced abnormally high levels of L-proline. The selection was for strains that have become resistant to L-azetidine-2-carboxylate, a toxic analogue of L-proline. We have found a number of mutants with this phenotype, due to mutations linked to *proAB*, the genes coding for the first two enzymes of the L-proline biosynthetic pathway. The mutations have not been characterized extensively, but we have presumed that they are alterations in the normal regulation of the overall activity of the L-proline pathway. Our rationale for the selection of mutants with increased osmotolerance was confirmed because we have found at least two which grow faster on high concentrations of NaCl than the parental strain, one of which, TC88, has a 1.8-fold increase in growth rate in the presence of 0.65 *M* NaCl.

Mutants which have become resistant to L-proline analogues because of defects in the normal regulation of L-proline biosynthesis have been isolated previously (Condamine, 1971; Ratzkin et al., 1978) and in some the alterations have been characterized to lead to loss of feedback inhibition by L-proline of L-glutamate kinase, the first enzyme of the L-proline biosynthetic pathway (Condamine, 1971). It is curious that of 34 (not necessarily independent) mutants having the property of L-azetidine-2-carboxylate resistance due to a mutation linked to the proAB genes, only two are more NaCl-tolerant than the parental strain, and pronounced NaCl tolerance is exhibited by only one of those two. It may be that all the mutations affect the same gene and the two which result in increased salt tolerance are extreme forms of the alteration; or it may be that several genes are involved in the regulation of L-proline biosynthesis and the differences in phenotypes might be due to the fact that different genes have been altered. Not much is known about the details of the regulation of L-proline biosynthesis, and regulatory mutants have not been characterized extensively. The proAB region present on the F'128 used for this work is derived from the *E. coli* K12 chromosome, and we intend to transfer the putative regulatory mutants back into their original host, to test if they confer NaCl tolerance to that organism also, and to characterize the mutations in terms of genetic complementation and mapping.

What Is The Function of L-Proline During Osmotic Stress?

The mechanism of the regulation of the intracellular osmotic tension in bacteria as a function of the extracellular osmolarity is not very well understood. There are reports that in salt stressed bacteria of a wide variety of species there is an enhanced accumulation of L-proline, in some cases and L-glutamate in others. Some of these studies were carried out with the bacteria grown in complex media (Measures, 1975; Koujima et al., 1978), thus making it impossible to interpret whether in these cases the enhancement of the L-proline concentration or L-glutamate is due to an increase in the net rate of biosynthesis of the two compounds or to an enhancement of transport. The levels of free amino acids were also examined in a number of species of bacteria grown in media of varying osmolarity in the absence of exogenous amino acids, but the results are complex. The one simple generalization that can be made is that as the external osmolarity is increased the concentration of one predominant amino acid is increased, which is L-glutamate in some species and L-proline in others (and in a limited number of examples, γ-amino butyric acid or a combination of all three). So, for example, with increased external osmolarity there are pronounced increases in the intracellular concentration of L-glutamate in *Beneckea harvey* (Makemson and Hastings, 1979), in *Klebsiella* (formerly *Aerobacter*) *aerogenes*, in *Bacillus megaterium* and *Bacillus polymyxa* (Tempest et al., 1972), whereas the increases of L-proline are most pronounced in several uncharacterized Gram-negative marine bacteria and in *Bacillus subtilis* var. Niger (Brown and Stanely, 1972). As we said earlier, one rationalization put forward to account for the increase in the concentration of these compounds is that they serve to balance the internal osmolarity against that of the exterior.

Since there are bacteria in which the intracellular levels of free L-proline do not vary greatly with increasing external osmolarity, L-proline does not seem to act universally as the osmotic balance, and in these organisms L-glutamate, or other substances, such as inorganic ions (Epstein and Schultz, 1965), might fill the role. There is an alternate explanation first suggested by Schobert (1977a,b,c) for the function of proline in osmoregulation: that it might have special effects on water structure and protein solubility under conditions of lowered water activity. The details of this proposed interaction have not been extensively described, but Schobert and Tschesche (1978) argued that because of the very high solubility of L-proline (1 mole per 3.94 moles of water at 25°) it must have a high affinity for water molecules and perhaps it might enhance the solubility of proteins under conditions of reduced water activity, by its effects on water structure. Consequently, the stimulation L-proline exerts on the growth rate of *Salmonella typhimurium* in media of high osmolarity might be interpreted to be mediated through this mechanism.

Prospects

The experimental results presented here are rather preliminary characterizations of the phenomenon that L-proline can counteract to a limited extent the inhibitory effect of high osmolarity in *Salmonella typhimurium*. Many important questions remain unanswered. For example, we need to determine how the concentration of free amino acids in *Salmonella typhimurium* depend on external osmolarity. There is one trivial possibility we have not yet mentioned that needs to be examined: perhaps there is a step in the L-proline biosynthetic pathway, somewhere between L-glutamate and charged L-prolyl-tRNA which is catalyzed by an enzyme which is the single most osmosensitive enzyme in the organism, and stimulation by L-proline is observed because high concentrations of the imino acid (whether concentrated from the medium or synthesized at abnormally high levels on account of regulatory mutations) could overcome or circumvent the block. We shall attempt to settle this point by measuring the extent of L-prolyl-tRNA charging in cells grown in media of high osmolarity. Since we do not know why all the mutations which resulted in L-azetidine-2-carboxylate resistance due to alterations linked to *proAB* did not all equally cause enhanced NaCl tolerance, it is necessary to isolate additional ones to see if it is possible to obtain strains with even greater NaCl tolerance, and also to elucidate the genetic differences that could account for the differences in phenotypes.

Perhaps, the most exciting, though further removed prospects, are to attempt to apply the methodology that is developed in the examination of the problem of osmoregulation in Salmonella typhimurium to other organisms of more practical value. It seems reasonable to select salt tolerant mutants of other bacteria of agricultural or industrial importance by selecting derivatives resistant to L-proline analogues. Another approach might be to transfer mutant pro alleles conferring analogue resistance from enteric bacteria into other species. Finally, the methodology of obtaining organisms with increased osmotolerance by selecting derivatives that are resistant to L-proline analogues and thus overproduce the imino acid might be applicable to cultured plant cells, which then might be regenerated into intact plants with increased salt tolerance.

ACKNOWLEDGMENTS

We thank Dr. R. C. Valentine for his encouragement and advice. This research was supported by the National Science Foundation under Grant No. PFR 77-07301. Any opinions, findings, and conclusions or recommendations expressed in this publication are those of the author and do not necessarily reflect the views of the National Science Foundation.

REFERENCES

Anderson, R. R., Menzel, R., and Wood, J. M., 1979, Biochemistry and regulation of a second L-proline transport system in Salmonella typhimurium, (manuscript in preparation).

Britten, R. J. and McClure, F. T., 1962, The amino acid pool in Escherichia coli, Bacteriol. Rev., 26:292.

Brown, C. M. and Stanley, S. O., 1972, Environment-mediated changes in the cellular content of the "pool" constituents and their associated changes in cell physiology, J. Appl. Chem. Biotechnol., 22:363.

Christian, J. H. B., 1955a, The influence of nutrition on the water relations of Salmonella orianeneburg, Aust. J. Biol. Sci., 8:75.

Christian, J. H. B., 1955b, The water relations of growth and respiration of Salmonella orianeneburg at 30°, Aust. J. Biol. Sci., 8:490.

Cohen, G. N. and Rickenberg, H. V., 1956, Concentration specifique reverisble des amino acides chez E. coli, Ann. Inst. Pasteur, Paris, 91:693.

Condamine, H., 1971, Sur la régulation de la production de proline chez E. coli, Ann. Inst. Pasteur, Paris, 120:126.

Epstein, W. and Schultz, S. G., 1965, Cation transport in Escherichia coli. V. Regulation of cation content, J. Gen. Physiol., 49:221.

Filutowicz, M., Ciésla, Z., and Klopotowski, T., 1979, Interference of azide with cysteine biosynthesis in Salmonella typhimurium, J. Gen. Microbiol., 113:45.

Grant, M. M., Brown, A. S., Corwin, L. M., Troxler, R. F., and

Franzblau, C., 1975, Effect of L-azetidine-2-carboxylic acid on growth and proline metabolism in *Escherichia coli*, *Biochem. Biophys. Acta*, 404:180.

Johnston, H. M. and Roth, J. R., 1979, Histidine mutants requiring adenine: selection of mutants with reduced *hisG* expression in *Salmonella typhimurium*, *Genetics*, 91:1.

Kohno, T. and Roth, J., 1979, Electrolyte effects on the activity of mutant enzymes *in vivo* and *in vitro*, *Biochem.*, 18:1386.

Koujima, I., Hayashi, H., Tomochika, K., Okabe, A., and Kanemasa, Y., 1978, Adaptional change in proline and water content of *Staphylococcus aureus* after alteration of environmental salt concentration, *Appl. and Env. Microbiol.*, 35:467.

Makemson, J. C. and Hastings, J. W., 1979, Glutamate functions in osmoregulation in a marine bacterium, *Appl. and Env. Microbiol.*, 38:178.

Measures, J. C., 1975, Role of amino acids in osmoregulation of nonhalophylic bacteria, *Nature*, 257:398.

Ratzkin, B., Grabnar, M., and Roth, J., 1978, Regulation of the major proline permease gene of *Salmonella typhimurium*, *J. Bacteriol.*, 133:737.

Ratzkin, B. and Roth, J., 1978, Cluster of genes controlling proline degradation in *Salmonella typhimurium*, *J. Bacteriol.*, 133:744.

Sanderson, K. E. and Hartman, P. E., 1978, Linkage map of *Salmonella typhimurium*, *Microbiol. Rev.*, 43:471.

Schobert, B., 1977a, The influence of water stress on the metabolism of diatoms. II. Proline accumulation under different conditions of stress and light, *Z. Pflanzenphysiol.*, 85:451.

Schobert, B., 1977b, The influence of water stress on the metabolism of diatoms. III. The effect of different nitrogen sources on proline accumulation, *Z. Pflanzenphysiol.*, 85:463.

Schobert, B., 1977c, Is there an osmotic regulatory mechanism in algae and higher plants?, *J. Theor. Biol.*, 68:17.

Schobert, B. and Tschesche, H., 1978, Unusual properties of proline and its interaction with proteins, *Biochem. Biophys. Acta*, 241:270.

Tempest, D. W., Meers, J. L., and Brown, C., 1970, Influence of environment on the content and composition of microbial free amino acid pools, *J. Gen. Microbiol.*, 64:171.

Umbarger, H. E., 1978, Amino acid bioxynthesis and its regulation, *Ann. Rev. Biochem.*, 47:533.

EFFECT OF ELECTROLYTES ON GROWTH OF MUTANT BACTERIA

T. Kohno, M. Schmid, and J. R. Roth

Department of Biology
University of Utah
Salt Lake City, Utah 84112

In many microbial systems, it has been observed that the phenotypes of certain mutants can be corrected by increasing the electrolyte concentration (or in some cases merely the osmotic strength) of their growth medium. The basis of these effects is not fully understood, but an understanding of these phenomena promises to shed light on how cells respond to such enviornmental stress.

We have studied this phenomenon using histidine-requiring (*his*) auxotrophs of the bacterium *Salmonella typhimurium* (Kohno and Roth, 1979). This study led us to some general conclusions on the reasons for salt correctability. More recently we have begun to study the effect of salt on some regulatory mutants of the *his* operon (Schmid and Roth, unpublished results). These new results suggest an alternative interpretation of the earlier conclusions. Here we would like to review both sets of results and discuss in a speculative way what they might suggest about effects of salt on bacteria. We will suggest the possibility that some of the effects of salt on cells may be due to reduction in the rate of intracellular proteolysis.

Work on *his* auxotrophs (Kohno and Roth, 1979) led to the following basic observations:

(1) All temperature-sensitive mutants tested show salt-correctability.

(2) Both heat- and cold-sensitive mutants are correctable.

(3) Several neutral salts were effective with NaCl and KCl showing the strongest effect (0.2 M was optimal).

(4) Neither divalent cations ($MgCl_2$, $CaCl_2$) nor purely osmotic agents (e.g., sucrose) showed correcting activity.

(5) Salt correctable, temperature-sensitive mutations were found in many different genes.

(6) Correction does not seem to be due to induction of misreading of the mutant codons (informational correction).

(7) Mutant enzyme (from cells grown at a permissive temperature) is temperature-sensitive *in vitro*.

(8) The temperature-sensitivity of mutant enzyme can be corrected *in vitro* by exposure to electrolytes.

These results led to the conclusion that salt correction of mutant phenotypes occurred by direct interaction of an electrolyte and the mutant protein. It was assumed that intracellular electrolyte levels were being altered and that these electrolytes permitted the mutant enzyme to assume an active conformation.

However, this conclusion did not easily accommodate several other bits of data. Each problem could be rationalized away but in aggregate they left some nagging doubts. The problems are as follows:

(1) If increased intracellular levels affect protein folding, one would expect to find some mutant enzymes whose folding is impaired by salts. This type of mutant would be auxotrophic only in the presence of high salt medium. Despite vigorous efforts, no such mutants were recovered (Hoppe and Roth, unpublished results).

(2) Both heat-sensitive and cold-sensitive mutants are corrected. One would expect these types of mutant enzymes to have very different structural defects, yet both show salt correctability.

(3) Although salt-correctable, heat-sensitive mutations were found in many different *his* genes, all the mutations in any one gene fall into one or two small regions of the gene ("hot-spots"). This is in contrast to temperature-sensitive mutations in systems other than *his*; usually heat-sensitive mutations are scattered widely throughout the gene.

(4) A higher temperature was required to inactivate the mutant proteins *in vitro* and *in vivo*.

(5) A higher salt concentration was required to correct the mutant enzymes in vitro than was needed in vivo.

(6) Some salts (e.g., $(NH_4)_2SO_4$) correct the mutant defect well in vitro but poorly in vivo. This could be due to permeability differences or it could suggest that the in vitro and in vivo results have different bases. These problems will be discussed later. First, we would like to consider the very different effects of salt on a different sort of bacterial mutant.

Recently we noticed that salt is extremely toxic to histidine regulatory mutants. These strains grow with high constitutive levels of all of the histidine biosynthetic enzymes (Johnston et al., 1980). It has been shown earlier that such mutants (due to their high level of the hisH and hisF enzymes) show a complex array of phenotypes including temperature sensitivity and growth inhibition by adenine (Murray and Hartman, 1972). While these defects are not understood in detail they seem to involve an interaction of the hisH and F proteins with adenine metabolism. (Constitutive strains may be made resistant to the salt toxicity by mutations which lower the hisH or F levels or by mutations in the purine genes purB and purH.) Nothing in the known physiology of the constitutive regulatory mutants suggests a reason for their acute sensitivity to salt.

Another sort of mutant with acute sensitivity to salt has recently been described by Sherman and co-workers (Singh, 1977; pers. comm.). Recently they have found that growth of yeast strains carrying nonsense suppressors is strongly inhibited by addition of salt to the growth medium. These suppressor mutations are known to act by allowing the cell to read through certain termination codons. Nothing in their mode of action suggests a direct involvement of these mutations with osmotic regulation.

Thus, three separate phenomena involving salt effects on microbial mutants are at hand: correction of auxotrophic phenotypes, toxicity to his regulatory mutants and toxicity to strains carrying nonsense supressor mutations. We have tried to find a unified explanation for these three distinct phenomena. One possibility now seems attractive.

The basic suggestion is that high concentrations of exogenous salt may inhibit intracellular protein degradation. This could occur in a variety of ways including inhibition of membrane-associated proteolytic enzymes such as those studied by Goldberg and co-workers (reviewed by Miller, 1975). Below we discuss how this suggestion might account for the three basic phenomena under consideration.

Temperature-sensitive mutant enzymes must undergo some structural rearrangement at the non-permissive temperature. This could make them vulnerable to proteolysis. Considerable data already support this rapid degradation of mutant enzymes (reviewed by Miller, 1975). Blocking proteolysis might extend the half-life of mutant enzymes and result in phenotypic correction. The possibility is especially attractive for *his* mutants. These mutants must lose virtually all function at high temperature; leaky *his* mutants escape detection since operon derepression compensates for slight losses of function when the supply of histidine limits growth. Thus we suggest that many temperature sensitive mutants owe their loss of function, at least in part, to proteolysis. Salt may slow proteolysis and thereby restore some enzyme function.

The *his* constitutive mutants have a marginal growth defect due to overproduction of the *hisH* and *F* enzymes. Proteolysis may relieve the growth defect somewhat by limiting the levels of these enzymes in the cell. If salt slows proteolysis, the levels of *hisH* and *F* may increase to a toxic level. This may be the cause of the acute salt toxicity seen in *his* regulatory mutants.

Nonsense suppressors cause the production of "read through" proteins when they permit protein synthesis to continue across the normal termination codons. These longer read-through proteins will be made for many genes in cells carrying a suppressor. Proteolysis may be essential to survival of such cells. Proteolysis can degrade these deformed and possibly toxic read-through proteins; proteolysis may also trim the excess amino acids from these proteins and restore a functional gene product. If this is so, it is clear that inhibition of proteolysis would be deleterious to such cells. Therefore we suggest that the salt sensitivity of suppressor strains found by Sherman may be due to salt inhibition of proteolysis.

In summary, three phenomena suggest that exogenous salt can affect the conditions prevailing inside cells. These salt effects occur despite the cellular mechanisms for osmoregulation (or perhaps because of it). The inhibition of proteolysis, which we propose speculatively, may even be a part of the cellular defenses against salt stress. If changes of intracellular conditions cause minor unfolding of many proteins, proteolysis may be a dangerous activity. Shutting down protein degradation may allow proteins to persist and even function under extreme conditions which affect their conformation. Thus, mutants such as those described here should be useful tools in a genetic approach to osmoregulation since they provide a means of scoring the cell's response to salt stress.

REFERENCES

Johnston, H. M., Barnes, W., Chumley, F., Bossi, L., and Roth, J. R., 1980, Model for regulation of the histidine operon of *Salmonella typhimurium*, *Proc. Natl. Acad. Sci. USA*, 77:508.

Kohno, T. and Roth, J. R., 1979, Electrolyte effects on the activity of mutant enzymes *in vivo* and *in vitro*, *Biochemistry*, 18:1386.

Miller, C. G., 1975, Proteases and peptidases of *E. coli* and *Salmonella typhimurium*, *Ann. Rev. Microbiol.*, 29:485.

Murray, M. L. and Hartman, P. E., 1972, Overproduction of *hisH* and *hisF* gene products leads to inhibition of cell division in *Salmonella*, *Can. J. Microbiol.*, 18:671.

Provtz, W. F. and Goldberg, A. L., 1972, Effects of protease inhibitors on protein breakdown in *E. coli*, *J. Biol. Chem.*, 247:3341.

Singh, A., 1977, Nonsense suppressors of yeast cause osmotic-sensitive growth, *Proc. Natl. Acad. Sci. USA*, 74:305.

β-GALACTOSIDASE FROM OSMOTIC REMEDIAL LACTOSE UTILIZATION MUTANTS OF *E. COLI*

R. T. Vinopal, S. A. Wartell, and K. S. Kolowsky

Microbiology Section, U-44
Biological Sciences Group
University of Connecticut
Storrs, Connecticut 06268

INTRODUCTION

Retrieval of genetic information for synthesis of an enzyme is usually thought of in terms of transfer of sequence information -- transcription of the base sequence of the gene into mRNA and translation of mRNA into the amino acid sequence of the polypeptide chain. The final expression of the gene, folding of the polypeptide to form the biologically active globular protein, is considered to result automatically from the sequence of amino acids. An information specialist will point out that the amino acid sequence does not possess enough information to specify the precisely folded shape of the active enzyme. The additional information comes from the specification of the environment in which the polypeptide folds. Anfinsen (1973) describes the thermodymanic hypothesis for protein folding in this way: "This hypothesis states that the three-dimensional structure of a native protein in its normal physiological milieu (solvent, pH, ionic strength, presence of other components such as metal ions or prosthetic groups, temperature, and other) is the one in which the Gibbs free energy of the whole system is the lowest; that is, that the native conformation is determined by the totality of interatomic interactions and hence by the amino acid sequence in a given environment. In terms of natural selection through the 'design' of macromolecules during evolution, this idea emphasizes the fact that a protein molecule only makes stable structural sense when it exists under conditions similar to those for which it was selected -- the so-called physiological state."

If a protein makes sense only within a certain range of folding environments, then, in the case of a cytoplasmic enzyme, the organism producing the enzyme must maintain a cytoplasm within this range to give meaning to the gene coding for the enzyme. This does not necessarily mean that the composition of the cytoplasm is regulated in the context of making genes meaningful. Regulation of the cytoplasm is required in any event for coordinated metabolism (Atkinson, 1969), and denatured proteins will often fold to active conformation _in vitro_ in environments differing in a number of ways from the native cytoplasm (Anfinsen, 1973). It then seems possible that any conceivable cytoplasm -- so long as it is aqueous, with a few ions and not too extreme a pH -- would provide for correct folding. Indeed, a surprisingly large proportion of mutations resulting in a single amino acid substitution are "neutral", without obvious effect on phenotype (Langridge, 1974a), suggesting that the cytoplasmic environment provides so much additional information for polypeptide folding that much amino acid sequence information is redundant.

In a special class of mutant microorganisms, however, the properties of the cytoplasm do seem to affect the meaning of the altered gene. As a result of a lesion in the gene for a cytoplasmic enzyme, the ability of these mutants to catalyze a reaction -- typically to synthesize an amino acid or utilize a sugar -- depends on where within the normal environmental range of the parent strain they are grown, in terms of the temperature (Ingraham, 1973; Langridge, 1968), or pH (Colb and Shapiro, 1977; Hawthorne and Friis, 1964), or partial pressure of carbon dioxide (Roberts and Charles, 1970), or ionic or osmotic concentration (Hawthorne and Friis, 1964; Kohno and Roth, 1979) of the medium. For at least some of these conditional mutants the effect of growth conditions may be inferred to be due to an effect on folding of the mutant polypeptide chain and/or on self-assembly of folded polypeptide subunits into quaternary structure. A number of examples have been reported among osmotic remedial mutants, strains having a mutant phenotype that is corrected by growth media with an osmotic concentration near the maximum tolerated by the parent.

Osmotic remedial mutants were first described in yeast (Hawthorne and Friis, 1964), and have since been studied in other fungi (Martin and DeBusk, 1975) and in bacteria (Kohno and Roth, 1979; Russell, 1972). More than a tenth of randomly tested amino acid auxotrophic mutants in yeast (Hawthorne and Friis, 1964) and in _Escherichia coli_ (unpublished) are osmotic remedial. The molecular basis of osmotic repair is not known for many osmotic remedial mutants (Fincham and Baron, 1977), but a number of generalizations have been made (Figure 1): osmotic remedial strains carry missense mutations (rarely nonsense mutations mapping close to the end of the gene or affecting one enzyme in a multienzyme complex) (Singh and Sherman, 1975), many of which also result in a temperature-sensitive phenotype (Hawthorne and Friis, 1964; Kohno and Roth, 1979). As any osmotically active solute increases in

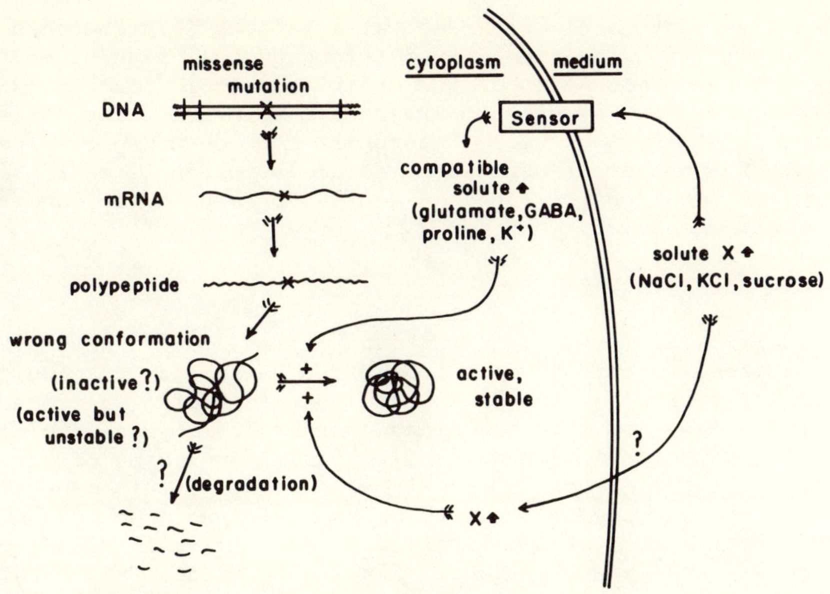

Fig. 1. Hypothetical osmotic remedial mutation in an enteric bacterium.

concentration in the growth medium (Hawthorne and Friis, 1964; Russell, 1972), the cell compensates to prevent plasmolysis by increasing the cytoplasmic concentration of a "compatible" solute or solutes (Brown, 1976) which do not grossly affect normal macromolecular conformation but somehow correct the folding (or subunit assembly -- most osmotic remedial mutations occur in enzymes with subunit structure) (Hawthorne and Friis, 1964; Fincham and Baron, 1977) of the mutant enzyme. In bacteria the compatible solutes are various combinations of proline, glutamic acid, γ-amino butyric acid and K^+ (Measures, 1973), while proline or various polyhydric alcohols serve this function in eukaryotic microorganisms (Brown, 1976; Schobert, 1977). In some cases permeation of a solute from the medium and direct interaction with the enzyme may occur (Kohno and Roth, 1979; Bassel and Douglas, 1970), but usually no enzyme activation in vitro by the osmotically active solutes used in permissive growth medium can be shown. Some osmotic remedial mutants lack the affected gene product in growth media of low osmotic concentration. This has been interpreted as being due to a block in enzyme synthesis (Bassel and Douglas, 1970), but because proteins of abnormal conformation, including some unassembled subunits, are specifically degraded in many organisms (Goldberg and St. John, 1976), it is possible that an altered enzyme is made and degraded (Kohno and Roth, 1979; Martin and DeBusk, 1975).

We are studying osmotic remedial mutations in *Escherchia coli* with the plan of using them in understanding cell mechanisms for regulating the composition of the cytoplasm and in learning whether any aspects of this regulation are needed to provide for correct folding of wild-type enzymes. We report here on the preliminary analysis of a number of osmotic remedial lactose utilization mutants, and provide both confirmations of and exceptions to the generalizations outlined above.

MATERIALS AND METHODS

Minimal medium was 007 (Clark and Manlow, 1967), with 0.2% carbon source, solidified with 1.5% Difco Bacto agar as required. Osmotically active solutes were added to this medium in the amounts indicated. Assay of β-galactosidase was by hydrolysis of o-nitrophenol-β-D-galactoside (ONPG) (Miller, 1972) and assay of lactose permease was by transport of ^{14}C-labeled methyl-1-thio-β-D-galactoside. SDS-polyacrylamide gel electrophoresis was done by a modification of the system of Laemli (Laemli, 1970; Truman and Berquist, 1976; Zipser and Bhavser, 1976). Genetic techniques and strains used for fine-structure mapping at the *lacZ* locus are those described by Miller (1972). Isopropyl-β-D-thiogalactoside (IPTG) was used to induce expression of the *lac* operon (Miller, 1972). Assay for the N-terminal portion of β-galactosidase was by use of the auto-α complementation reaction (Truman and Berquist, 1976).

RESULTS AND DISCUSSION

We expected that if osmotic remedial (OR) mutations occurred in *lacZ*, the gene for β-galactosidase, they would be relatively easy to characterize. A number of lactose non-utilizing mutants isolated after ethyl methane sulfonate mutagenesis of RT1000, a *proC* derivative of the *E. coli* K-12 strain W3110, were tested for being able to grow on lactose minimal agar with added NaCl, KCl, or sucrose. Strains able to grow on lactose agar with any of these solutes were common (Table I). Good suppression of the lactose phenotype was obtained at 1% NaCl (0.17 M) or KCl (0.13 M) or 12% sucrose (0.35 M; controls ruled out growth on contaminating glucose as the basis for stimulation). 1.5% NaCl or KCl in agar stimulated growth of OR mutants on lactose but was somewhat inhibitory to the wild-type strain.

We chose 33 independent lactose OR strains for analysis. All mutations were mapped at the lactose operon by cotransduction with *proC*. By deletion mapping the mutations of 31 strains were located within the *lacZ* gene, while two strains were lactose permease mutants (Figure 4).

The strength of osmotic repair varied with the strain. Optimal NaCl, KCl or sucrose concentrations stimulated an increase in colony

Table I. Osmotically corrected lactose utilization mutants.

	OR[a]	not OR
TS[b]	21	19
not TS[c]	40	127
	61	146

[a]Osmotic remedial; colony diameters on lactose minimal agar with added NaCl (1% or 1.5%), KCl (1% or 1.5%) or sucrose (12%) are at least two-fold greater than on lactose minimal agar without additions.

[b]Temperature sensitive; unable to utilize lactose at 42 C, but able to utilize it at 30 C.

[c]Unable to utilize lactose at any temperature.

diameter on lactose agar of from 2-fold to 8-fold. Most strains were not OR in lactose liquid minimal medium with any concentration of solute. It may be that adsorption at the agar surface lowers the thermodynamic activity of water. Selection for growth of lacZ OR strains in lactose liquid medium with salts or sucrose yielded derivatives with second-site mutations at lacZ that were strongly OR in liquid medium. For example, strain SW1035 was not OR in lactose liquid medium, but its derivatives SW1080 and SW1081 were strongly suppressed by NaCl (Table II) or by KCl or sucrose. The best osmotic suppression was seen at ca. 0.3 M NaCl or KCl and, surprisingly, at only a slightly greater concentration of sucrose, which should be only half as osmotically active on a molar basis. Liquid culture OR derivatives were much more responsive than their parent to osmotic suppression on lactose agar; SW1035 showed a 2-fold increase in colony diameter with 1% NaCl, while its derivative SW1081 showed an increase of 10-fold.

All 33 lactose OR strains were induced for the lactose operon with IPTG and assayed for β-galactosidase and lactose permease. When grown at 42 C, 26 of the strains had much less than 1% of wild-type β-galactosidase activity, as expected from the lactose negative phenotype, since even 1% activity gives appreciable growth on lactose. Two strains had more than 10% of wild-type β-galactosidase activity. These were lactose permease mutants, by mapping (Figure 4) and assay (ca. 4% wild-type permease activity). Surprisingly,

Table II. Lactose minimal liquid medium osmotic suppression; doubling time, in hours.

	(NaCl)[a]			
Strain[b]	0	1%	2%	3%
SW1035	∞	∞	∞	∞
SW1080	∞	8.8	4.4	6.4
SW1081	∞	3.6	3.5	4.4
RT1000[b]	1.0	1.0	1.5	2.5

[a]NaCl w/v added to lactose minimal medium.

[b]SW1035 is a lacZ OR mutant. SW1080 and SW1081 are second-site lacZ mutant derivatives of SW1035 selected for osmotic suppressibility in liquid medium. RT1000 is the $lacZ^+$ parent of SW1035.

five strains had between 2% and 10% of wild-type β-galactosidase activity, although they were completely negative for growth on lactose due to mutations within *lacZ*. In assays of cells induced at 30 C additional strains with too much β-galactosidase activity to account for their inability to utilize lactose were found. Our assays were for ONPG hydrolysis in the standard Z buffer, which contains a high concentration of monovalent cations (Table III). A monovalent cation is required for activity of wild-type β-galactosidase; Na^+ is most effective for hydrolysis of ONPG, K^+ for hydrolysis of lactose (Wallenfels and Weil, 1972). Osmotically stressed *E. coli* has high cellular levels of K^+ (Table III), up to 300 mM (Measures, 1973), and it was possible that some OR β-galactosidase enzymes required higher levels of monovalent cation for activity than the wild-type enzyme, and showed unexpectedly high activities *in vitro* because of activation of Z buffer. We dialyzed extracts against a Tris-based buffer to remove enough monovalent cation to reduce the activity of wild-type β-galactosidase greatly, then tested the effect of added K^+ and Na^+ (Table IV). SW1035, completely unable to grow on lactose yet having 2% of wild-type β-galactosidase activity in the standard assay, was found to have an enzyme more dependent than the wild-type on monovalent cations. The liquid medium OR derivatives of this strain, SW1080 and SW1081, are even more highly activated, with K^+, the compatible solute, being more effective than Na^+, opposite to the preference of the

Table III. Compatible solutes of E. coli and components of Z buffer.

Compatible solutes	
K^+	KCl 0.01 M
Glutamic acid	Na_2HPO_4 0.06 M
Proline	NaH_2PO_4 0.04 M
γ-amino butyric acid	$MgSO_4$ 0.001 M
	β-mercaptoethanol 0.05 M

Table IV. β-galactosidase activity of dialyzed extracts.[a]

	RT1000	SW1035	SW1080	SW1081
(Z buffer)	(100%)	(2%)	(3%)	(10%)
Tris-based[c] + addition	Activity:	1000x (ΔOD_{420}/min/mg protein)		
-	752	0	1	7
Na^+ 10^{-3} M	1811	2	2	1
10^{-2} M	2333	44	12	22
10^{-1} M	1778	96	382	194
K^+ 10^{-3} M	707	1	1	1
10^{-2} M	1264	1	20	10
10^{-1} M	1305	114	543	2280

[a]Cells grown in 1% casein hydrolysate 007 medium at 30 C, and induced with IPTG.

[b]Strain derivation in footnote b, Table II.

[c]0.1 M Tris-HCl pH 7, 1 mM $MgSO_4$, 1 mM β-mercaptoethanol; Na^+ and K^+ added as the chlorides.

of the wild-type enzyme (Table IV). The enzyme from SW1081 is even more active than the wild-type enzyme in assays with 0.1 M KCl (Table IV). Nine of the 31 lacZ mutants had β-galactosidase more than ten-fold more dependent on monovalent cations in vitro than the wild type enzyme (indicated by A in Figure 4). Some OR strains have a β-galactosidase not activated by K^+ or Na^+ alone but active in vitro in a buffer containing all of the compatible solutes of E. coli.

OR mutants described in the literature are typically altered in subunit enzymes. High salt concentrations can stabilize hydrophobic bonds, which are involved in subunit interactions. We tested the lacZ OR strains for having altered subunit interactions in the β-galactosidase tetramer by measuring the loss of activity after two hours in the presence of 1 M urea, at 37 C, a treatment not affecting the activity of the wild-type enzyme (Langridge, 1974b). Six mutants had enzymes losing from 80% to 99.9% of initial activity (indicated by U- in Figure 4), while two had enzymes with activity reproducibly more than two-fold higher after this treatment (U+ in Figure 4).

We expected lactose OR mutations to be missense, but anticipated that some might result in a gene product with a conformation that would be recognized as abnormal and lead to degradation (Zipser and Bhavser, 1976). Physical absence of β-galactosidase in OR mutants grown under non-suppressing conditions could also be due to failure to initiate synthesis, or to premature release or lability during synthesis. OR strains were tested for producing full sized β-galactosidase monomer when induced under non-suppressing conditions by subjecting cell extracts to SDS-polyacrylamide gel electrophoresis; the 135,000 MW β-galactosidase subunit makes a distinct, well-separated band on gels (Zipser and Bhavser, 1976). We found three classes of lacZ OR mutants: strains with wild-type amounts of monomer at both 30 C and 42 C, strains with no detectable monomer at one temperature (usually 42 C) and greater although usually not wild-type amounts at the other temperature, and strains with no detectable monomer at either temperature. To determine whether strains lacking monomer made full sized subunits which were rapidly degraded or failed to make subunits, we exposed cells induced with IPTG to a short pulse of high specific activity ^{14}C-leucine, then chased with unlabeled leucine and took samples for electrophoresis and autoradiography immediately and after 30 minutes. Results for representative strains of the three classes noted above are shown in Figure 2. The first class (e.g., SW1057-1) makes stable monomer, the second (e.g., SW1055-1) makes unstable monomer, and the third (e.g., SW1003-1) makes little or no monomer.

Strains producing unstable monomer (cross-hatched in Figure 4) show much larger amounts of monomer when induced in osmotically suppressing growth medium; in those strains with a temperature-

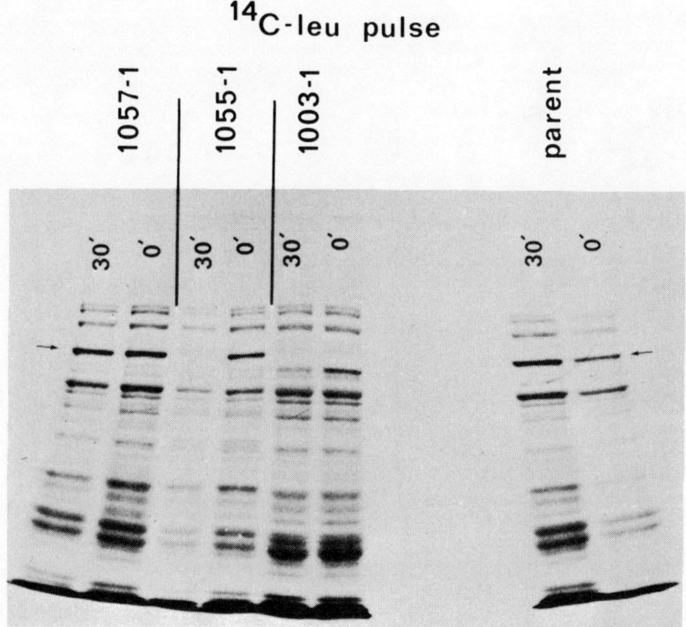

Fig. 2. Autoradiograph of an SDS-polyacrylamide gel of extracts of pulse labeled lacZ OR mutants. Cells growing in glycerol minimal medium at 42 C were induced with IPTG for 10 min, exposed to ^{14}C-leucine for 3 min, then chased with a 100-fold excess of unlabeled leucine. Samples were taken at the time of addition of unlabeled leucine (0') and after 30 min.

sensitive phenotype the relatively stable monomer produced at 30 C in non-suppressing medium is unstable after shift to 42 C, as may be shown on gels (Figure 3) or followed by enzyme assay. One temperature-sensitive strain produced β-galactosidase that lost activity and was degraded rapidly at 42 C *in vivo* but was stable and active at 42 C *in vitro*. This enzyme may be enzymatically active but abnormally shaped at 42 C.

Seven lacZ OR strains made no detectable monomer in media of low osmotic concentration at 30 C or 42 C (Table V). These mutants were strongly osmotic remedial on lactose agar, but even in growth medium with high concentrations of suppressing solutes β-galactosidase activity were only a few percent of wild-type, and only a faint band of β-galactosidase monomer was seen on gels. Synthesis of β-galactosidase was initiated in non-suppressing medium, as shown by assay of IPTG-induced cells for the N-terminal portion of the enzyme, using the "auto-α" *in vitro* complementation reaction (Table V; auto-α levels were expected to be low for SW1060-3,

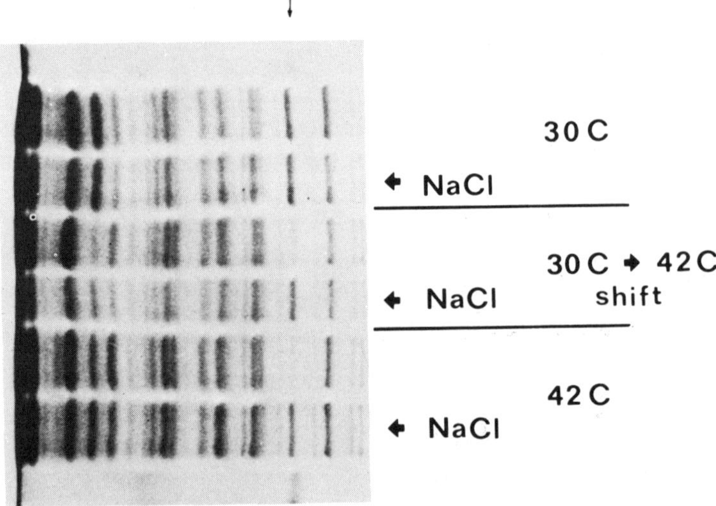

Fig. 3. Osmotic protection of unstable β-galactosidase against degradation. A culture of SW1002-3, a temperature-sensitive lacZ OR mutant, was spread to solid medium with casein hydrolysate as carbon source, IPTG as inducer, and with or without 1.0% NaCl. Agar plates were incubated for 48 hr at 30 C or at 42 C, or for 48 hr at 30 C followed by 24 hr at 42 C, during which no significant increase in cell mass occurred. Cells were scraped from plates; solubilized, and subjected to electrophoresis.

because of its map position, and levels for the other strains were probably low due to degradation of an incomplete polypeptide) (Bukhari and Zipser, 1973). The lacZ mutations in these strains were polar on the expression of lactose permease (Table V), indicating premature release of nascent polypeptide. Surprisingly, three of these strains were found to carry ochre (UAA) mutations in lacZ; the strains reverted to lactose utilization by mutation to ochre suppression, introduction of known ochre suppressors restored growth on lactose, and when the lacZ alleles were moved into an amber (UAG) suppressing genetic background reversion to growth on lactose occurred by mutation at lacZ of UAA to UAG. Known ochre mutations of bacterial genes and of bacteriophage T_4 were not osmotic remedial in our strains. The ochre lacZ OR mutations were not as polar on expression of lactose permease as expected; at

Table V. Osmotic remedial lacZ mutants lacking β-galactosidase monomer in non-suppressing medium.

Strain	Map region	Auto-α, 30 C (% wild-type)	Lactose permease, 37 C (% wild-type)	Mutation type
SW1060-3	1	8	6	?
SW1003-1	15-19	30	55	UAA
SW1012-4	15-19	21	66	UAA
SW1013-3	15-19	56	36	?
SW1018-1	15-19	10	46	?
SW1058	15-19	11	70	UAA
SW1055-2	20-26	64	49	?

their map positions typical UAA lesions reduce permease activity to 20% or less of wild-type strains (Zipser et al., 1970), while the OR strains had between 55% and 70% of wild-type permease activity (Table V). The OR ochre mutations may then occur at sites in the gene where context (neighboring base sequence) increases ambiguity in translation (Fluck et al., 1977). Ambiguity-restricting streptomycin resistance alleles (Zengel et al., 1977) reduced or eliminated the suppressing effects of salts and sucrose when introduced into the ochre lacZ OR strains.

The remaining four lacZ OR strains listed in Table V carry mutations of unknown type that apparently also cause premature termination of polypeptide synthesis. These mutations were more polar on lactose permease expression than the UAA mutations (Table V), although most were not as polar as expected for nonsense mutations at their map locations. Spontaneous revertants of these strains to growth on lactose were often temperature-sensitive lethal strains, unable to grow on any medium at 42 C. These temperature-sensitive revertants suppressed amber, ochre and UGA mutations of bacterial genes and of bacteriophage T_4, and could possibly carry ribosomal ambiguity mutations (Zimmerman et al., 1971).

It is not certain whether osmotic suppression of the ochre lacZ OR mutations is due to informational suppression resulting from osmotically-induced ambiguity of translation or is due to

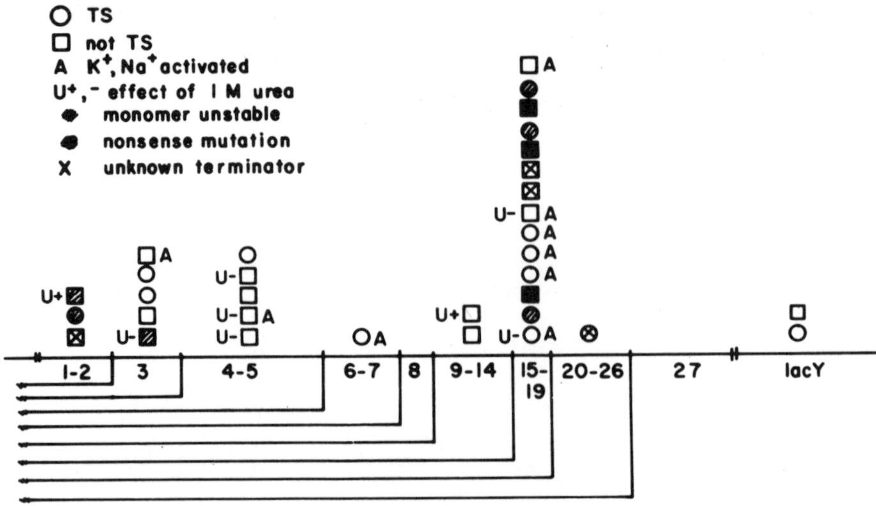

Fig. 4. Fine-structure mapping of lacZ OR mutations. Map intervals are as described (Miller, 1972; Zipser et al., 1970). TS indicates that the mutation results in a temperature-sensitive lactose utilization phenotype.

osmotic stabilization of low amounts of labile β-galactosidase produced as the result of spontaneous ambiguity of translation not resulting from osmotic effects. Preliminary evidence supports the first mechanism. We are testing to see if growth in medium of high osmotic concentration results in generalized misreading during protein synthesis. Osmotically-affected premature termination of polypeptides need not involve nonsense codons. Some of the unidentified mutations in Table V might affect the sites where wild-type β-galactosidase is often prematurely released (Manley, 1978), sites at present not understood.

SUMMARY

Mutants of Escherichia coli able to utilize lactose only in a medium of high osmotic concentration are described. Most of these mutants are altered in the lacZ gene, the structural gene for β-galactosidase. The basis of osmotic suppression of the lactose utilization phenotype differs among strains. Some osmotic remedial lacZ strains produce an enzyme with an increased requirement for monovalent cations, some an enzyme with altered subunit interactions, some an enzyme which is degraded in vivo in medium of low osmotic concentration, and some fail to produce a full-sized monomer in dilute medium, due to premature termination of the polypeptide chain.

ACKNOWLEDGMENTS

Work reported here was supported by NSF Grant PCM 77-07114. Heartfelt thanks to Jean Winters for typing this at the last moment.

REFERENCES

Anfinsen, C. B., 1973, Principles that govern the folding of protein chains, Science, 181:223.

Atkinson, D. E., 1969, Limitation of metabolic concentrations and the conservation of solvent capacity in the living cell, in: "Current Topics in Cellular Regulation," Vol. I, B. L. Horecker and E. R. Stadtman, eds., Academic Press, Inc., New York.

Bassel, J. and Douglas, H. C., 1970, Relationship between solute permeability and osmotic remediability in a galactose-negative strain of Saccharomyces cerevisiae, J. Bacteriol., 104:707.

Brown, A. D., 1976, Microbial water stress, Bacteriol. Rev., 40: 803.

Bukhari, A. I. and Zipser, D., 1973, Mutants of Escherichia coli with a defect in the degradation of nonsense fragments, Nature New Biology, 243:238.

Clark, D. J. and Maaloe, O., 1967, DNA replication and the division cycle in Escherichia coli, J. Mol. Biol., 23:99.

Colb, M. and Shapiro, A., pH-conditional mutant of Escherichia coli, Proc. Natl. Acad. Sci. USA, 74:5637.

Fincham, J. R. S., and Baron, A. J., 1977, The molecular basis of an osmotically repairable mutant of Neurospora crassa producing unstable glutamate dehydrogenase, J. Mol. Biol., 110: 627.

Fluck, M. M., Salser, W., and Epstein, R. H., 1977, The influence of the reading context upon suppression of nonsense codons, Molec. Gen. Genet., 151:137.

Goldberg, A. L. and St. John, A. C., 1976, Intracellular protein degradation in mammalian and bacterial cells: part 2, Ann. Rev. Biochem., 45:747.

Hawthorne, D. C. and Friis, J., 1964, Osmotic-remedial mutants. A new classification for nutritional mutants in yeast, Genetics, 50:829.

Ingraham, J. L., 1973, Genetic regulation of temperature responses, in: "Temperature and Life," Precht, Christopherson, Hensel, and Larcher, eds., Springer-Verlag, New York.

Kohno, T. and Roth, J., 1979, Electrolyte effects on the activity of mutant enzymes in vivo and in vitro, Biochemistry, 18:1386.

Laemli, U. K., 1970, Cleavage of structural proteins during the assembly of the head of bacteriophage T_4, Nature (London), 227:680.

Langridge, J., 1968, Thermal responses of mutant enzymes and temperature limits to growth, Molec. Gen. Genet., 103:116.

Langridge, J., 1974, Mutation spectra and the neutrality of mutations, Aust. J. Biol. Sci., 27:309.

Langridge, J., 1974, Genetic and enzymatic experiments relating to the quaternary structure of β-galactosidase, Aust. J. Biol. Sci., 27:321.

Manley, J. L., 1978, Synthesis and degradation of termination and premature-termination fragments of β-galactosidase in vitro and in vivo, J. Mol. Biol., 125:40.

Martin, C. E. and DeBusk, A. G., 1975, Temperature-sensitive, osmotic remedial mutants of Neurospora crassa: osmotic pressure induced alterations of enzyme stability, Molec. Gen. Genet., 136:31.

Measures, J. C., 1973, Role of amino acids in osmoregulation of non-halophilic bacteria, Nature (London), 257:398.

Miller, J. H., 1972, "Experiments in Molecular Genetics," Cold Spring Harbor Laboratory, New York.

Roberts, G. A. and Charles, H. P., 1970, Mutants of Neurospora crassa, Salmonella typhimurium, and Escherichia coli specifically inhibited by carbon dioxide, J. Gen. Microbiol., 63:21.

Russell, R. R. B., 1972, Temperature-sensitive osmotic remedial mutants of Escherichia coli, J. Bacteriol., 112:661.

Schobert, B., 1977, Is there an osmotic regulatory mechanism in algae and higher plants?, J. Theor. Biol., 68:17.

Singh, A. and Sherman, F., 1975, Genetic and physiological characterization of met15 mutants of Saccharomyces cerevisiae: a selective system for forward and reverse mutations, Genetics, 81:75.

Truman, P. and Berquist, P. L., 1976, Genetic and biochemical characterization of some missence mutations in the lacZ gene of Escherichia coli K-12, J. Bacteriol., 126:1063.

Wallenfels, K. and Weil, R., 1972, β-galactosidase, in: "The Enzymes," Vol. 7, P. Boyer, ed., Academic Press, New York.

Zengel, J. M., Young, R., Dennis, P. P., and Nomura, M., 1977, Role of ribosomal protein S12 in peptide chain elongation: analysis of pleiotropic, streptomycin-resistant mutants of Escherichia coli, J. Bacteriol., 129:1320.

Zimmerman, R. A., Garvin, R. T., and Gorini, L., 1971, Alteration of a 30S ribosomal protein accompanying the ram mutation in Escherichia coli, Proc. Natl. Acad. Sci. USA, 68:2263.

Zipser, D., Zabell, S., Rothman, J., Grodzicker, T., Wenk, M., and Novitski, M., 1970, Fine structure of the gradient of polarity in the Z gene of the lac operon of Escherichia coli, J. Mol. Biol., 49:251.

Zipser, D. and Bhavser, P., 1976, Missense mutations in the lacZ gene that result in degradation of β-galactosidase structural protein, J. Bacteriol., 127:1538.

SECTION II

OSMOREGULATION IN EUKARYOTIC MICROORGANISMS

OSMOREGULATION IN YEAST

A. D. Brown and Margaret Edgley

Department of Biology
University of Wollongong
Wollongong, N.S.W. 2500, Australia

INTRODUCTION

For the purposes of this article, osmoregulation will be defined as the maintenance of approximately constant cell volume and turgor pressure in the face of changing water potential. Microorganisms must accomodate the entire range of environmental water potential whereas the range is narrowed for the cells of healthy higher plants and animals by regulatory systems associated with their more complex anatomy and physiology. For this reason, cellular responses to extreme environmental conditions are best seen in microorganisms and, indeed, microorganisms are the only inhabitants of certain types of extreme environments such as hypersaline lakes.

Osmoregulation is achieved by varying the content of one or more specific solutes. At very low water potentials or water activities another factor in addition to volume or turgor maintenance becomes important. Under these conditions, solute concentrations are high and it is essential for the continuation of metabolism that the osmoregulator is not inhibitory at the prevailing concentrations. For reasons such as this the term "compatible solute" was introduced to describe those noninhibitory substances that function as osmoregulators under extreme conditions (Brown and Simpson, 1972; Brown, 1978).

The outstanding compatible solute in eukaryotes is glycerol. It has this function in yeasts, fungi and the halotolerant flagellate, Dunaliella; it has the related role of an anti-freeze in insects. Enzymological and physicochemical aspects of the interaction of glycerol and other non-electrolytes with proteins have been discussed elsewhere (Brown, 1978). We may start, then, with the premise that, even though some biological activities may persist at low water activities without an appropriate concentration of compatible solute, the full expression of biological activity is impossible without it.

In the ensuing discussion the terms "water activity" (a_w) and "water potential" (ψ) are used interchangeably when referring qualitatively to the amount of thermodynamically available water. When quantitative statements are made, water activity is used. The two are related by the equation,

$$\psi = \frac{RT \ln a_w}{\bar{V}_w} + P$$

where R is the gas constant, T the absolute temperature, $\bar{V}_w$ the partial molal volume of water, P is hydrostatic pressure and, of course, ψ is the water potential and a_w water activity. Water activity is independent of temperature in an ideal solution and, in a non-ideal solution, dependent on temperature only to the extent that temperature affects the activity coefficient of solute or solvent.

The yeasts used in the comparisons described below are *Saccharomyces rouxii*, a xerotolerant yeast and *S. cerevisiae* which is non-tolerant. *Saccharomyces cerevisiae* can grow at levels of water activity down to about 0.90 in sugars and over a narrower range in salts or polyethylene glycol (PEG, mol. wt. 200). *Saccharomyces rouxii* will span the range from approximately zero to $0.62a_w$ in some sugars but in salt its lower limit is about $0.85a_w$. In PEG the range is still smaller. Growth rates of xerotolerant strains are generally slower than those of the non-tolerant strains (Anand and Brown, 1968). For technical reasons the experimental results described in this paper were obtained, for the most part, with solutions adjusted with either NaCl or PEG 200.

Measurement and Interpretation of Osmoregulatory Response

Plant physiologists have tended to use biophysical methods for the study of cellular osmoregulation (see reviews by Cramm, 1976; Zimmermann, 1978). The techniques applied have commonly involved direct measurement of turgor pressure, cell volume and membrane potential. The volume of microorganisms, including yeast, can be determined, but not with the accuracy obtainable with larger cells. Neither turgor pressure nor membrane potential can yet be measured directly. There are good reasons, therefore, to adopt a biochemical rather than a biophysical approach to the study of osmoregulation in yeast and other microorganisms.

An understanding of the response of a microorganism to the water potential of its environment requires information in two distinct areas. They are:

1) The phenotypic characteristics of the organism when it is fully adapted to and growing at each of several levels of water potential;

2) The changes that occur in the organism when it is adapting from one water potential to another.

In a strict but rather narrow sense, the concept of osmoregulation might be confined to the second area, the transition from one environment to another. Nevertheless, there is a genuine homeostatic regulation necessary to maintain the compatible solute at a level appropriate to a particular water potential.

The transition has several distinct phases that can be readily identified. Phase 1 is the duration of the water flux that is the immediate and inevitable result of a water stress. It has a time scale of seconds and its end is marked by the cell's reaching equilibrium with the new environment, at least in respect of water potential. Phase 2 is the period during which volume, turgor pressure and normal biological activities are restored. This phase can be divided into 2a, volume/turgor recovery and 2b, biochemical readjustment. For organisms of our experience Phase 2a has a time scale of an hour or so whereas 2b is of the order of a generation time (Brown and Borowitzka, 1979). Phase 2a is the period of osmoregulation in the strict sense. Energy is needed during this phase and, indeed, must be marshalled very soon after the end of Phase 1 when intracellular conditions are very different from the final fully adapted condition and presumably are more inhibitory than at any other time during the period of adaptation. For that reason, a solute stress (decrease in water potential) is likely to be more difficult to deal with than a dilution stress (increase in water potential) since eukaryotes are generally xerotolerant rather than xerophilic. A solute stress is therefore likely to shift internal conditions further from the optimum for enzyme activity and, at the beginning of Phase 2, there is minimal protection from the compatible solute. Phase 3 is the new phenotype, the fully adapted steady-state organism. There is a significant amount of information about all phases of transition for the alga, <u>Dunaliella</u> (see review by Brown and Borowitzka, 1979) but most data available for the yeasts refer to the fully adapted, Phase 3 organisms. We are only now beginning to examine the transition sequence in yeasts.

Two general questions are applicable to osmoregulation. They are: 1) what is the site(s) of the regulation, and 2) what is the signal that triggers the osmoregulatory response? Identification of the site requires first the recognition of the physiological type of regulation. Some organisms regulate with K^+. This is conspicuously true of the Halobacteriaceae and less conspicuous but probably no less true for some other bacteria as well. The method, at least for the Halobacteriaceae, seems to involve a simple "pump and leak" system with active exclusion of Na^+. We shall not discuss this further. The flagellate, Dunaliella, regulates with glycerol and does so entirely by biochemical adjustments. That is to say, glycerol content is controlled entirely by synthesis and degradation, not to any significant extent by leakage from the cell (see Brown and Borowitzka, 1979). The salt-tolerant yeast, Debaryomyces hansenii maintains a relatively high K^+/Na^+ ratio, higher than in S. cerevisiae for example, but regulates with glycerol by both synthesizing more glycerol and retaining a greater proportion of it in response to increased salinity (Gustafsson and Norkrans, 1976; Gustafsson, 1979).

The two yeasts that are the subject of this paper, the tolerant S. rouxii and the non-tolerant S. cerevisiae each use components of the method employed by D. hansenii. Saccharomyces rouxii responds to diminished water potential by retaining an increased proportion of a constant total amount of glycerol whereas S. cerevisiae uses the "American method," the energetically wasteful system of synthesizing more glycerol in response to diminished water potential and retaining a constant low proportion of it (Brown, 1978). Thus these two species respond to water stress superficially in the same way, namely by regulating glycerol content. Although glycerol is essential for xerotolerance, its production is not in itself sufficient to confer more than a very limited tolerance; tolerance in the present sense requires the glycerol to be produced in a manner that does not exhaust the yeast's energy resources. In looking for a site of regulation we should therefore look initially to the plasma membrane of S. rouxii and to the intermediary metabolism of S. cerevisiae.

Biochemical and Physiological Components of the Osmoregulatory Mechanisms

There is little to say at this level about membrane function except to note that glycerol is apparently transported actively in S. rouxii but not in S. cerevisiae (Brown, 1974; Edgley, cited by Brown, 1978). Furthermore, there is a major difference in fatty acid composition of the two yeasts when grown in a basal medium (a_w 0.997); S. rouxii is rich in linoleic acid ($C_{18:2}$) whereas none could be detected in S. cerevisiae (Edgley, unpublished results). It is of some interest that Walker and Kummerow (1964) reported a correlation between the content of linoleic acid and impermeability to glycerol in erythrocytes. Dunaliella, whose plasma membrane is remarkably impermeable to glycerol, is also rich in linoleic and linolenic ($C_{18:3}$) acids (Ben-Amotz, personal communication).

The activities of several enzymes of carbohydrate metabolism and their response to water stress are different in the two species and correlate well with their respective methods of osmoregulation. The specific activity of phosphofructokinase, a key glycolytic enzyme, is perhaps a little lower in *S. rouxii* than in *S. cerevisiae* but, significantly, is much more inhibited by ATP in the first species. Growth of the yeasts in 1.8 molal NaCl suppressed formation of the enzyme in *S. rouxii* but enhanced it in *S. cerevisiae* (Table I).

Table I. Phosphofructokinase activity after growth of each yeast in media with and without added salt.

Nucleotide in Assay	Basal Medium		Medium + NaCl (1.8 molal)	
	S. rouxii	*S. cerevisiae*	*S. rouxii*	*S. cerevisiae*
ATP (1 mM)	54±8	157±37	36±5	174±23
(0.1 mM)	146±11	168±20	82±6	353±32
GTP (1 mM)	154±17	179±23	-	-

Enzyme activites (± S.E.M.) are expressed as nmole substrate changed/min/mg protein.

Thus, adaptation of *S. cerevisiae* enhances the activity of a major glycolytic enzyme but this type of effect is much more remarkable with enzymes closer to glycerol itself, most notably *sn*-glycerol-3-phosphate dehydrogenase. Table II shows this effect with the enzyme assayed in the direction of dihydroxyacetone phosphate (DHAP) reduction and includes comparative values for the reduction of dihydroxyacetone (DHA).

The levels of NAD-linked capacity for reducing DHAP were low in *S. rouxii* and, in contrast with the enormous response of *S. cerevisiae*, were unaffected by growth at higher salt concentrations. The apparent involvement of NADPH in glycerol production by *S. rouxii* is noteworthy, however, and is probably very significant for the physiology of that species. Reduction by NADPH of DHAP and DHA were each faster under the assay conditions than reductions dependent on NADH; together the two NADP-linked reactions probably represent the major route to glycerol in this species. Neither reaction was affected by growth at higher salinities. The NADP-linked reduction of DHAP in *S. cerevisiae* was enhanced by growth in salt to a point where it matched the analogous reaction in *S. rouxii* but it remained, nevertheless, an insignificant proportion of the total apparent capacity for reducing DHAP in *S. cerevisiae*.

Table II. The effects of salt concentration of the growth medium on the specific rates of reduction of dihydroxyacetone phosphate (DHAP) and dihydroxyacetone (DHA) by NADH and NADPH in cell-free extracts of each yeast.

[NaCl] in growth medium (molal)	*S. rouxii*		*S. cerevisiae*	
	NADH	NADPH	NADH	NADPH
	DHAP Reduction			
0	2.3-4.5	4.3-8.8	8.1-16	1.2-3.3
0.87	1.8-3.5	5.8-9.3	144-400	3.7-9.5
1.79	3.0-3.1	5.2-7.7	229-642	7.7-12
	DHA Reduction			
0	negligible	8-13	negligible	0.9-1.6
0.87	negligible	13-23	negligible	2.7-3.1
1.79	negligible	5-13	negligible	1.7-5.0

Enzyme activities are expressed as nmole substrate changed/min/mg protein.

We assume that *S. rouxii* produces its arabitol via the pentose phosphate pathway and, since arabitol content does not respond to water stress under the experimental conditions, it was also reasonable to assume that this cycle is not regulated to any significant extent in *S. rouxii* by water potential. Table III, which lists activities of four key pentose phosphate cycle enzymes in crude cell-free preparations from each yeast, supports this assumption.

Table III. Specific activities of key enzymes of the pentose phosphate cycle in cell-free extracts of both yeasts.

Enzyme	[NaCl] in growth medium (molal)	*S. rouxii*	*S. cerevisiae*
Glucose-6-P-Dehydrogenase	0	540±15	163±8
	1.8	451±64	316±34
6-P-gluconate-Dehydrogenase	0	218±4	46±4
	1.8	273±7	146±4
Transaldolase	0	234-284	92-94
Transketolase	0	79	69

Enzyme activities are expressed as nmole substrate changed/min/mg protein ± S.E.M. where there were enough replicates to justify statistical analysis.

When the yeasts were grown in basal medium the cycle was apparently more active in *S. rouxii* than in *S. cerevisiae* but this difference was diminished by increasing the salinity of the growth medium.

Thus there is clear and consistent evidence that complete adaptation of *S. cerevisiae* to Phase 3 involves regulation of gene expression in a manner that affects various enzymes of carbohydrate metabolism involved immediately or remotely in glycerol production. No such effect is evident in *S. rouxii*. This is not the only respect in which the two yeasts respond differently at the genetic level to environmental factors. For example, *S. rouxii* does not show a Crabtree effect whereas *S. cerevisiae* does (Brown, 1975). Xerotolerant yeasts in general do not form petite mutants (Kreger-van Rij, 1969) whereas the non-tolerant yeasts, or perhaps more relevantly, the vigorous fermenters do. In general, there is a high correlation between absence of a Crabtree effect and failure to produce petite mutants.

There are several reports suggesting that changes in the concentration of any of several non-electrolytes, including glycerol itself, might directly affect gene expression, in some cases at the level of transcription. For example, glycerol is reported to stimulate petite mutations in starved *S. cerevisiae*, to activate DNA transcription in a cell-free bacterial preparation and, in common with glucose and sucrose, to stimulate transcription by rat thymus DNA-dependent RNA polymerase (Buss and Stalter, 1978). Changes in water activity, adjusted with sucrose, have been reported to lower cyclic AMP concentrations in *Escherichia coli* to the point where patterns of enzyme induction were changed. Thus there is a real possibility that changes in glycerol concentration, initiated by some other mechanism during the transition from one water potential to another, might, by positive feedback, contribute to the changes in levels of enzyme activity in *S. cerevisiae*.

The transition. Our information about the yeasts in transition is meagre. Measurements of turbidimetric changes of yeast suspensions suggest that Phase 1 takes about 1 min for *S. cerevisiae* and substantially less than that for *S. rouxii* when stressed by transfer from $0.998a_w$ to $0.935a_w$ in PEG. We do not yet know the duration of Phase 2a (the period of readjustment of glycerol concentration) in our conventional batch situation but we have some relevant results for *S. rouxii* in continuous culture ($D = 0.12\ h^{-1}$, nitrogen limitation, 30 C). When transferred under these conditions from 1% to 20% glucose (1.1 M) a steady-state population (stable growth rate) was reestablished in about 20 h. Glycerol content reached an appropriate concentration in about 1 h but thereafter oscillated and did not stabilize until about 30 h after the transition (A. J. Markides, unpublished results). Markides also showed that during this transition there was a surge of ethanol production that lasted about 10 h. No ethanol was detected before the transition nor after the surge; ethanol disappearance was caused predominantly by metabolism rather than wash-out. Similar results were obtained by transfer from the basal medium ($0.997a_w$) to medium adjusted with PEG 200 to $0.977a_w$ but, in this case, the ethanol took longer to disappear.

The signal(s) that triggers the osmoregulatory response is most likely to be identified during the transition, especially early in Phase 2a. Osmoregulatory signals are still very much a matter of conjecture although accumulating evidence is beginning to narrow the range of possibilities. There is a fairly general consensus that turgor pressure is regulated in walled cells and volume is regulated in wall-less cells (Cram, 1976; Zimmermann, 1978) although we have suggested (Brown and Borowitzka, 1979) that turgor might also be involved in the fine tuning, that is the homeostatic maintenance of wall-less cells such as Dunaliella. Be that as it may, volume and turgor are reference parameters, not signals.

Zimmermann (1978) has discounted the involvement of chemical reactions directly in the primary step of turgor sensing because of the very high hydrostatic pressures needed to reverse a reaction. He commented, "We are thus driven to the conclusion that the basic steps of the turgor sensing mechanism must be of mechanical or electrochemical nature." He argued elsewhere in the same review that osmoregulation is dependent on membrane potential.

Energy metabolism affects the membrane potential of cells of organelles by causing ion fluxes; the ions exchanged always include protons. The direction of the fluxes depends on the nature of the cell or organelle and the type of energy transduction that initiates them. Ion fluxes are often mutually coupled and also coupled to the transport of metabolites, including non-electrolytes. Yeast is no exception to this generalization. A supply of an energy-yielding substrate to washed yeast normally produces a proton efflux accompanied among other things, by the uptake of K^+.

Under our experimental conditions (suspension of washed yeast in dilute phosphate buffer, pH 5.9, 0.5 mM containing K^+, Na^+ and Mg^{2+} at 30 C) there is initially a loss of K^+ and an uptake of H^+. Presumably the leaking of K^+ drives the proton uptake. Adding glucose reverses this process and causes a rapid efflux of H^+ accompanied by a K^+ uptake. There is a lag of about 30 sec before S. rouxii responds to glucose whereas the response of S. cerevisiae is virtually instantaneous. During the first 30 sec after glucose addition, however, there is a shock efflux of arabitol from S. rouxii (Brown, 1974). Each yeast ejects five times as many protons as K^+ taken up but S. cerevisiae exchanges twice as many ions/mole glucose consumed as does S. rouxii (Table IV).

When the yeasts are stressed by transfer to a solution of PEG 200 (0.935a_w) containing the same dilute phosphate buffer, the glucose response, as measured by ion fluxes is delayed for some 11.5 min in S. rouxii and about 5.5 min in S. cerevisiae, in both cases well past the end of Phase 1.

There is another difference in the response of the two yeasts, at present best described only qualitatively. When the yeasts are transferred to the PEG solution, the pre-glucose efflux of K^+ is greatly enhanced, especially in S. cerevisiae, and is accompanied by a rapid efflux of H^+. The apparent counter coupling of H^+ and K^+ fluxes is thus destroyed. After 2-2.5 min the proton efflux reverses in both species but the K^+ efflux proceeds unchanged. Glucose has no effect on these early events. After the lag already described, glucose can reestablish a proton efflux but at a much lower rate than in the unstressed yeast. The rate is normally several times greater in *S. cerevisiae* than in *S. rouxii*. Under these conditions K^+ efflux is scarcely affected in *S. cerevisiae* and is usually slightly retarded in *S. rouxii*. The effect of glucose is influenced by the time of its addition. When added an hour after stress it causes a slight reversal of K^+ efflux in both yeasts but, by that time the fluxes have greatly diminished anyway.

Table IV. Cations exchanged after adding glucose to washed suspensions of yeast in dilute phosphate buffer.

Yeast	Cation Transported (mole/mole glucose)	
	H^+	K^+
S. cerevisiae	+0.30±0.05	-0.06±0.02
S. rouxii	+0.15±0.04	-0.03±0.02

Results are given ± S.D.; +, efflux; -, uptake.

The results are variable and incomplete but some simple facts can be consistently identified. Salient among them is the destruction by solute stress of the apparent 5:1 coupling between proton efflux and K^+ uptake. Thus the internal ion composition of each yeast is quite different in the two sets of circumstances, the immediately obvious difference being a lower K^+ concentration in the early stages of Phase 2 adaptation. The effect is slightly more marked in *S. cerevisiae* than in *S. rouxii*.

Some Implications of Glycerol Production by Glycolysis and Via The Pentose Phosphate Cycle

When a yeast such as *S. cerevisiae* produces glycerol in response to a solute stress and does so within the framework of the Emden-Meyerhof scheme, there are various implications for its overall energy metabolism. First, degradation of glucose to the level of triose phosphate does not generate any reducing potential; the sequence must continue past glyceraldehyde-3-phosphate to 1,3, diphosphoglycerate in order to produce NADH (Scheme 1).

Scheme 1. The glycolytic pathway.

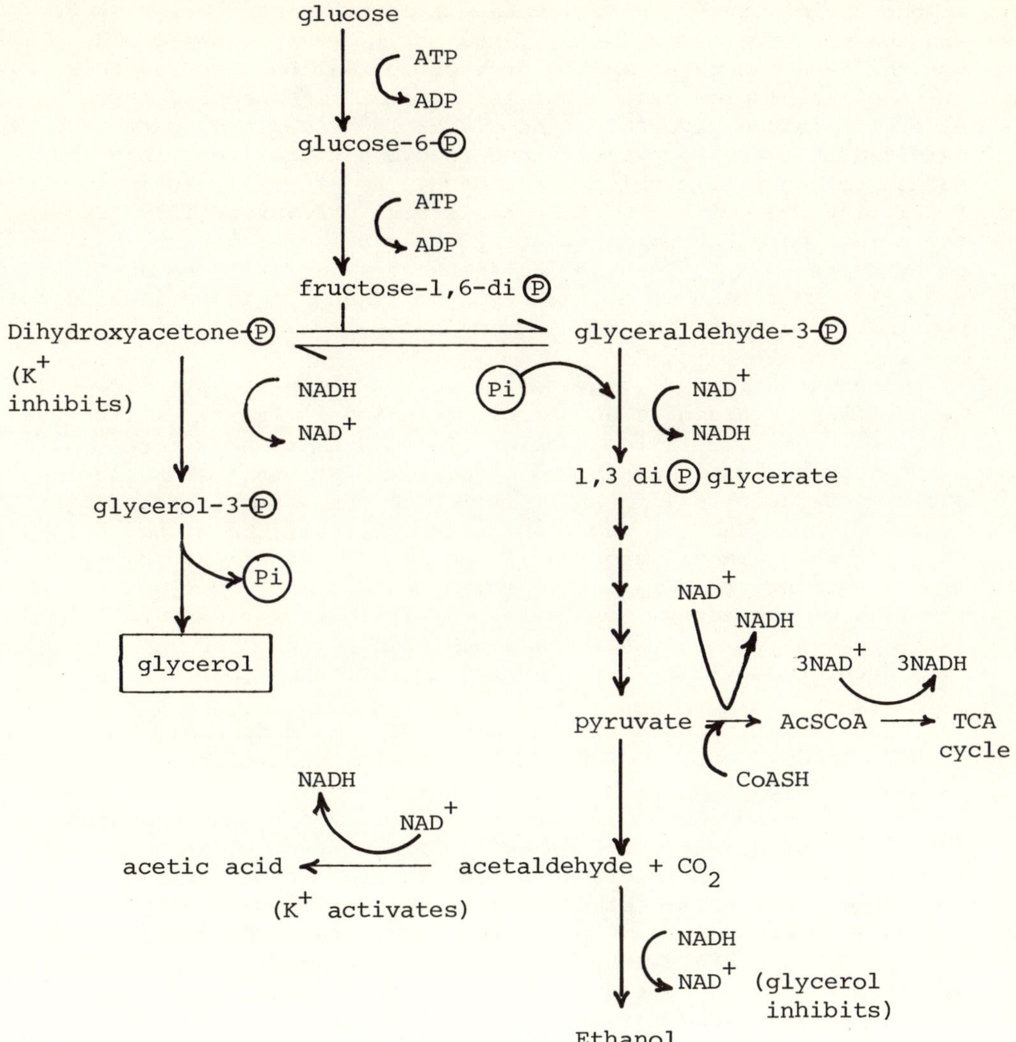

This scheme shows the normal glycolytic sequence as it occurs in S. cerevisiae, the branch pathway from dihydroxyacetone phosphate to glycerol and all reactions, including side reactions, which use NAD^+/NADH. Reduction of dihydroxyacetone phosphate to glycerol is inhibited by K^+; oxidation of acetaldehyde to acetic acid is catalyzed by a dehydrogenase, at least one form of which is activated by K^+; reduction of acetaldehyde to ethanol is inhibited by glycerol (see text).

The question of what metabolite(s) accumulates as a result of this then arises since a) the process is essentially a fermentation in the strict sense, b) all metabolites up to and including pyruvate are too reactive to accumulate, and c) if ethanol is formed it will use the NADH generated in the previous oxidation step and thus deny it to glycerol production. It is therefore inescapable that, in order to produce glycerol, some glyceraldehyde-3-phosphate must be oxidized at least to pyruvate and some additional reactions are needed to generate supplementary NADH. Additional reactions potentially involved with NAD^+/NADH start with pyruvate. They are i) the oxidation of pyruvate to acetyl CoA, ii) the decarboxylation of pyruvate to acetaldehyde and thence its oxidation to acetic acid and iii) reduction of acetaldehyde to ethanol. There is also the possible involvement of the citrate cycle.

The oxidation of pyruvate to acetyl CoA in the presence of NAD^+ is normally regarded as an aerobic reaction in yeast. All that is required, however, is reoxidation of the NADH that is formed. Since respiration is inhibited by a solute stress (Brown, 1975; and recent unpublished results from our laboratory), the rate of reoxidation of NADH by the respiratory chain is diminished but acetyl CoA production should proceed if the NADH is reoxidized by reduction of DHAP. Since respiration is inhibited one must then wonder about the fate of the acetyl CoA so formed; it is significant that growth of _S. cerevisiae_ in a saline medium leads to enhanced lipid production (Hunter and Rose, 1971; Edgley, unpublished observation).

The oxidation of pyruvate to acetyl CoA is probably the major "supplementary" source of NADH for glycerol production. An additional contribution by the citrate cycle is also possible. This cycle needs aerobic conditions strictly only for the oxidation of succinate which is coupled directly to the respiratory chain. The three moles of NADH generated by one turn of the cycle could be reoxidized by any fermentation sequence, in this case the reduction of DHAP. Since respiration continues at a reduced rate under solute stress it is possible that some of the NADH generated in the citrate cycle is also used to reduce DHAP. If that does happen, less ATP will be generated by oxidative phosphorylation which, in turn, as a manifestation of the Pasteur effect, should increase glucose consumption. This does happen with _S. cerevisiae_ but not _S. rouxii_ growing in a saline medium (see Brown, 1978).

The oxidation of acetaldehyde to acetic acid also generates NADH but is not likely to be of major significance in solute-stressed yeast. We have not specifically analyzed for acetic acid but solute-stressed yeast does not lower the pH of a suspending solution containing glucose as much as unstressed yeast (Brown, unpublished results). Furthermore at least one form of yeast aldehyde dehydrogenase is activated by K^+ (Black, 1958), which is depleted in the early stages of solute stress (with PEG; see above).

Reduction of acetaldehyde to ethanol is competitive for NADH but the reaction is subject to inhibition by glycerol. Glycerol at concentrations in the order of 30% is an effector of alcohol dehydrogenase; it lowers the K_m (EtOH) and K_m (NAD^+) and raises the K_m (NADH) (Meyers and Jakoby, 1975). Thus it favors reoxidation of ethanol and exerts a positive feedback control at the point of alcohol dehydrogenase.

The primary site of regulation in Phase 2, however, is probably the reduction of DHAP to glycerol-3-phosphate. The enzyme responsible, glycerol-3-phosphate dehydrogenase, is inhibited by K^+ at physiological concentrations (Gancedo et al., 1958) and, as already emphasized, the ion is rapidly depleted by solute stress.

It is apparent, therefore, that there are several possible regulatory sites responsible for enhanced glycerol synthesis. The reduction of DHAP should be stimulated by the K^+ depletion that follows the stress (at least in PEG). Respiration is inhibited by the stress but in spite of that, oxidation of pyruvate to acetyl CoA should proceed by virtue of the regeneration of NAD^+ in the reduction of DHAP. In due course glycerol accumulation should inhibit ethanol formation and shift the fermentation more in favor of glycerol. A logical prediction therefore is that glycerol production during Phase 2 should be non-linear, the rate increasing with time in the early to mid stages of the response. Ultimately the process should be accelerated further by the great increase in the levels of relevant enzymes but we have no information yet about the mechanism responsible for this increased enzyme synthesis. It is also obvious that a complete justification of the metabolic changes outlined above will require a comprehensive carbon budget under several sets of conditions.

A yeast dependent on the pentose phosphate cycle rather than glycolysis does not have the same need of supplementary reactions to generate reducing capacity, especially if glycerol synthesis is achieved with NADPH rather than NADH. The pentose phosphate cycle produces triose phosphate as one of its intermediates but, by the time that happens, it has already produced NADPH. (In Scheme 2, withdrawal of intermediates such as arabitol and triose phosphate will change the stoichimetry from that shown. The extent of the change will, of course, depend on the proportion of carbon abstracted from the cycle.) This being so there is no need to oxidize glyceraldehyde-3-phosphate in order to generate NADH. The pathway from glyceraldehyde-3-phosphate to pyruvate should be needed only for subsequent metabolism in biosynthesis and for generation of ATP by oxidative phosphorylation. It is noteworthy that *S. rouxii* does not produce significant quantities of ethanol except transiently after a solute stress when respiration is impaired.

Scheme 2. Glycerol production and the pentose phosphate pathway and modified glycolysis.

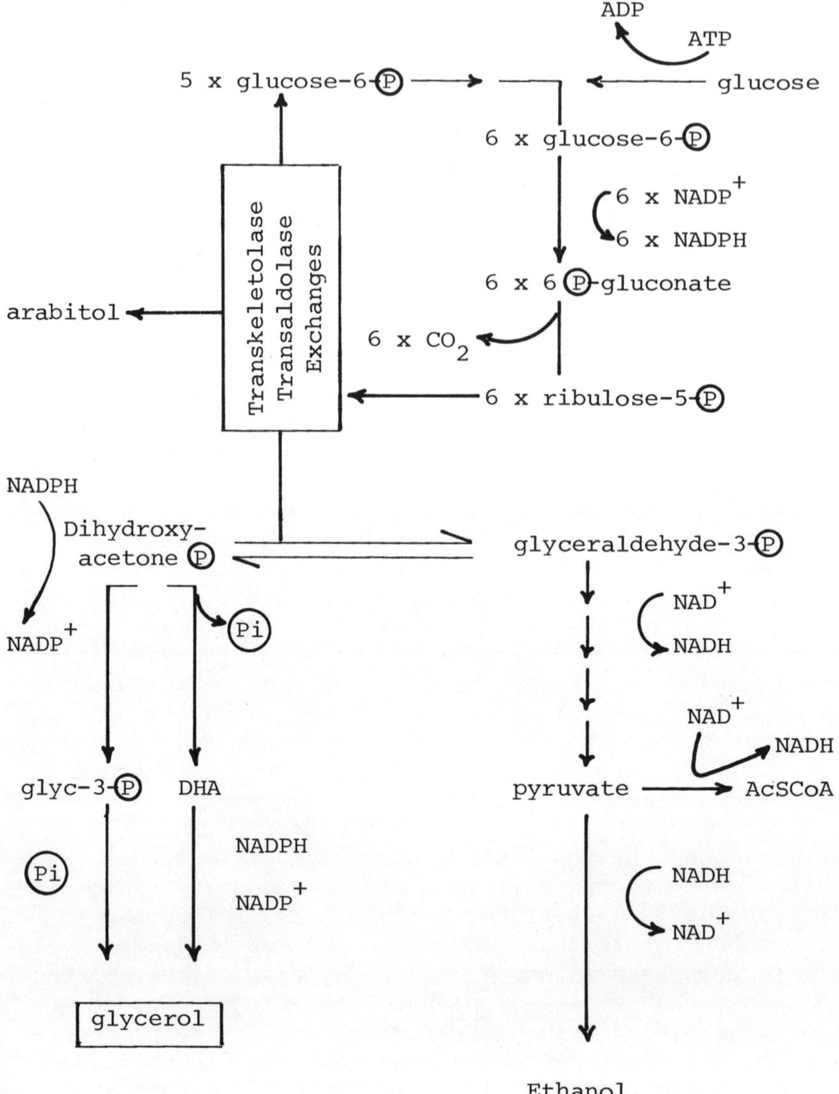

Two sequences to glycerol production as they are thought to occur in S. rouxii. The reduction of DHAP to sn-glycerol-3-phosphate is also achieved to a limited extent with NADH (see text).

It is presumably the reserve reducing capacity produced as NADPH by the pentose phosphate cycle that enables S. rouxii to synthesize glycerol in considerable quantity even without stress. Since it does not depend predominantly on glycolysis and NADH it is not obliged to respond to K^+ depletion by increased reduction of DHAP as we have suggested happens with S. cerevisiae. When coupled to the control of glycerol retention this apparent immunity from metabolic stimulus can be seen as a basis of a very economical stress-resistant physiology. (We expect that S. rouxii might be obliged to synthesize more glycerol at levels of water activity lower than we have used experimentally.)

It is of some comparative interest that Dunaliella depends primarily on NADPH for glycerol production and that it responds to salinity stress in the dark apparently by starch degradation via the pentose phosphate cycle (Brown and Borowitzka, 1979). In that case the process is regulated at the level of preformed enzymes since inhibitors of protein synthesis do not impair the response.

SUMMARY

Both xerotolerant and non-tolerant yeasts osmoregulate by accumulating glycerol but use different physiological methods to do so. The tolerant species, S. rouxii, regulates biophysically; it responds to water stress by regulating the proportion of its total glycerol production that is retained within the cell. The non-tolerant S. cerevisiae, on the other hand, regulates biochemically; it responds by controlling the amount of glycerol synthesized while retaining a constant proportion of it. This is energetically wasteful and eventually limits tolerance of low water potential by diverting an unacceptably large proportion of metabolic activity to glycerol production. When fully adapted to various levels of water potential, a number of major relevant phenotypic differences become evident. There is a major difference in fatty acid composition. Saccharomyces rouxii apparently has an active transport system for glycerol whereas S. cerevisiae does not. Saccharomyces cerevisiae depends almost exclusively on glycolysis and NADH generation for glycerol synthesis whereas S. rouxii appears to depend predominantly on NADPH generated by the pentose phosphate cycle. The specific activities of some relevant glycolytic enzymes, most notably glycerol-3-phosphate dehydrogenase, are greatly increased by solute stress in S. cerevisiae but are virtually unaffected in S. rouxii. The transition from high to low water potential produces various transient responses in the yeasts including a depletion of cellular K^+. This is possibly a primary osmoregulatory signal in S. cerevisiae and is discussed in relation to the broader physiological implication of dependence on glycolysis on the one hand and on the pentose phosphate cycle on the other.

REFERENCES

Anand, J. C. and Brown, A. D., 1968, Growth rate patterns of the so-called osmophilic and non-osmophilic yeasts in solutions of polyethylene glycol, J. Gen. Microbiol., 52:205.

Black, S., 1955, Potassium-activated yeast aldehyde dehydrogenase, in: "Methods in Enzymology," Vol. 1, S. P. Colowick and N. O. Kaplan, eds., p. 508, Academic Press, New York.

Brown, A. D., 1974, Microbial water relations; features of the intracellular composition of sugar-tolerant yeasts, J. Bacteriol., 118:769.

Brown, A. D., 1975, Microbial water relations. Effects of solute concentration on the respiratory activity of sugar-tolerant and non-tolerant yeasts, J. Gen. Microbiol., 86:241.

Brown, A. D., 1978, Compatible solutes and extreme water stress in eukaryotic micro-organisms, Adv. Microbial Physiol., 17:181.

Brown, A. D. and Borowitzka, L. J., 1979, Halotolerance of Dunaliella, in: "Biochemistry and Physiology of Protozoa," Vol. 1, 2nd Ed., M. Levandowski and S. H. Hunter, eds., p. 139, Academic Press, New York.

Brown, A. D. and Simpson, J. R., 1972, Water relations of sugar-tolerant yeasts: the role of intracellular polyols, J. Gen. Microbiol., 72:589.

Buss, W. C. and Stalter, K., 1978, Stimulation of eukaryotic transcription by glycerol and polyhydroxylic compounds, Biochem., 17:4825.

Cram, J., 1976, Negative feedback regulation of transport in cells. The maintenance of turgor, volume and nutrient supply, in: "Encyclopedia of Plant Physiology," New Series, Vol. 2, Part A, U. Lüttge and M. G. Pitman, eds., Springer-Verlag, Berlin.

Gancedo, C., Gancedo, J. M., and Sols, A., 1968, Glycerol metabolism in yeasts - pathways of utilization and production, Europ. J. Biochem., 5:165.

Gustaffson, L., 1979, The ATP pool in relation to the production of glycerol and heat during growth of the halotolerant yeast, Debaryomyces hansenii, Arch. Microbiol., 120:15.

Gustaffson, L. and Norkrans, B., 1976, On the mechanism of salt tolerance. Production of glycerol and heat during growth of Debaryomyces hansenii, Arch. Microbiol., 110:17.

Hunter, K. and Rose, A. H., 1979, Yeast lipids and membranes, in: "The Yeasts," Vol., 2, A. H. Rose and J. S. Harrison, eds., p. 211, Academic Press, New York.

Myers, J. S. and Jakoby, W. B., 1975, Glycerol as an agent eliciting small conformational charges in alcohol dehydrogenase, J. Biol. Chem., 250:3785.

Walker, B. L. and Kummerow, F. A., 1964, Erythrocyte fatty acid composition and apparent permeability to non-electrolytes, Proc. Soc. Exptl. Biol. Med., 115:1099.

Zimmermann, U., 1978, Physics of turgor and osmoregulation, Ann. Rev. Plant Physiol., 29:121.

OSMOREGULATION IN THE HALOPHILIC ALGAE DUNALIELLA AND ASTEROMONAS

Ami Ben-Amotz* and Mordhay Avron**

*Israel Oceanographic & Limnological Research
Tel-Shikomna, P.O.B. 8030, Haifa, Israel

**The Weizmann Institute of Science
Biochemistry Department, Rehovot, Israel

INTRODUCTION

The genus Dunaliella of the order Volvocales includes a variety of ill-defined species of unicellular green microscopic algae (Butcher, 1959). Members of the genus Dunaliella are generally ovoid in shape, 4-15 μm wide and 10-25 μm long. The cells are motile, due to the presence of two equal long flagellae in each cell, and contain one large cup-shaped chloroplast which occupies about half of the cell volume. The chloroplast contains a large pyrenoid surrounded by polysaccharide granules, the storage product.

Asteromonas gracilis (Peterfi and Manton, 1968), a green flagellate of the order Prasinophyceae, is a unicellular microscopic alga present in salt marshes and small saline pools. Members of the genus are generally spindle in shape, bluntly pointed at both ends, widest a little forward of center, 8-16 μm wide and 15-22 μm long. The cells are motile, with two flagellae of 1.5-2 times body length in each cell. The cell shape is characterized by six keel-like ridges running longitudinally. Each cell contains a basin-shaped chloroplast which occupies about half of the cell volume, with a big single asymmetric pyrenoid situated between the chloroplast and the nucleus.

Distinguishing features of Dunaliella and Asteromonas are as follows: (a) the cells lack a cell wall; the cells are enclosed only by a thin elastic cell envelope. This permits rapid changes in cell shape, makes the cell highly responsive to osmotic changes, and causes it to burst if subjected to extreme hypotonic osmotic stress; (b) Asexual flagellates may produce thick-walled cysts of

12-20 μm, with 2-5 μm wall thickness, which allows survival of the algae under environmental stress; (c) natural habitats of the halotolerant algae include saltpans, brine lakes, small pools and salt water ditches near the sea in media containing salt in concentrations ranging from dilute to concentrate. These unique characteristics of Dunaliella and Asteromonas attracted the attention of scientists interested in studying and understanding the metabolic mechanism responsible for osmoregulation and the adaptation of the algae to low water activity under water stress.

MATERIALS AND METHODS

Dunaliella salina was obtained from the culture collection of Dr. Thomas, La Jolla, CA, USA. Asteromonas gracilis was obtained from the Culture Centre of Algae and Protozoa, Cambridge, England. Dunaliella bardawil was a local isolated species. Asteromonas gracilis was cultivated on seawater medium enriched with 5 mM KNO_3, 0.2 mM KH_2PO_4, 1.5 μM Fe, 30 μM EDTA, 20 mM HEPES, pH 7.5, 5 mM $NaHCO_3$, NaCl at the indicated concentration, and trace metal mix. Growth conditions, glycerol assay and other methods have been described previously in full (Ben-Amotz and Avron, 1973; Ben-Amotz, 1975; Ben-Amotz and Avron, 1978).

RESULTS AND DISCUSSION

Growth and Adaptability to Salinity

Table I illustrates the specific growth rates of Asteromonas gracilis, Dunaliella bardawil and Dunaliella salina at several concentrations of NaCl. Optimal growth is around 1 M NaCl which approximates the optimum for photosynthesis in Dunaliella parva (Ben-Amotz and Avron, 1972). Table I shows, however, that both Dunaliella strains and A. gracilis grow over a full spectrum of salinities, from the NaCl concentration of seawater to saturated solutions. It is apparently the halotolerance adaptability of the organism over a wide range of salinities, rather than its obligatory halophilism, which gives Asteromonas and Dunaliella competitive advantage in saline environments. Further evidence of the adaptability of Dunaliella to salinity has been illustrated previously (Borowitzka and Brown, 1974; Borowitzka et al., 1977; Avron and Ben-Amotz, 1979) through continuous transfers of the algae into low or high osmotic media until a new growth rate was achieved.

Table II illustrates the effect of salt concentrations on maximal concentration of cells per ml of A. gracilis, D. bardawil and D. salina. The maximal population in the medium seems to be related to both salt concentration and algal size. The large cell-size species A. gracilis and D. bardawil do not exceed a density of 2 x 10^6 cells ml^{-1}, while the smaller cells of D. salina can reach densities of 2 x 10^7 cells ml^{-1}. It should be noted, however, that

Table I. The specific growth rates (μ day^{-1}) of A. gracilis, D. bardawil and D. salina at different salt concentrations.*

NaCl concentration	Specific growth rate		
	A. gracilis	D. bardawil	D. salina
M	μ day^{-1}	μ day^{-1}	μ day^{-1}
0.5	2.05	-	-
1.0	2.20	2.40	3.40
1.5	2.10	-	-
2.0	-	2.40	2.70
2.5	1.53	-	-
3.0	-	1.76	2.10
3.5	1.35	-	-
4.0	-	1.45	1.80
4.5	1.20	-	-

*Growth conditions were as previously described (Ben-Amotz and Avron, 1974). The specific growth rate was calculated from the logarithmic phase of growth curves. A specific growth rate of 2 represents doubling of the algae every 24 hours.

Table II. Effect of salt concentration on the maximal concentration of A. gracilis, D. bardawil and D. salina.*

NaCl concentration	Maximal cell concentration		
	A. gracilis	D. bardawil	D. salina
M	cell ml^{-1}	cell ml^{-1}	cell ml^{-1}
0.5	1.2×10^6	-	-
1.0	1.2×10^6	1.5×10^6	2.0×10^7
1.5	8.5×10^5	-	-
2.0	-	7.0×10^5	5.0×10^6
2.5	6.0×10^5	-	-
3.0	-	5.5×10^5	3.0×10^6
3.5	4.0×10^5	-	-
4.0	-	3.5×10^5	1.5×10^6
4.5	2.0×10^5	-	-
5.0	-	1.5×10^5	6.0×10^5

*Growth conditions were as described previously (Ben-Amotz and Avron, 1974). Cells were counted in a Coulter Counter at the stationary phase of the growth cycle.

maximal populations of small and large cells are both equivalent on a dry weight basis, both reaching about 1 g dry weight per liter.

Intracellular Composition

In contrast to the situation encountered with halophilic bacteria (Larsen, 1967), the intracellular salt content of *Dunaliella* is too low to support iso-osmotic equilibrium with the extracellular salt concentration (Okamoto and Suzuki, 1964; Gimmler and Schirling 1978). A generally accepted value of the intracellular salt concentration is as yet unavailable because of the technical difficulties of obtaining reliable analyses. However, major evidence against high salt accumulation in the cells lies in the salt sensitivity of cytoplasmic and chloroplastic enzymes of Dunaliella (Johnson et al., 1968; Ben-Amotz and Avron, 1972; Borowitzka and Brown, 1974). The solute responsible for most of the intracellular osmotic pressure under high salinity has been clearly shown to be glycerol (Wegmann, 1971; Ben-Amotz and Avron, 1973; Borowitzka and Brown, 1974; Borowitzka et al., 1977; Ben-Amotz and Avron, 1978). Furthermore, glycerol has been shown to be the major osmoregulator of *D. salina* and other strains of *Dunaliella* (Ben-Amotz and Avron, 1973; Borowitzka et al., 1977); its *in vivo* concentration changes in direct proportion to the salt concentration to which the algae are adapted.

In analogy to glycerol accumulation in *Dunaliella*, *A. gracilis* produces and accumulates a high content of intracellular glycerol. The response of glycerol concentration to salinity in *A. gracilis*, *D. bardawil* and *D. salina* is shown in Figure 1. The glycerol content values in *A. gracilis* and in *D. bardawil* of between 50 and 400 pg glycerol per cell are higher than values of glycerol per cell in *D. salina*, due to the larger volume of these two flagellates.

The cell volume, glycerol per cell and chlorophyll per cell of these three algae grown in 4 M NaCl are shown in Table III. It is clear that irrespective of cell volume, these three algae, when grown in 4 M NaCl, reach similar high glycerol-to-chlorophyll ratios of about 60-70 g g^{-1}. The response of glycerol-to-chlorophyll ratio to salinity in the three algae confirms previous conclusions that internal glycerol serves as the major solute which osmotically balances the external salt concentration (Ben-Amotz and Avron, 1973, 1978; Borowitzka et al., 1977). In contrast to the strong inhibition of chloroplastic and cytoplasmic enzymes of *Dunaliella* by NaCl, glycerol does not inhibit cytoplasmic and membrane-bound enzymes at physiological concentrations (Borowitzka and Brown, 1974; Ben-Amotz, 1975; Brown, 1976). Glycerol, therefore, may serve as an osmoregulator and, at its elevated concentrations, as a compatible solute under conditions of water dehydration (Brown, 1977; Schobert, 1977).

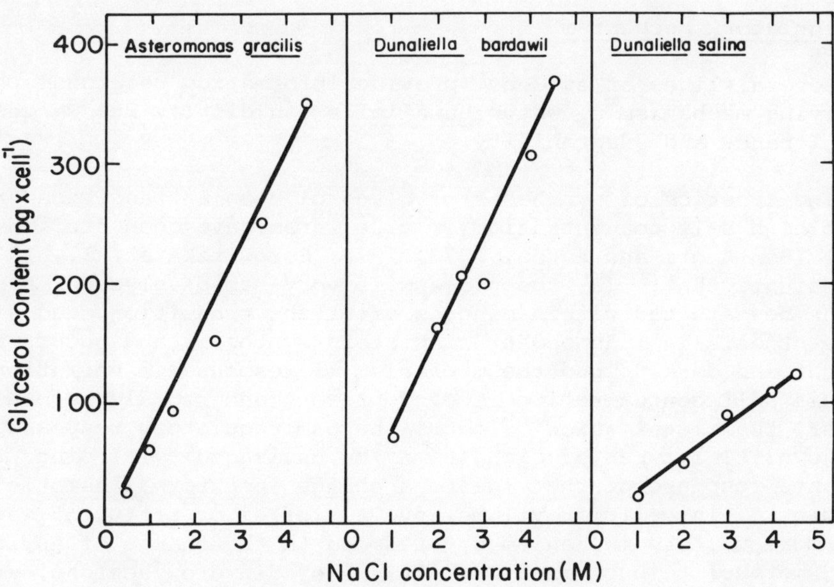

Fig. 1. Effect of extracellular NaCl concentration on the intracellular glycerol content in A. gracilis, D. bardawil and D. salina. At the late logarithmic phase, algae were counted and samples were assayed for glycerol as previously described (Ben-Amotz and Avron, 1978).

Table III. Intracellular volume, glycerol content and chlorophyll content of algae maintained at 4 M NaCl.*

	A. gracilis	D. bardawil	D. salina
Cell volume, μm^3	400	400	100
Chlorophyll content, pg $cell^{-1}$	5.5	5.0	1.8
Glycerol content, pg $cell^{-1}$	350	350	110
Glycerol-to-chlorophyll ratio, g g^{-1}	64	70	60

*Cell volume was estimated microscopically; chlorophyll, glycerol and cell counting were assayed as previously described (Ben-Amotz and Avron, 1973, 1978).

Osmoregulatory Mechanism

Several lines of evidence provide information regarding the underlying mechanism by which *Dunaliella* can display its unique halotolerance and adaptability.

The kinetics of synthesis of glycerol upon a transition from low to high salt concentration or vice versa have been studied in detail (Ben-Amotz and Avron, 1973, 1978; Borowitzka et al., 1977), and indicate that: (a) the process is very rapid; glycerol synthesis can be detected within minutes after the transition, and (b) such synthesis is independent of protein synthesis and occurs both in light and dark. Thus the mechanism of response to variations in media salt concentration is ever present and rapidly responding. However, the signal which triggers the osmoregulatory response is not known. Two types of signals may be envisaged: (a) since the immediate response of the alga to a change in external osmotic pressure is, in cellular volume, due to influx or efflux of water, the osmoregulatory mechanism may respond to the change of an involved metabolite such as NADP, ATP, H^+ or dihydroxyacetone, etc. (see below); (b) since such volume changes bring about changes in the tension within the plasma membrane, a membrane-bound enzyme may respond to such a change in tension (or thickness). The possibility that the signal involves direct detection of the extracellular salt concentration or the transmembrane salt concentration difference seems to be ruled out by the observation that the osmoregulatory system responds equally well to sucrose or salt (Borowitzka et al., 1977).

Two unique enzymes have been described for *Dunaliella* which in all probability are involved in its osmoregulatory response. The first is dihydroxyacetone reductase (also referred to as glycerol dehydrogenase) which catalyzes the reaction (Ben-Amotz and Avron, 1974; Borowitzka and Brown, 1974):

$$\text{Dihydroxyacetone} + \text{NADPH} + H^+ \rightleftarrows \text{glycerol} + \text{NADP}^+ \quad K'\text{eq (pH 7)} = 2 \times 10^4.$$

The enzyme is $NADP^+$ specific, with a low affinity for glycerol (apparent K_m = 1.5 M). The analogous enzyme, which catalyzes the reaction of glycerol-phosphate and dihydroxyacetone-phosphate and utilizes NAD^+, has also been described for *Dunaliella* (Wegman, 1971).

The second unique enzyme is a dihydroxyacetone kinase, which is highly specific toward dihydroxyacetone (Lerner and Avron, 1977).

Taking into consideration the described characteristics of glycerol synthesis, particularly during an increase in the salt concentration of the medium (rapidity of response, occurrence in dark), it is clear that glycerol synthesis must, in the first instance, depend on supply of its carbon skeleton from stored non-osmotically active substances. Polysaccharides, such as starch, are obvious candidates. Thus, the immediate osmoregulatory response to an increase in salt concentration in the medium must involve the conversion of storage polysaccharides to glycerol. Such conversion requires supply of ATP and NADPH, both of which are products of the photosynthetic electron transport chain:

1. $H_2O + NADP^+ + ADP + Pi \rightarrow 1/2\ O_2 + NADPH + H^+ + ATP$

2. $3\ (CH_2O) + NADPH + ATP + H^+ \rightarrow \text{glycerol} + NADP^+ + ADP + Pi$

This mechanism predicts that an increase in salt concentration in the medium should result in a CO_2-independent O_2 evolution in *Dunaliella*, which has been recently demonstrated (Kaplan and Avron, in press). Conversely, the immediate response to a decrease in salt concentration in the medium would be expected to involve the conversion of glycerol into storage polysaccharides.

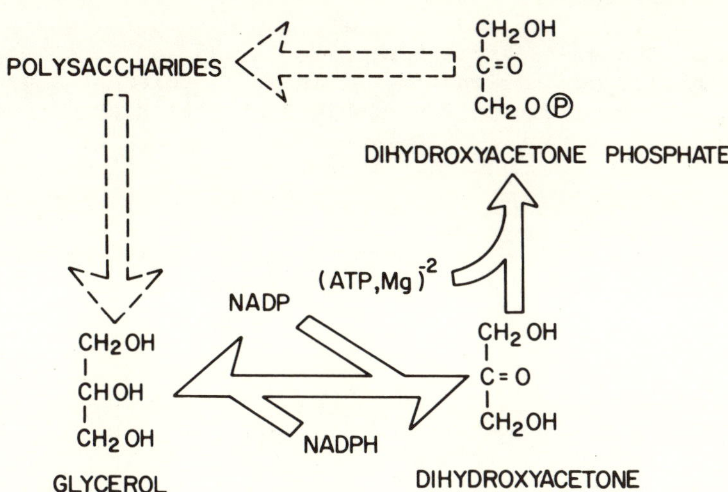

Fig. 2. Hypothetical scheme of the osmoregulatory metabolism of *Dunaliella*

Considering this overall metabolic framework and inserting the known unique enzymes of *Dunaliella*, a reasonable hypothesis of the metabolic events that take place during osmoregulatory responses in *Dunaliella* is presented in Figure 2. Glycerol may accumulate by production of triose-phosphate from polysaccharides via the glycolytic pathway, followed by reduction to glycerol-phosphate and dephosphorylation. Conversion of glycerol to polysaccharides may proceed via oxidation to dihydroxyacetone and phosphorylation to dihydroxyacetone phosphate. It should be noted, however, that the metabolic pathway in *Asteromonas*, the complete metabolic pathway in *Dunaliella*, and the signal triggering the osmoregulatory responses remain uncertain and warrant continued investigation.

SUMMARY

The unicellular wall-less green algae *Asteromonas gracilis*, *Dunaliella bardawil* and *Dunaliella salina* can adapt to and grow in media containing salt in concentrations ranging from seawater level to Dead Sea level. The specific growth rate of these three species is not affected by salt concentrations between 0.5 M and 2.0 M and only gradually decreases in salt concentrations above 2.0 M. These halotolerant algae were found to accumulate very large amounts of glycerol, the concentration of which varies in direct proportion to the extracellular salt concentration. In high salt concentrations of around 4 M, *Asteromonas gracilis* and *Dunaliella bardawil*, two large flagellates of about 400 μm^3, accumulate glycerol to 350 pg per cell, while the smaller flagellate *Dunaliella salina*, of about 100 μm^3, accumulates glycerol to 100 pg per cell. The glycerol-to-chlorophyll ratio in 4 M NaCl is about 65 g g^{-1} for the three species. The above results support and extend previous information (Ben-Amotz and Avron, 1978; Avron and Ben-Amotz, 1979) on the metabolic adaptation of wall-less halotolerant algae to low water activity by accumulation of high intracellular concentrations of glycerol.

ACKNOWLEDGMENT

The authors wish to thank Mrs. Tamar Grunwald for her help in carrying out various aspects of the research.

REFERENCES

Avron, M. and Ben-Amotz, A., 1979, Metabolic adaptation of the alga *Dunaliella* to low water activity, *in*: "Strategies of Microbial Life in Extreme Environments," M. Shilo, ed., Verlag Chemie, Weinheim, pp. 83-91.

Ben-Amotz, A., 1975, Adaptation of the unicellular alga *Dunaliella parva* to saline environment, *J. Phycol.*, 11:44.

Ben-Amotz, A. and Avron, M., 1972, Photosynthetic activities of the halophilic alga *Dunaliella parva*, *Plant Physiol.*, 49:240.

Ben-Amotz, A. and Avron, M., 1973, The role of glycerol in osmotic regulation of the halophilic alga Dunaliella parva, Plant Physiol., 51:875.

Ben-Amotz, A. and Avron, M., 1974, Isolation, characterization and partial purification of an $NADP^+$-dependent dihydroxyacetone reductase from the halophilic alga Dunaliella parva, Plant Physiol., 53:628.

Ben-Amotz, A. and Avron, M., 1978, On the mechanism of osmoregulation in Dunaliella, in: "Energetics and Structure of Halophilic Microorganisms," S. R. Caplan and M. Ginzburg, eds., Elsevier, Amsterdam, pp. 529-541.

Borowitzka, L. and Brown, A. D., 1974, The salt relations of marine and halophilic species of the unicellular green alga Dunaliella, Arch. Microbiol., 96:37.

Borowitzka, L., Kessly, D. S. and Brown, A. D., 1977, The salt relations of Dunaliella. Further observations on glycerol production and its regulation, Arch. Microbiol., 113:131.

Brown, A. D., 1976, Microbial water stress, Bacteriol., Rev., 40: 803.

Brown, A. D., 1977, Compatible solutes and extreme water stress in eukaryotic microorganisms, Adv. Microbiol. Physiol., 17:181.

Butcher, R. W., 1959, An introductory account of the smaller algae of British coastal waters. I. Introduction and Chlorophyceae, Fish. Invest. Ser., 4:1.

Gimmler, H. and Schirling, R., 1978, Cation permeability of the plasmalemma of the halotolerant alga Dunaliella parva. II. Cation content and glycerol concentration of the cells as dependent upon external NaCl concentration, Z. Pflanzenphysiol., 87:487.

Johnson, M. K., Johnson, E. J., McElory, R. D., Speer, H. L., and Bruff, B. S., 1968, Effects of salts on the halophilic alga Dunaliella viridis, J. Bacteriol., 95:1461.

Kaplan, A. and Avron, M., 1980, Salt induced metabolic changes in D. salina, Plant Physiol., (in press).

Larsen, H., 1967, Biochemical aspects of extreme halophilism, Adv. Microbiol. Physiol., 1:97.

Lerner, H. R. and Avron, M., 1977, Dihydroxyacetone kinase activity in Dunaliella parva, Plant Physiol., 59:15.

Okamoto, H. and Suzuki, I., 1964, Intracellular concentration of ions in the halophilic strains of Chlamydomonas. I. Concentration of Na, K and Cl in the cell, Z. Allg. Microbiol., 4:350.

Peterfi, L. S. and Manton, I., 1968, Observations with the electron microscope on Asteromonas gracilis Artari (Stephanoptera gracilis Artari) with some comparative observations on Dunaliella sp., Br. Phycol. Bull., 3:423.

Schobert, B., 1977, Is there an osmotic regulatory mechanism in algae and higher plants?, J. Theor. Biol., 68:17.

Wegmann, K., 1971, Osmotic regulation of the photosynthetic glycerol production in Dunaliella, Biochim. Biophys. Acta, 234:323.

IONS AND OSMOREGULATION

J. A. Raven, F. A. Smith, and S. E. Smith

Department of Environmental Biology, Research School of Biological Sciences, A.N.U., Canberra City, A.C.T. 2601, Australia; Permanent Address: Department of Biological Sciences, University of Dundee, Dundee DD14HN, Scotland, United Kingdom

and

Departments of Botany and Agricultural Biochemistry, The University of Adelaide, South Australia 5001, Australia

INTRODUCTION

Various aspects of the involvement of ions in osmoregulation (turgor regulation, volume regulation) have been reviewed recently by Bisson and Gutknecht (1979), Cram (1976), Flowers et al. (1977), Hellebust (1976), Osmond (1979) and by Pitman and Cram (1977). These reviews show that, while we know something of the phenomena of ionic involvement in osmoregulation, little is known of the molecular mechanisms of ion transport, or of the regulation of this transport in response to (e.g.) turgor-related signals. We shall discuss the ways in which ions are involved in osmoregulation in the context of the evolution of phototrophic plants; discussion of the various strategies will, we hope, provide a background for consideration of the genetics of osmoregulation in a particular plant. Our discussion will necessarily deal not only with the constraints on the ionic composition of the cell and its various compartments, but with possible mechanisms of ion transport at the cell and the whole-plant level and the regulation of this transport.

IN THE BEGINNING: SELECTIVE PRESSURES WHICH MAY HAVE BEEN INVOLVED IN SHAPING THE IONIC REQUIREMENTS AND IONIC TOLERANCE OF CYTOPLASM

We start from the premise that the earliest cells were prokaryotic in their organization, with no wall and with the plasmalemma as their only membrane. A number of selective pressures may have acted to 'imprint' certain cytoplasmic ionic requirements.

The total ionic activity tolerated by catalytic and informational macro-molecules is a function of their chemical nature (Wyn Jones et al., 1977) which in turn reflects the monomers available in the "primeval soup". It is possible that the earliest cytoplasm had an ionic concentration similar to that of the ambient 'sea', which was more dilute than present-day seawater. The early wall-less prokaryote would have had a volume-regulation problem due to the presence of impermeant macro-molecules in the cytoplasm; this was met by a membrane of low Na^+ permeability, and subsequently by an Na^+ extrusion pump to give 'dynamic' volume regulation (Gutknecht and Dainty, 1968; Raven, 1976). Considerations of pH regulation, and of the properties of Ca^{2+}, provide selective pressures for the occurrence of regulated H^+ and Ca^{2+} extrusion pumps (Smith and Raven, 1978, 1979; Halman, 1974; Raven and Smith, 1979).

These intrinsic chemical considerations, together with extrinsic considerations (e.g., volume regulation) can be used to rationalize the "generalized" ionic composition of cytoplasm (Raven, 1977a), i.e., a total inorganic ionic concentration of 100-500 mOsm, a higher internal K^+ than Na^+ concentration, and free H^+ and Ca^{2+} of 0.1-10 μM. Anions, aside from macromolecules and organic metabolites, include relatively low concentrations of $SO_4{}^{2-}$ and $NO_3{}^-$ (probably not available to the earliest cells in a reducing environment) $HPO_4{}^{2-}$ and Cl^-. Acquisition of walls (and hence of hypertonicity), and the invasion of fresh-water (with consequent lower internal tonicity for a given turgor pressure) led to modifications in cytoplasmic ion content similar to that described below for eukaryotes.

PRIMARILY AQUATIC EUKARYOTES

Accepting (albeit with some reluctance) the endosymbiotic hypothesis for the evolution of mitochondria and chloroplasts in eukaryotes, we must postulate a wall-less 'pre-eukaryote' host which acquired its organelles by a phagocytotic mechanism. The wall-less eukaryotic organism (Figure 1) would presumably be similar in its ion transport processes and in the constraints on its cytoplasmic composition to the prokaryotic mentioned in the previous section. Such an organism would consist of a number of 'tight' compartments (e.g., stroma of plastids, thylakoids of plastids, matrix of mitochondria, lysosomes) whose membranes have a low (and regulated) permeability to low-molecular weight metabolites (including inorganic ions). There could also be 'leaky' compartments (intermembrane spaces of mitochondria and plastids; microbodies) which are bound, at least in part, by membranes permeable to compounds with a M_r less than 10^3 or so (Raven, 1980). Regulation of the volume of 'tight' compartments involves mediated or active solute (including ion) transport, while regulation of the

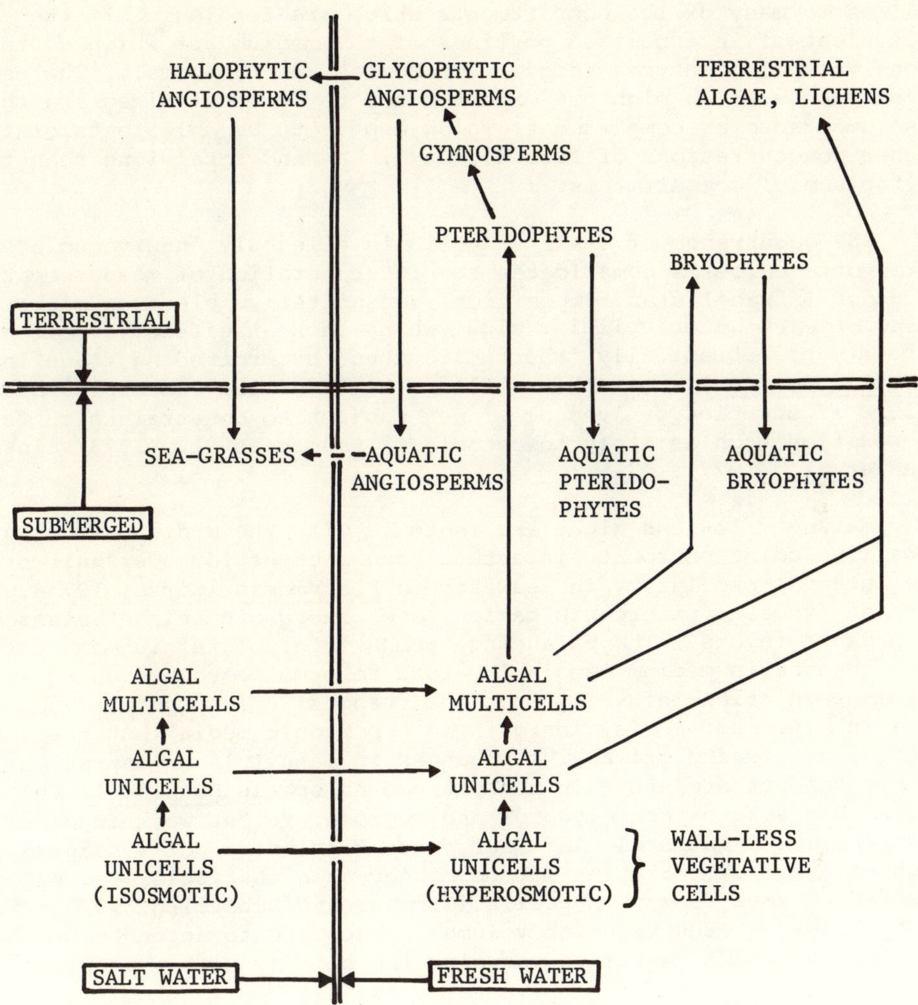

Fig. 1. Possible evolutionary scheme for phototrophic eukaryotic plants. Unless otherwise stated, vegetative cells are walled and hypertonic. All taxa of terrestrial plants (except gymnosperms) contain 'resurrection plants'.

shape and volume of 'leaky' compartments relates to the area of bounding membrane and the content and net charge of impermeant macromolecules. It appears that the constraints on ionic composition of the cytosol apply to the 'endosymbiotic' organelles as well as to many of the compartments which are topologically the equivalent of internalized portions of the medium but which contain 'non-lysosomal' enzymes (e.g., microbodies, dictyosomes). The only compartments which might be expected _not_ to be constrained are the lysosome-vacuolar compartments which appear to be able to tolerate higher concentrations of free H^+, Ca^{2+}, Na^+ and total ions than the 'cytoplasmic' compartments.

The occurrence of small vacuoles in allegedly 'non-vacuolate' eukaryotic unicells complicates the interpretation of measurements of gross intracellular composition. Quantitative electron microscopy of various unicellular algae which lack an effective cell wall (absent, or mechanically 'short-circuited' by protruding flagella as in _Chlamydomonas_) shows that at least 5-10% of the intra-plasmalemma volume is occupied by 'vacuoles' not subject to constraints on their composition such as apply to cytoplasm (Schotz et al., 1973; Sicko-Goad et al., 1977).

Marine wall-less algae are isotonic with the medium; inorganic ions (including phosphate in esters) can account for over half of the internal osmolarity in seawater in _Platymomas_ (Kirst, 1977a,b) with an excess of inorganic cations over inorganic anions balanced by organic anions. The balancing 'compatible solute' is mannitol. Step changes in medium osmolarity lead to osmotic volume changes; volume regulation initially involves changes in ion content (ion loss in hypotonic media, ion gain in hypertonic media) but these ion changes are traded off against changes in mannitol concentration in longer-term treatments. In the extreme halophile _Dunaliella_ the compatible solutes are glycerol and sucrose; recent work suggests that the gross intracellular NaCl concentration in cells adapted to high external osmolarities may be higher than that tolerated by extracted enzymes from these cells (Gimmler and Schirling, 1978); it is not clear if the vacuolar volume is adequate to account for the bulk of this NaCl, assuming vacuolar isotonicity with the cytosol.

Freshwater wall-less algae are hypertonic to the medium, and volume regulation involves expulsion of a hypotonic solution _via_ contractile vacuoles (see Raven, 1976). The total osmolarity is usually low, perhaps as a means of reducing the energetic load of osmoregulation. In such cells the cytoplasmic osmolarity appears to reflect a value close to the lowest ionic concentration compatible with life (see Raven, 1976). The inorganic anion concentration in such cells is very low (Raven, 1980), and in many cases the measured concentration of organic anions is not balanced by the measured content of inorganic cations; the reason for this is not clear (Raven, 1980).

Many unicellular algae, and all multicellular algae, have cell walls; such cells are hypertonic to the medium. Their possible phylogenetic relationship to the non-walled algae and to higher plants is shown in Figure 1. A cell wall seems to be a necessary (but not a sufficient) condition for the presence of a large vacuole (i.e., one with a larger volume than the rest of the cytoplasm; see Raven, 1976, 1980; Atkinson et al., 1974; Sicko-Goad et al., 1977; Fagerburg et al., 1979). Many marine algae with cell walls are able to regulate their internal osmotic pressure so as to maintain constant turgor over a wide range of external osmolarities (Hellebust, 1976; Bisson and Gutknecht, 1979; Bisson and Kirst, 1979a; Kirst and Bisson, 1979). In those algae with large vacuoles the major fraction of the internal osmolarity is always generated by inorganic ions, and in many cases there is a large enough change in some 'compatible solute' content on a whole cell basis to account for much of the cytoplasmic solute increase in cells adapted to high external osmolarities. This requires that all of the 'compatible solute' is in the cytoplasm; the evidence for such a location is generally indirect (Bisson and Kirst, 1979a; Kirst and Bisson, 1979). Furthermore, as pointed out by Wyn Jones et al. (1977), there is a substantial body of data on giant-celled marine algae (Raven, 1976; Findlay et al., 1978) suggesting higher cytoplasmic ion concentrations than expected, on comparative grounds, to be consistent with optimal cytoplasmic activity in the cytoplasm. However, it is noteworthy that a direct estimate of K^+ activity in the cytoplasm of *Griffithsia* gives a lower value than for the vacuole (Vorobiev, 1967).

Most 'vacuolate' marine algae use KCl, NaCl or a mixture of KCl and NaCl as their major osmotica (Bisson and Gutknecht, 1979; Bisson and Kirst, 1979a; Kirst and Bisson, 1979); in the case of algae with high 'average' intracellular Na^+ levels there is some evidence (Raven, 1976; Findlay et al., 1978) for relatively high K^+/Na^+ in the cytoplasm. In most algae the major ion pump involved in turgor generation is active Cl^- influx, although in some cases the major role is played by active K^+ influx (Bisson and Kirst, 1979a); hypersomotic adaptation generally involves a major effect on active ion uptake, while in hypotonic adjustment a major role is played by increased passive efflux of the pumped ion.

In addition to roles in increasing the cell surface area per unit of cytoplasm volume, and as repositories for inimical compounds, the vacuoles of algae can act as storage compartments for nutrients which could not be tolerated at high concentrations in the cytoplasm. The algae provide some of the best direct evidence (as opposed to the presumptive evidence more commonly available) for inorganic nutrient storage and subsequent use. *Laminaria* and *Alaria* of Northern seas accumulate NO_3^- at up to 150 mM (28,000 times the external concentration) in winter when NO_3^- is available, and use it for spring growth as photosynthate becomes available (Chapman and Craigie, 1977; Buggeln, 1978). What happens to the osmoregulation of the organism as NO_3^- is stored and then used is not clear, although it is of interest that mannitol shows changes approximately anti-parallel to NO_3^- (Chapman and Craigie, 1978).

In the marine walled algae with a small fraction of the cell occupied by vacuole the whole-cell content of inorganic cations generally exceeds that of intracellular anions, the balance being organic acid and amino-acid anions (Raven, 1976; Bisson and Kirst, 1979a), the total ionic concentration is less than that in the seawater medium.

Fresh-water algae with cell walls, like their marine ancestors, can either be multicellular or unicellular, and can have large or small vacuoles. Those with a small vacuolar fraction (e.g., *Chlorella*; Atkinson et al., 1974) have an excess of inorganic cations over inorganic anions (Raven, 1980). In many cases considerable turgor adaptation occurs under hyperosmotic stress, with 'compatible solutes' playing the major role in generating a higher intracellular osmolarity (Greenway and Setter, 1979). In vacuolate fresh-water algae (as with their marine relatives) the 'whole-cell' balance of inorganic cations and inorganic anions is much closer (Raven, 1976, 1980; Raven and De Michelis, 1979a). The extent of turgor regulation in response to altered external osmotic pressure is not clear. Cram (1976) discusses the work of Tazawa and co-workers on *Nitella flexilis* which suggests that internal osmolarity rather than turgor is the regulated parameter, and that generation of increased internal osmolarity can occur in the absence of exogenous ions, although the major osmotica are usually ions. Cram (1976) suggests that osmotic pressure regulation and turgor regulation are ecologically identical for submerged fresh-water organisms. While many Charophytes can grow in brackish water or at even higher osmolarities, it is only recently that 'phenotypic' rather than 'ecotypic' turgor regulation has been demonstrated (Bisson and Kirst, 1979b).

For the walled eukaryotic algae as a whole, much of the data is compatible with the cytoplasm having 50-300 mOsm inorganic ions and 20-100 mOsm 'metabolic anions', with additional osmolarity contributed by neutral (or anionic, digeneaside; Kirst and Bisson, 1979) 'compatible solutes'. Vacuolar osmotica are generally mainly inorganic. There is little evidence for major changes in the relative contents of organic and inorganic anions as part of a 'biochemical pH stat' depending on the form in which N is supplied to algae (Raven and Smith, 1974; Raven and De Michelis, 1979a,b).

TERRESTRIAL EUKARYOTES

Figure 1 shows the considerable taxonomic diversity of terrestrial phototrophic eukaryotes. Starting with those most closely related to the aquatic algae, the terrestrial algae and the lichens are poikilohydric plants (i.e., able to withstand desiccation in the vegetative state but with little ability to regulate their water content; Walter and Stadelmann, 1968). These algal cells generally have a relatively low extent of vacuolation, and by analogy with aquatic algae of similar structure we might expect them to have an excess of inorganic cations over inorganic anions, with the balance of anions being organic; relatively few direct analyses are available. Walter and Stadelmann (1968) consider that the low degree of vacuolation is an adaptation permitting large water losses without damage to the cytoplasm in these and other poikilohydric plants; however, it does not seem to be possible to generalize about structural correlates of resistance to large changes in water content (Tucker, Costerton and Bewley, 1975).

The osmotic constituents of the archeogoniates (bryophytes and pteriodophytes) were investigated by Walland and Kinzel (1964); these workers provide data for the (poikilohydric) gametophytes of bryophytes, and the sporophytes of pteridophytes; the latter are potentially homoiohydric (Raven, 1976; Walter and Stadelmann, 1968). In all cases investigated ions represented a major component (usually more than 50%) of the osmolarity of the expressed sap, and in all cases more than half of the anions were organic. While in some cases the cells only had a small fraction of vacuole, in others the vacuolation was as great as in terrestrial spermatophytes. Archegoniates thus contrast with their algal ancestors in that vacuolate cells contain a preponderance of organic anions rather than having mainly inorganic anions. Two possible contributory factors may be cited in relation to this dominance of organic anions. One relates to the possibility that understorey bryophytes may obtain a considerable fraction of their inorganic nutrients as leachate from the leaves of trees above them, and thus perforce reflect the composition of the leachate. The other possibility relates to the use of organic acid synthesis as a means of neutralizing excess OH^- generated in NO_3^- assimilation in cells remote from an extracellular sink for OH^- (e.g., the soil solution; Raven and Smith, 1974, 1976). Further work is needed to determine the involvement of these two factors in the dominance of organic anions in archegoniates.

Data on the ionic contribution to the intracellular osmolarity of spermatophytes are provided by Albert and Kinzel (1973), Ashcroft and Wallace (1976), Cram (1976), Nierhaus and Kinzel (1971), Oechssler (1968), Pitman and Cram (1977), Wallace et al. (1974) and Winter et al. (1976). Data expressed on a dry weight basis must be corrected for cell wall uronates, and for insoluble oxalates, before comparison with 'expressed sap' data. As in the archegoniates, ions contribute the bulk of the osmolarity, and organic anions exceed inorganic anions. Plants growing at low water potentials (halophytes and drought-enduring xerophytes) have higher intercellular osmolarity than glycophytes and mesophytes, and often have a higher proportion of inorganic anions (Cl^-, SO_4^{2-}). The major intracellular cation is usually K^+, although in many halophytes the Na^+ concentration is high. There is considerable variation in the phenotypic ability to regulate turgor with changes in external osmolarity.

If averaged over the whole cell, the ionic concentration in the halophytes would substantially inhibit enzyme activities (Flowers et al., 1977). Evidence for a cytoplasmic ionic composition consistent with little inhibition of enzyme activity comes from a variety of sources; the analogy between cytoplasm and phloem sap; compartmental (flux) analysis; organelle isolation and analysis; histochemistry; and the analysis of shoot and root apices which quantitative electron microscopy shows have a low degree of vacuolation (Raven, 1977c; Flowers et al., 1977; Wyn Jones et al., 1977; Gerson and Poole, 1972; Jeschke and Stelter, 1976; Lyndon and Robertson, 1979). We may provisionally conclude that the use of ions to generate osmolarities in excess of 500 mOsm does not occur in the cytoplasm, but may occur frequently in the vacuole of halophytes and drought-tolerating xerophytes. While we are dealing here only with solutes bearing a net charge, we must point out that sucrose and proline (for example) can contribute several hundred mOsM to vacuolar solutes.

Before turning to whole-plant aspects of the involvement of ions in osmoregulation in terrestrial spermatophytes, some factors which relate to the organic anion content of higher plant cells will be considered (cf. Cram, 1976; Raven and Smith, 1974, 1976). Higher plants grown with NH_4^+ as their N-source usually have considerably lower concentrations of organic anions in their tissues than do plants grown on NO_3^-. The higher organic anion concentration in NO_3^--grown plants reflects the operation of the biochemical pH-stat which generates a strong organic acid from neutral or weakly acid precursors as a means of neutralizing the excess OH^- resulting from NO_3^- assimilation; the salts of the organic acid(s) represent a 'slag' resulting from this process. However, when NH_4^+ is the N-source the production of organic N leads to a net generation of H^+ within the cell; even so, there is a considerable content of organic anions within the cells grown on NH_4^+, although the synthesis of this organic acid involves further excess H^+ production within the cell which must be excreted if cellular acid-base balance is to be maintained. We may conclude that although organic acid synthesis has a role (often a major one) in pH regulation during NO_3^- assimilation, there is a 'basal' level of organic anions regardless of the nature of the N-source. In those plants which neutralize essentially all of their excess OH^- generated in NO_3^- assimilation by synthesis of organic acid, the osmolarity of the 'slag' (inorganic cation plus organic anion) may exceed that required for turgor generation if all of the 'slag' is soluble; Raven and Smith (1976) have suggested that the production of oxalic acid as the 'pH-statting' acid, and the subsequent precipitation of oxalate as the Ca salt, may be a commonly used resolution of the osmotic regulation/pH regulation antithesis. Despite this precipitation of slag, it is commonly observed that both the ion content and the total osmolarity is higher in NO_3^--grown than in NH_4^+-grown plants (Kirkby, 1968; Huber and Watson, 1974; Chambers et al., 1980).

The osmotic problem created by pH-regulation during NO_3^- assimilation is exacerbated in those plants which synthesize more organic acid than is needed to neutralize the OH^- generated in NO_3^- assimilation; these plants excrete H^+ while growing on NO_3^-. This situation is found in Lycopersicon and Fagopyrum growing at low NO_3^- concentrations (Kirkby and Knight, 1977; Argyriadis et al., 1976), in a number of desert halophytes even at high external NO_3^- concentrations (Wallace et al., 1974; Ashcroft and Wallace, 1976), and in certain plants under Fe-deficient conditions (Aktas and van Egmond, 1979). While the halophytes may not produce excess osmotic potential by producing this excess of organic acids, the NO_3^--deficient and Fe-deficient plants may well do so; it is not clear why NO_3^- deficiency leads to this excess organic acid synthesis; in the case of Fe deficiency the availability of Fe is increased at the low rhizosphere pH values generated by net H^+ efflux. Further aspects of the involvement of organic acid synthesis and N-assimilation in acid-base balance will be discussed in relation to whole-plant aspects of osmoregulation.

At the whole cell level, the terrestrial plants differ from the aquatic plants in that at least some of their cells are at a considerable distance from the source of ions for generation of turgor pressure, and the sink for any excess ions which have been taken up or generated in metabolism. This means that the osmoregulation in shoot cells is closely related to the relationship between the rate of growth of the shoot and the rate of supply of compounds which generate osmotic potential in the xylem stream. Any excess solutes must be rendered osmotically inactive *in situ* (e.g., by precipitation -- a strategy already mentioned in relation to oxalate), excreted *via* salt glands (e.g., in some halophytes) or re-exported to the root in the phloem. The lack of universal occurrence of the mechanism (salt glands), or its quantitative inadequacy (e.g., NaCl transport in the phloem; Greenway et al., 1966; Robinson, 1971; Pitman and Cram, 1977; Raven, 1977b,c) suggests that very important regulatory properties must be sought at the level of solute loading into the xylem in the roots.

One important aspect of this regulation is clearly that of total salt loading into the xylem. For halophytes, the problem is largely one of *exclusion* of salt, since calculations based on transpiration per unit fresh weight increment, and the salt concentration in the medium and the shoot, show that the salt concentration in the xylem must only be of the order of a few percent of that in the surrounding seawater. The extent to which the endodermis constitutes a complete barrier to non-mediated salt entry is not clearly established. This exclusion problem is exacerbated by the fact that transpiratory convective fluxes of NaCl up to the root surface commonly exceed the net influx of NaCl into the root even in glycophytes (Nye and Tinker, 1977); in halophytes the much higher external NaCl concentration, together with the generally low ion diffusivity in saline soils, means that substantial accumulations of NaCl at the halophyte root surface are to be expected. Despite these problems, the required exclusion is achieved (see Pitman and Cram, 1977).

Another important aspect of the regulation of xylem loading in the root is that of insuring acid-base balance in the shoot. This has been discussed by Raven and Smith (1976), Cram (1976), Pitamn and Cram (1977) and by Raven (1977b,c). Shoot cells cannot deal with net excess H^+ either by metabolic neutralization or by retranslocation to the root whence excess H^+ can be excreted. This suggests that the organic acid anions found in the shoots of all plants, except those where NO_3^- reduction in the shoots is equivalent to the organic anions found there, must come from root metabolism (see Raven and Smith, 1976). This does not rule out interconversions of carboxylates in the shoot (we are not suggesting that cell wall uronates are transported up the xylem!) but does require that no _net_ organate synthesis shall occur in the shoots. The nature of the control signal from the shoot which determines how much organic anion is loaded into the xylem (and hence how much total organic acid is synthesized in the root) is not known. A case which would repay further study is that of _Capsicum_; most of the NO_3^- reduction in this plant occurs in the shoot (Steer, 1974), and it appears that the organic anion (probably malate plus oxalate) generated in neutralizing the excess OH^- produced in NO_3^- assimilation is a major osmoticum in leaves of _Capsicum_ grown under non-saline conditions (Bernstein, 1961, 1963). However, in the leaves of plants grown under saline conditions this organic anion is replaced by Cl^- (thus increasing the osmotic effectiveness of unit cation; Bernstein, 1961, 1963); what is not clear is how NO_3^- assimilation changes to accommodate this. Is NO_3^- reduction transferred to the roots of salinized plants, or is the organate moved down the phloem if NO_3^- reduction is retained in the leaves?

SECONDARILY AQUATIC EUKARYOTES

Figure 1 shows possible lines of evolution leading to secondarily aquatic phototrophic eukaryotes. The submerged aquatic mosses, liverworts, pteridophytes and angiosperms were probably derived from glycophytic terrestrial ancestors, while the seagrasses (secondarily aquatic marine angiosperms which normally live fully submerged) are thought by Den Hartog (1970) to have evolved from halophytic terrestrial ancestors rather than from freshwater submerged ancestors.

The ionic osmotica of the submerged aquatic bryophytes and tracheophytes are listed by Walland and Kinzel (1964), Lehtoranta (1956), Cram (1976) and Jeschke (1976). In all cases the preponderance of inorganic cations over inorganic anions, and the significant role of organic anions, resembles that of their immediate terrestrial ancestors (i.e., terrestrial bryophytes and tracheophytes; previous Section) rather than of their more distant ancestors (the algae) with which they now cohabit. It should be emphasized that these secondarily aquatic plants generally have vacuolate mature cells, so that their high content of organic anions cannot be attributed to a very high fraction of the cell being occupied by cytoplasm. The spatial problems related to disposal of H^+ or OH^- resulting from N-assimilation in terrestrial plants are absent from these plants living in an aqueous medium; the extent to which the land-plant phenomenon of biochemical neutralization of excess OH^- resulting from NO_3^- reduction is retained in their aquatic descendents is not known; nor is the extent of phenotypic turgor regulation in these organisms (cf. freshwater algae, Primarily Aquatic Eukaryotes Section).

The sea-grasses clearly show turgor regulation when exposed to hypo- or hypertonic media (Cram and Tyerman, 1978); regulation appears to involve changes in the whole-cell content of ions (mainly Na^+ and Cl^-) with no significant changes in the whole-cell content of possible 'compatible solutes' which might be involved in cytoplasmic changes of osmoticum (Cram and Tyerman, 1978; Drew, 1978). The growing zone at the leaf base is exposed to a lower osmotic pressure due to the characteristics of the leaf sheath (Tyerman, 1979). Preliminary results (Tyerman, cited by Pitman and Cram, 1977) suggest that the acid-base regulation related to NO_3^- assimilation in *Zostera* involves OH^- excretion rather than organic acid synthesis.

SYMBIOSIS AND TURGOR REGULATION

The phototrophs which form the chief subject matter of this article form a variety of symbiotic associations with heterotrophs (and with other phototrophs). There are important but little-known implications for osmoregulation in these symbioses. Marine (Dinyophyceae and *Platymonas*) and fresh-water (*Chlorella*) algae form intracellular associations with invertebrates; the marine algae are probably isotonic with their hosts, while the (walled) *Chlorella* is probably hypertonic. The nature of the immediate ionic environment of the algal cells, and its influence on algal ionic relations, deserves further investigation.

Phototroph associations with fungi include lichens and mycorrhizal and biotrophically parasitized higher plants; phototrophs also associate with nitrogen-fixing prokaryotes. Endomycorrhizal fungi (and other biotrophic symbionts) presumably maintain a turgor pressure greater than that in the host cells during the formation of intracellular structures involving invagination of the host plasmalemma (arbuscules, vesicles, haustoria, and infection threads in the Rhizobium-nodule system). Such excess pressure would be required to effect invagination, even though the host cell plasmalemma itself may be considerably increased before the fungus actually penetrates the cell (Cox and Sanders, 1974; Ingram et al., 1976). Although increased H_2O fluxes into mycorrhizal plants have been reported (Safir et al., 1971), the overall importance of the fungal compartment in water uptake has not been established. Net water flux from fungus to phototroph would involve a reversal of the water potential difference which produced the original arbuscules and vesicles.

The presence of mycorrhizal fungi in roots can alter the ionic composition of the host plant and the distribution of ions between root and shoot, with repercussions for turgor regulation (Smith, 1979; F. A. Smith, unpublished; Bowen and Smith, 1980; Chambers et al., 1980). In Trifolium subterrraneum, mycorrhizal infection decreases K^+/Na^+ in roots but increases K^+/Na^+ in shoots (F. A. Smith and S. E. Smith, unpublished).

Finally, we must mention angiosperm parasites on angiosperms; generally "xylem-feeding" parasites have similar ionic compositions to the bulk shoot of the host, while "phloem-feeders" have less Na^+, Ca^{2+} and other relatively phloem-immobile elements (Ziegler, 1975; Wallace et al., 1978). Both types require a more negative osmotic potential than the host to obtain water from it; the phycobionts of lichens are in a similar position with respect to water supply (from the mycobiont) and loss (to the atmosphere).

SUMMARY AND CONCLUSIONS

1) The nature of the organic compounds produced in the period of prebiotic synthesis, together with various ionic requirements 'imprinted' on cytoplasm by early selective pressures, can account for many of the present-day constraints on the ionic composition of cytoplasm.

2) The total ion concentration in eukaryotic cytoplasm ranges from 50-60 mOsm in freshwater flagellates to 500-600 mOsm (or perhaps more) in halophytes. K^+ generally exceeds Na^+ in concentration; free Ca^{2+} and free H^+ are about 0.1 μM; and the major anions are organic (carboxylates and phosphates) with relatively low concentrations of free inorganic anions (including Cl^-). The shortfall of osmolarity between that accounted for as ions and the total required for volume or turgor regulation is made up with 'compatible solutes'.

3) Constraints on vacuolar composition are less severe than those on the cytoplasm; in many primary aquatic organisms (algae)

the vacuolar contents are almost all inorganic, while in terrestrial glycophytes and secondarily aquatic freshwater plants there is a large content of organic anions. Halophytes (terrestrial and secondarily marine) show a more diverse ionic composition of their vacuoles. K^+ generally exceeds Na^+ in the vacuole, except in certain halophytes (both primary and secondary).

4) The extent of phenotypic turgor regulation is variable; most halophytes can acclimate to a range of external osmolarities of at least 300 mOsM, with the major vacuolar osmoticum which alters during adaptation being inorganic ions. Glycophytes may regulate turgor to a limited extent; the upper limit on regulation may be imposed by the inability to regulate the ionic content of the cytoplasm.

5) Turgor regulation in the shoots of terrestrial vascular plants, as well as the closely related phenomenon of acid-base regulation in the shoot, is mainly achieved by regulation of the quantity and nature of solutes loaded into the xylem in the root. Halophytes with salt glands can effect major regulation of shoot osmoticum content subsequent to the xylem-loading step.

6) Symbiotic associations pose important but little investigated problems of ionic involvement in turgor generation. Economically important examples include nitrogen-fixing symbioses, and mycorrhizal plants, in saline habitats.

7) Osmotic and turgor regulation involving ions is a very complex phenomenon, with repercussions for pH regulation as a function of N source and for Fe nutrition. It would be over-optimistic to expect very significant short-term economic results from genetic engineering of ionic osmoregulation, given the breadth of the metabolic events that are involved.

ACKNOWLEDGMENTS

We wish to thank Drs. W. G. Allaway, M. Bisson, G. O. Kirst and S. Tyerman (University of Sydney) for stimulating discussions and for providing material prior to publication.

REFERENCES

Aktas, M. and van Egmond, F., 1979, Effect of nitrate nutrition on iron utilization by an Fe-efficient and an Fe-inefficient soybean cultivar, Pl. Soil, 51:257.

Albert, A. and Kinzel, H., 1973, Unterscheidung von Physiotypen bei Halophyten der Neusiederseegebeites (Osterreich), Z. Pflanzenphysiol., 70:138.

Argyriadis, G. A., Dijkshoorn, W., and Lampe, J. E. M., 1976, Level and origin of carboxylate in buckwheat, Pl. Soil., 44:669.

Ashcroft, R. T. and Wallace, A., 1976, Sodium relations in desert plants. 5. Cation balance when grown in solution culture and in the field of three species of *Lycium* from the Northern Mojave Desert, *Soil Sci.*, 122:48.

Atkinson, A. W., Jr., Gunning, B. E.S., John, P. C. L., and McCullough, W., 1972, Dual isotherms of ion absorption, *Science*, 176:694.

Bernstein, L., 1961, Osmotic adjustment of plants to saline media. II. Steady State, *Am. J. Bot.*, 48:909.

Bernstein, L., 1963, Osmotic adjustment of plants to saline media. II. Dynamic phase, *Am. J. Bot.*, 50:360.

Bisson, M. A. and Gutknecht, J., 1979, Osmotic regulation in algae, *in*: "Membrane Transport in Plants," R. M. Spanswick and W. J. Lucas, eds., North-Holland, Amsterdam.

Bisson, M. A. and Kirst, G. O., 1979a, Osmotic adaptation in the marine alga *Griffithsia monilis* (Rhodophyceae): the role of ions and organic compounds, *Aust. J. Pl. Physiol.*, 6:523.

Bisson, M. A. and Kirst, G. O., 1979b, The brackish-water charophyte, *Lamprothamnium* : membrane potentials and osmotic responses, *in*: "Membrane Transport in Plants," R. M. Spanswick and W. J. Lucas, eds., in press, North-Holland, Amsterdam.

Bowen, G. D. and Smith, S. E., 1980, The effects of mycorrhizas on nitrogen uptake by plants, *Ecol. Bull.*, in press.

Buggeln, R. G., 1978, Physiological investigations on *Alaria esculenta* (Laminariates, Phaeophyceae). IV. Inorganic and organic nitrogen in the blade, *J. Phycol.*, 14:156.

Chambers, C. A., Smith, S. E., and Smith, F. A., 1980, Effects of ammonium and nitrate ions on mycorrhizal infection, nodulation and growth of *Trifolium subterraneum*, New Phytol., 85:(in press).

Chapman, A. R. O. and Craigie, J. S., 1977, Seasonal growth in *Laminaria longicruris* : relations with dissolved inorganic nutrients and internal reserves of nitrogen, *Mar. Biol.*, 40:197.

Chapman, A. R. O. and Craigie, J. S., 1978, Seasonal growth in *Laminaria longicruris* : relations with reserve carbohydrate storage and production, *Mar. Biol.*, 46:209.

Cox, G. C. and Sanders, F. E. T., 1974, Ultrastructure of the host-fungus interface in a V-A mycorrhiza, *New Phytol.*, 73:901.

Cram, W. J., 1976, Negative feedback regulation of transport in cells. The maintenance of turgor, volume and nutrient supply, *in*: *Encyclopedia of Plant Physiology, New Series, Vol. IIA*, U. Lüttge and M. G. Pitman, eds., Springer-Verlag, Berlin.

Cram, W. J. and Tyerman, S., 1978, Turgor regulation in sea grasses, *Proc. Inaugural Meeting F.E.S.P.P.*, Edinburgh, July, 1978.

Drew, E. A., 1978, Carbohydrate and inositol metabolism in the seagrass *Cymodocea nodosa*, *New Phytol.*, 81:249.

Fagerburg, W. R., Moon, R., and Truby, E., 1979, Studies on *Sargassum*. III. A quantitative ultrastructural and correlated physiological study of the blade and stipe organs of *S. filipendula*, *Protoplasma*, 99:247.

Findlay, G. O., Hope, A. B., Pitman, M. G., Smith, F. A., and Walker, F. A., 1978, Ionic relations of the marine alga Valoniopsis pachynema, Aust. J. Pl. Physiol., 5:675.

Flowers, T. J., Troke, P. F., and Yeo, A. R., 1977, The mechanism of salt tolerance in halophytes, Ann. Rev. Pl. Physiol., 28:89.

Gerson, D. F. and Poole, R. J., 1972, Chloride accumulation by mung bean root tips. A low affinity active transport system at the plasmalemma, Pl. Physiol., 50:603.

Gimmler, H. and Schirling, R., 1978, Cation permeability of the plasmalemma of halotolerant alga Dunaliella parva. II. Cation content and glycerol concentration of the cells as dependent upon NaCl concentration, Z. Pflanzenphysiol., 87:435.

Greenway, H. and Setter, T. L., 1979, Na^+, Cl^- and K^+ concentrations in Chlorella emersonii exposed to 100 and 335 mM NaCl. Aust. J. Pl. Physiol., 6:61 (cf. Aust. J. Pl. Physiol. 6:571).

Greenway, H., Gunn, A., and Thomas, D. A., 1966, Plant responses to saline substrates. VIII. Regulation of ion concentrations in salt-sensitive and halophytic species, Aust. J. Biol. Sci., 19:741.

Gutknecht, J. and Dainty, J., 1968, Ionic relations of marine algae, Oceanogr. Marine Biol. Ann. Rev., 6:163.

Halman, M., 1974, Models of prebiological phosphorylation, in: "Cosmochemical Evolution and the Origins of Life," J. Oro, S. L. Miller, C. Ponnamperuma, and R. S. Young, eds., D. Reidel Publ. Co., Dardrecht-Holland.

Hartog, C. den, 1970, "The Sea-Grasses of the World," North-Holland, Amsterdam.

Hellebust, J. A., 1976, Osmoregulation, Ann. Rev. Pl. Physiol., 27:485.

Huber, D. M. and Watson, R. D., 1974, Nitrogen form and plant disease, Ann. Rev. Phytopathol., 12:319.

Ingram, D. S., Sargent, J. A., and Tommerup, I. C., 1976, Structural aspects of infection by biotrophic fungi, in: "Biochemical Aspects of Plant-Parasite Relationships," J. Friend and J. D. R. Threfall, eds., Academic Press, London.

Jeschke, W. D., 1976, Ionic relations of leaf cells, in: "Encyclopedia of Plant Physiology, New Series, Vol. IIB, U. Lüttge and M. G. Pitman, eds., Springer-Verlag, Berlin.

Jeschke, W. D. and Stelter, W., 1976, Measurement of longitudinal ion profiles in single roots of Hordeum and Atriplex by use of flameless atomic absorption spectroscopy, Planta, 128:107.

Kirkby, E. A., 1969, Ion uptake and ionic balance in plants in relation to the form of nitrogen nutrition, Symp. Ecol. Soc., 9:215.

Kirkby, E. A. and Knight, A. H., 1977, Influence of the level of nitrate nutrition on ion uptake and assimilation, organic acid accumulation and cation-anion balance in whole tomato plants, Pl. Physiol., 60:349.

Kirst, G. O., 1977a, Coordination of ionic relations and mannitol concentrations in the euryhaline alga Platymonas subcordiformis (Hazen) after osmotic shocks, Planta, 135:69.

Kirst, G. O., 1977b, Ion composition of unicellular marine and fresh-water algae, with special reference to Platymonas subcordiformis cultivated in media with different ionic strengths, Oecologia, 28:177.

Kirst, G. O. and Bisson, M. A., 1979, Turgor pressure regulation in marine algae. Ions and low molecular weight compounds, Aust. J. Pl. Physiol., 6:539.

Lehtoranta, L., 1956, The cation and chloride content of cells of aquatic plants, Ann. Botan. Soc. Zool. Botan. Fennicae (Vanamo), 29:1.

Lyndon, R. F. and Robertson, E. S., 1979, The quantitative ultra-structure of the pea shoot apex in relation to leaf initiation, Protoplasma, 87:387.

Nierhaus, D. and Kinzel, H., 1971, Vergleichende Untersuchungen über die organischen Säuren in Blättern höheher Pflanzen, Z. Pflanzenphysiol., 64:107.

Nye, P. H. and Tinker, P. B., 1977, "Solute movement in the soil-root system," Blackwells Scientific Publications, Oxford.

Oechssler, G., 1968, Jahreszeitliche Schwankungen des Gehaltes an organischen Säuren in den Nadeln von Pseudotsuga menziesii (Mirb.) Franco, Picea abies (L) H. Karsten und Larix decidua Mill., Z. Pflanzenphysiol., 59:213.

Osmond, C. B., 1979, Ion uptake, transport and excretion, in: "Arid-Land Ecosystems: Structure, Functioning and Management," Vol. 1, R. A. Perry and D. W. Goodall, eds., Cambridge University Press, Cambridge.

Pitman, M. G. and Cram, W. J., 1977, Regulation of ion content in whole plants, Symb. Soc. Exp. Biol., 31:391.

Raven, J. A., 1976, Transport in algal cells, in: "Encyclopedia of Plant Physiology, New Series, Vol. IIA," U. Lüttge and M. P. Pitman, eds., Springer-Verlag, Berlin.

Raven, J. A., 1977a, Regulation of solute transport at the cell level, Symp. Soc. Exp. Biol., 31:73.

Raven, J. A., 1977b, The evolution of vascular land plants in relation to supracellular transport processes, Adv. Bot. Res., 5:153.

Raven, J. A., 1977c, H^+ and Ca^{2+} in phloem and symplast: relation of the relative immobility of the ions to the cytoplasmic nature of the transport paths, New Phytol., 79:465.

Raven, J. A., 1980, Nutrient transport in microalgae, Adv. Microb. Physiol., in press.

Raven, J. A. and De Michelis, M. I., 1979a, Acid-base regulation during nitrate assimilation in Hydrodictyon africanum, Plant Cell and Environment, 2:245.

Raven, J. A. and De Michelis, M. I., 1979b, Acid-base regulation during NH_4^+ and NO_3^- assimilation in Hydrodictyon africanum, in: "Membrane Transport in Plants," R. M. Spanswick and W. J. Lucas, eds., North-Holland, Amsterdam, in press.

Raven, J. A. and Smith, F. A., 1974, Significance of hydrogen ion transport in plant cells, Can. J. Bot., 52:1035.

Raven, J. A. and Smith, F. A., 1976, Nitrogen assimilation and transport in vascular plants in relation to intracellular pH regulation, New Phytol., 76:415.

Raven, J. A. and Smith, F. A., 1979, The chemiosmotic viewpoint, in: "Membrane Transport in Plants," R. M. Spanswick and W. J. Lucas, eds., North-Holland, Amsterdam, in press.

Robinson, J. B., 1971, Salinity and the whole plant, in: "Salinity and Water Use," T. Talsma and J. R. Philip, eds., Macmillan, London.

Safir, G. R., Boyer, J. S., and Gerdemann, J. W., 1971, Mycorrhizal enhançement of water transport in soybean, Science, 172:581.

Schötz, F., Bathelt, H., Arnold, C-G., and Schimmer, A., 1972, Die Architektur und Organisation der Chlamydomonas-Zelle: Ergebnisse der Elektronemikroscopie von Serialschlnitten und der daraus resultierenden dreidimensionalen Rekonstruktion, Protoplasma, 75:229.

Sicko-Goad, L., Stoermer, E. F., and Lademski, B. G., 1977, A morphometric method for correcting phytoplankton cell volume estimates, Protoplasma, 93:147.

Smith, F. A. and Raven, J. A., 1978, The evolution of H^+ transport and its role in photosynthetic energy transduction, in: "Light Transducing Membranes: Structure, Function and Evolution," D. W. Deamer, ed., Academic Press, New York.

Smith, F. A. and Raven, J. A., 1979, Intracellular pH and its regulation, Ann. Rev. Pl. Physiol., 30:289.

Smith, S. E., 1979, Mycorrhizas of higher plants, in: "Handbook of Food and Nutrition," M. Rechcigl, Jr., ed., C.R.C. Press, Inc., Palm Beach, FL, in press.

Steer, B. T., 1973, Diurnal variations in photosynthetic products and nitrogen metabolism in expanding leaves, Pl. Physiol., 51:744.

Tucker, E. G., Costerton, J. W., and Bewley, J. D., 1975, The ultrastructure of the moss Tortula ruralis on recovery from desiccation, Can. J. Bot., 53:94.

Tyerman, S., 1979, Turgor regulation and the development of water potential gradients in Posidonia, in: "Membrane Transport in Plants," R. M. Spanswick and W. J. Lucas, eds., North-Holland, Amsterdam, in press.

Vorobiev, L. N., 1967, Potassium ion activity in the cytoplasm and the vacuole of cells of Chara and Griffithsia, Nature, 216:1325.

Wallace, A., Romney, E. M., and Alexander, G. C., 1978, Mineral composition of Cuscuta nevadensis Johnston (dodder) in relation to its hosts, Pl. Soil, 50:227.

Wallace, A., Romney, E. M., Cha, J. W., and Alexander, G. V., 1974, Sodium relations in desert plants: 3. Cation-anion relationships in three species which accumulate high levels of cations in leaves, Soil Sci., 118:397.

Walland, A. and Kinzel, H., 1966. Über die Zusammensetzung der Zellsäfte bei Archegoniaten, Flora, 156A:597.

Walter, H. and Stadelmann, E. J., 1968, The physiological prerequisites for the transition of autotrophic plants from water to terrestrial life, Biosci., 18:694.

Winter, K., Troughton, J. H., Evenari, M., Lauchli, A., and Lüttge, U., 1976, Mineral ion composition and occurrence of CAM-like diurnal malate fluctuations in plants of coastal and marine habitats of Israel and the Sinai, Oecologia, 25:125.

Wyn Jones, R. G., Storey, R., Leigh, R. A., Ahmad, N., and Pollard, A., 1977, in: "Regulation of Cell Membrane Activities in Plants," E. Marré and O. Cifferi, eds., North Holland, Amsterdam.

Ziegler, H., 1975, Nature of transported substances, in: "Encyclopedia of Plant Physiology, New Series, Vol. I," M. H. Zimmerman and J. A. Milburn, eds., Springer-Verlag, Berlin.

PANEL DISCUSSION ON MOLECULAR BIOLOGY OF OSMOREGULATION BY MICROORGANISMS

Discussants: R. C. Valentine (Chair), J. Roth, R. Vinopal, L. Csonka, A. D. Brown, J. Raven, and A. Ben-Amotz

R. C. Valentine: In a nutshell, where are the frontiers and what are the exciting experiments to come? Please keep in mind that the orientation of this meeting emphasizes higher plants as well as microorganisms.

J. Roth: If I were going to speculate as to how one might go about studying osmoregulation in bacteria in a systematic way, I would utilize a sort of standard genetic approach that has been fruitful in many areas of molecular biology in the past. To start with, I'd choose an organism for which there is some genetic background. The enteric bacteria (e.g., E. coli, Salmonella) and yeast would be the two microbes that I would explore once it is established that they exhibit some kind of osmoregulation. I'd then proceed to isolate mutants that are unable to osmoregulate. Such mutants would probably appear as some sort of salt-sensitive mutants. This general approach involving isolation of mutants that are defective in the particular property you are interested in studying and then analyzing each of these mutants and in turn classifying them and trying to get some feeling for the complexity of the phenomenon has been very successful for a variety of systems and a variety of processes. Now it's no guarantee that it would work in this case. One just might get into horrible complexities but at least it is a systematic approach that one could take as a first step. This approach has not yet been taken extensively though some workers are starting to do it a bit now, but it seems to me that that might show some promise.

G. Brosseau: What organisms would you choose? Could you set out your agenda?

J. Roth: Well, to do that with a heavy genetic slant, which is my prejudice, you need an organism for which there is a background of genetic data. If you picked an organism at random then there are a lot of background information that has to be accumulated before you can attack the property you are really interested in. At the moment the genetics of E. coli and Salmonella are in pretty good shape and the same is true for the genetics of a couple of yeasts, also Neurospora. So I think even though they may not be the nicest organisms for your particular interest that to rough out the phenomenon these relatively simple systems might give a picture that can be applied to other organisms. The advice again is that you don't have to work out the whole thing in a system where the genetics is difficult.

G. Brosseau: That makes sense. Do you have reason to believe that in the organisms that are most suited for experimental attack, that they have the kinds of genetic variability needed, for instance salt tolerance? Is there any reason to believe that this exists?

J. Roth: That is a complication; obviously if they don't do the phenomenon they are not going to be much good to analyze. But the Enterics, for example, grow over a range of salt, say from no addition to between 0.5 to 1 molar sodium chloride.

G. Brosseau: Is the basic point that I am hearing from the panel that there are already certain organisms that are well suited for this kind of genetic analysis. What I am concerned about goes back to basic principles. The first thing that a geneticist works with is a difference and so what I was curious about is are there reasons to believe that the kinds of differences that are of importance and interest here already exist in the organisms discussed at this meeting? Or do we have to make some investments in the spade work in another group? On the other hand, if the organisms you've discussed are suitable, then I agree with you. Let's go all out and plow the ground that we know.

J. Roth: What do you mean by differences?

G. Brosseau: Do they show or can they tolerate a range of, let's say, salinity? Or alternately, are you aware whether there are strains that differ from one another in some kind of definable trait?

J. Roth: To start out, you need to have an organism called wild type, that can regulate its composition or physiology so as to adapt quickly to a range of exogenous salt concentrations. If the wild type organism adapts over a range that is wide enough to really convince you that there is an osmoregulatory phenomenon then you can select for mutants that are unable to adapt. In this way you generate the differences by the mutational procedure.

G. Brosseau: That will be fine.

R. C. Valentine: I had hoped from this morning's session that perhaps the least information that we would have gotten across is that we are now truly satisfied that the osmoregulatory (Osm) trait is present in many bacteria. That is the Osm gene or traits, however complicated it turns out to be, is indeed present and is carried by the wild type strains of E. coli, Salmonella and yeast. We are totally convinced of that. Just to add to what John Roth had to say, it seems to me it's very logical which organisms you are going to go after for the really basic aspects of the work.

G. Brosseau: Okay, I guess one of the things that I was struck by at the Boyce Thompson Conference a couple of weeks ago was that researchers reported on efforts to develop the genetics of Halobacterium. This turns out to be a very complex genetic system; for example, mutational rates of one in a thousand, to one in ten thousand were described. Harking back to my, I am afraid, outdated days in the field of genetics, this is a signal that you are dealing with an unstable genetic system.

R. C. Valentine: I would follow George's comments by saying that we believe a basic, core system of osmoregulation is seen throughout the world of microbes, extending on up through algae, yeast and on into the plant world. Halobacterium and even certain marine plants (e.g., mangrove) have evolved rather specialized mechanisms of salt tolerance, mechanisms which may turn out to be atypical of our current crop plants. I think that the organisms that are salt tolerant are the ones you should be looking at. Organisms, such as many gram-negative bacteria, that can grow from zero to about one molar salt fit that category. The salt tolerant organisms are the ones that we feel we should concentrate on rather than Halobacterium, although this organism has been found to possess many exiting biochemical properties. I hope there are some people here who will defend Halobacterium. Although a genetic system for Halobacterium is of enormous basic interest you are going to have to invest a great deal of effort to get this system up to par compared to the more extensively studied bacterial systems.

G. Brosseau: We need this kind of information spelled out as specifically as it can be; because it is only when we have these kind of specific recommendations that they can be translated into action programs.

R. Vinopal: It might be worth pointing out that mutants of E. coli with restrictive osmotic range already exist. Strains that are deficient in making putracine are unable to grow in dilute media. They are able to grow in normal medium but if you decrease the concentration, they slow up and finally require being fed putracine from the outside in order to grow in dilute medium. It's thought that it works this way. When you're growing a bacterium in a concentrated medium, there is lots of potassium on the inside. As the medium gets more and more dilute, the potassium content drops and the ionic strength of the cytoplasm drops as well. It's important for interactions of macromolecules to preserve ionic strength inside the cytoplasm. As the potassium drops, the putracine increases in concentration. Ionic strength is a function of the square of the charge so as potassium with a single charge comes down in dilute medium, putracine with two positive charges increases. It's four times as active in terms of ionic strength, but particle for particle it has the same osmotic effect. So there has been quite a considerable literature on the requirement for polyamines for growth in dilute medium. So far half of the mutant analysis has already been done, at least a start covering one half of the range. What we need to find now are mutants that are specifically salt sensitive, unable to adjust at the high osmotic concentration end of the range.

R. C. Valentine: Bob, could you briefly outline a scheme for how we are going to isolate osmoregulatory muants? What classes might you expect?

R. Vinopal: As shown in the introductory slide of my talk earlier today, it's known that for at least Enteric bacteria, that when any osmotically active substance increases in concentration in the growth medium, what happens is that the cells increase the osmotic concentration of their cytoplasm. They are osmotic conformers, but they are ionic regulators. The things that go up outside are not the same things that are increased inside. And once again, the things that increase in bacteria inside are proline, glutamic acid, and gamma aminobutyric acid, which is simply the decarboxylation product of glutamic acid. If you take a look at most bacteria, one or more of these substances increases in concentration in the cytoplasm to match the cells' surroundings. It's often hard to know what the exact concentrations are, and even these values aren't what we really want. We want the osmotic activities. But there's a pretty close match between the increase in osmotic concentration in the growth medium and the sum of these responses on the inside. In this regard, there's an interesting paper in NATURE, published two or three years ago, by John Measures, in which he used an amino acid analyzer to examine the cytoplasm of a whole range of bacteria, both gram-negative and gram-positive under osmotic stress. Measures arrived at the generalization mentioned above. Now there has to be some sort of a transducer, something to sense the increase of of osmotic concentration in the medium and direct the increase in the cell of compensating substances. It would be of considerable interest to test this hypothesis by mutant analysis. Here's Measures' idea. Glutamic dehydrogenase is potassium activated in all gram-negative bacteria that have been examined. So with that fact, here is the scenario. Up goes the salt concentration. The bacterium is caught by surprise, water is withdrawn by osmosis. As the cell shrinks with the volume getting smaller on the inside, the potassium that is already there becomes more concentrated. Potassium activates glutamic dehydrogenase; more glutamate is made that's converted on to proline and so forth. The potassium then is pumped in by active transport until it just balances the negative charge on glutamic acid. So, here's the hypothesis already. There's no question that an increase in the osmotic concentration of the medium will trigger this response. If you want to test this notion that is based on glutamic dehydrogenase, then I suppose you look for mutants that are sensitive to salt, that can't respond osmotically. These should be damaged somewhere in the system. It might be a little tricky, because an organism that can't make glutamate is at least under some conditions unable to grow at all, because it needs glutamate directly for protein synthesis and indirectly for proline which is obtained from glutamate. But there is at least a possibility here.

R. C. Valentine: There are lots of glutamate mutants available already?

R. Vinopal: Yes there are lots of glutamate mutants available. You have to feed them glutamate. It looks as though such an organism will use proline and glutamate from the medium preferentially for protein synthesis, so will cells use proline and glutamate from the medium preferentially for osmotic regulation. So there's probably two different kinds of regulation, not only a response like this, to make more proline and glutamate endogenously, in minimal medium, but if proline and glutamate are available in the medium, it looks as though there's likely to be some sort of regulation of the transport system so that during osmotic stress they're turned on allowing the concentration of these substances for osmotic balance. It's worth saying that the proline transport system in *Salmonella* and *E. coli* are activated by sodium. It's not the common situation for transport systems in Enteric bacteria. This may be trying to tell us that these transport systems are regulated not only to bring in proline and glutamate for protein synthesis, but also for osmoregulation. So a standard sort of mutant analysis could be brought to bear for selection of salt sensitive mutants.

J. Radin: You're listening to a plant physiologist talking here, but what you fellows are proposing is that we take an organism which already has the capability to adapt to high salt and look at mutants which have lost that ability. What we really want with plants is a plant which does not adapt to high salt and make it adapt in some way. So you fellows have it "bass ackwards". What guarantee do we have that the mechanisms you're looking at are the most appropriate ones for when the situation is reversed?

R. Vinopal: No guarantee at all, except in the past, in every case when you have basic information about how one organism does it in detail, then it has always helped in understanding other systems. I think it's misguided to try to take something like *E. coli* and make it able to grow in two molar sodium chloride. It's not going to be able to. If you want an organism that does beautifully in high salt, you should go out into nature and look in high salt environments and find an organism there. What you can do with *E. coli* is: 1) you know that it responds to osmotic stress; 2) it has all the machinery; 3) use mutant analysis to find out how it does it. The best way to do it is by loss of function. In the past it's always been useful to have basic information about one system and try to extrapolate it to another. I think that's what *E. coli* and other genetically tractable organisms are good for.

J. Roth: Yes. It's just a basic difference in thrust between your goal and what we're suggesting. You're talking about solving the immediate problem that needs solving. We're proposing ways of analyzing how cells approach the problem of osmoregulation. And our faith is that if you understand how cells adapt to high salt, then you could devise ways of solving your problem. That is, of helping them do that or improving how they do that. But I think generally, it's a lot like strains that have been developed for industrial purposes. It's been largely empirical that you select and you look for strains which do a little bit better in the direction you want to go and you keep doing that. And that works to a degree, but ultimately, at least with bacterial systems, you get to a place where that empirical method fails and now you need to have some quantum jump in understanding how to do the next big engineering trick. And the idea is if you understood exactly how they adapt to salt, then maybe you could play some sneaky trick on them to put together some combination of super mutations to fool them. But you need to understand the mechanism. That's sort of the article of faith.

G. Brosseau: Well, I guess my question is really why does this relate to the very well understood genetics of only a few organisms? Why can't this be studied in other organisms as well?

R. Vinopal: The basic approach of doing genetics involves being able to isolate a lot of mutants, grouping them by either genetic mapping or by complementation tests into functional groups, and then studying each of these functions individually. In order to carry out these fairly standard genetic manipulations it is essential to have an organism where there are other known genetic markers for mapping, complementation testing, etc. There simply aren't very many organisms to choose from without a lot of background work. I emphasize again that all that ground work has already been done in the Enterics. It just saves an enormous amount of work.

J. Raven: I agree that this is a very nice model. What I would like to point out is that it appears that most plants don't use the GDH mechanism for making glutamate anyway, at least in organisms like Chlorella grown in potassium deficient conditions. If you add potassium, they do indeed increase their internal osmolarity by synthesizing potassium glutamate, but it seems highly likely they'll be making this through the glutamine synthetase (GS)-glutamate synthase (GOGAT) pathway. Does anyone know if GS or GOGAT is potassium stimulated in eucaryotes or would one be relying on some pH dependence of the GS-GOGAT pathway? The potassium deprived organisms seem to have a relatively low internal pH and I know that it's been proposed for some bacteria that the regulation of GS at least is based on changes in the pH. This isn't knocking the fundamental approach, but I think one has to be a little wary if in fact the eukaryotic organisms is using an entirely different system for synthesizing glutamate from the prokaryote.

R. C. Valentine: As you know the glutamate synthase (GOGAT) pathway was first described in Enteric bacteria; the plant system was found later. So we have a lot of information about the GS-GOGAT in bacteria.

R. W. Breidenbach: I think I could not agree more in what's been said about using the bacterial systems as a way to solve parts of the problem. I think it's an important direction that has to be taken. However, one word of caution is that plants do many things that bacteria don't (vacuoles, mitochondria, etc.) and in this sense are far more complicated than E. coli. In short, we have to realize that the translation of bacterial information is going to go up to some point and then one has to really have information on the kinds of higher organisms that you're interested in applying the problem to. We need to start now in building a suitable background of genetic understanding of plants. One of the things that was really interesting about Laszlo's talk this morning was his talking about the simple application of the Osm genes for plants. Unfortunately, we cannot yet select for mutants in plants.

J. Roth: Going on with the discussion of the choice of organism. The halophiles are extremely interesting, but when you look inside you find that every enzyme is profoundly different from the enzymes of other organisms, many more acidic amino acid residues and so forth. Obviously in developing salt tolerant plants, you can't very well change every single enzyme. It's conceivable that there may be an organism out there, perhaps a moderate halophile which, instead of changing every enzyme, has some sort of osmotic stabilizing compound floating around in the cytoplasm. In fact, you might be lucky, and find the whole business on a plasmid. This kind of thing has certainly been seen in other situations. You look in nature for an organism that utilizes kemp, well the kemp utilization genes may be are on a plasmid. So it's worthwhile looking, I think, for halophiles with enzymes with roughly the same kind of amino acid composition that normal creatures have, and then asking what is it about these organisms that makes them relatively resistant to high salt. You may find a protective agent like proline, or something much more effective that could be moved around with genetic engineering techniques. So that going to nature is always a smart thing to do. It's not a substitute for genetic analysis of genetically tractable organisms, but it's something that should be done.

R. C. Valentine: I would like to ask Laszlo Csonka to comment about his approach of selecting salt tolerant mutants.

L. Csonka: Well, I'm not sure I can really add much more to John Roth's and Bob Vinopal's comments about the importance for developing classical genetic approaches. I think the interesting thing is the response, rather than the actual amounts that various organisms can tolerate. For instance, the fact that you can grow *Salmonella* in either zero concentrations of sodium chloride or very little, or up to, say, approximately one molar, the range of concentration difference it has to respond to is perhaps as profound as for halophiles. So, really to learn about how one can engineer plants that can tolerate more salt, one has to look at the response rather than the actual concentrations that they can tolerate. And as far as the approaches, it's just a philosophical difference. I kind of think, although I will give very serious consideration to John's suggestion about looking for mutants that are salt sensitive, that there's probably many more ways that you can take a system apart than to make it better. In other words, the chances of picking up mutants that are uninteresting are much more likely using the negative mutant approach. If you go the route of getting salt sensitive mutants, though, it may be productive. There also may be more fishing around to do, whereas if you can go the other way, of taking a *Salmonella* or *E. coli* or yeast, and getting a mutant that can grow in more salt, slightly faster, that may perhaps be a more direct route. And again, the point is not to look necessarily for a dramatic improvement, such as asking yeast to grow in two molar salt, but maybe asking yeast to grow faster in some inhibitory salt concentration.

R. Vinopal: The trouble with trying to improve bacteria more and more for any given trait is that in practice you very often run up against a stone wall where there are many things becoming growth limiting at the same time. For example, people who are studying the temperature range for growth of wild type bacteria have found that it's impossible to increase the maximum temperature for growth or to decrease the minimum temperature for growth with a single mutation. The interpretation has been that there are very many enzymes ready to become sensitive at roughly the same temperature.

So the change in no one of them will confer no appreciable or discernible increase in the maximum temperature for growth. It may be that looking for improved mutants, although Laszlo has done it with proline for osmotic stress, he may now be stopped at the point where no further improvement will be possible. Especially if you need two mutations, then the chances are that you won't see them. Each mutation at one in 10^8; two is one in 10^{16}; if you have three that's one in 10^{24} or whatever. If you understand what's going on you may be able to construct or put together the three and not need to come upon them spontaneously.

L. Csonka: Well, I'm not sure that I'm that pessimistic because, for instance, if one were to look at mutants that grow faster in salt by direct selection and come across another class such as, perhaps, the class Dr. Brown suggested earlier that have some alteration in potassium uptake, then maybe they will add a little improvement. If you put that together with the proline mutants, you might get a much more dramatic sum total effect.

R. Vinopal: Right, but how did you find the proline mutant? You didn't do that by direct selection at all. You had a preconceived notion which came from analytic work that other people did with that background. You tried it and it worked.

L. Csonka: Although I don't understand why I couldn't have gotten it by direct selection either.

R. Vinopal: Did you try by direct selection?

L. Csonka: Yes, it didn't work (laughter).

R. C. Valentine: We haven't discussed the role of plasmids much. I'm excited about the bacterial systems because as far as I am aware, the first salt tolerant plasmid has been described at this meeting. Laszlo constructed that plasmid using conventional genetic engineering techniques. It is a conjugative plasmid that is movable by the standard conjugation systems from one bacterium to another though in this case plasmid movement is restricted because it is a so-called narrow host range or F-plasmid which doesn't permit sex with a very wide number of female hosts. But it's well known now that there are more promiscuous and even super promiscuous systems such as discussed this morning. The promiscuous system for conjugation of bacteria is widely used and is becoming one of the key tools in moving genes from one organism to another. It's a marvelous system for moving DNA among distinct organisms. I think it would be very interesting to construct promiscuous salt tolerant systems so you could move the Osm genes into any gram-negative bacterium, including organisms such as Rhizobium, which of course, is important in symbiotic nitrogen fixation. So really, if I don't miss my guess, we're essentially one step away from having transferable, promiscuous salt tolerant plasmids that can be widely moved at least among the bacterial world.

Dr. D. Brown: I would like to start by commenting on something Dr. Roth said in the beginning. It was subsequently clarified in a later comment, but maybe it should be spelled out just for the reason of being sure what we're talking about. He did say in his opening comments that it might be worthwhile trying to convert by genetic manipulation, an organism which could not osmoregulate into one which could. To the best of my knowledge, all cells can osmoregulate and the question at issue is the range of water activities. We ought to be fairly clear about what we're aiming at. And the next thing is that if one is to pursue the game of genetic manipulation, then clarity of the target surely has got to be at least as important as a background of genetic information about the organisms you're talking about. For example, does one want to aim at an organism which is salt tolerant within a narrow range, or adaptable to a wide range of salt concentrations? Now, in a sense, that question has already been answered, but I suppose it can best be exemplified with halobacterium and *Dunaliella*. Now these two genera, one a prokaryote and the other a eukaryote, have equal tolerance of high salt concentrations. They are bedfellows in hypersaline lakes and they are pretty well unique in that respect. But halobacterium has a very limited range of tolerance. *Dunaliella* will span the whole salt concentration range from saturated right down to something less than 0.5 molar, and that's pretty well true of every species of it as far as I know, even those which are called halophilic. They might have to be trained to get down there, but they can do it. So I think that's one of the questions which needs to be asked and answered. To be sure, there are very specific reasons for not setting your sights on halobacterium, and they have already been mentioned. But the next thing, I think if you're going to change an organism into something else, then you want to know what it is that you are trying to do with it, and you need to know more about the target situation than simply that it might fill itself up with some appropriate compatible solute. I went to some lengths this afternoon to compare these two yeasts, each of which filled themselves up with glycerol, but one was tolerant and one was not. And the reason that they had this difference was they had different means of accumulating the glycerol and I can identify at least five methods, five physiological methods by which cells of one kind or another, osmoregulate. One is that they can take up a solute from their environment and control that, and this is what happens with halobacterium. Another is that they can synthesize a solute, a metabolite, and regulate, not control the synthesis, but regulate its retention. Now that in a sense is similar to what I've just been saying. In the first instance when a solute is taken up from the environment, there's no control on its production, all that's controlled is the permeation-transport situation. In the second one which I've just mentioned, there's no control on the synthesis; again there is control on the permeation-transport situation. Now that's another physiological method. Then you could have regulation of synthesis and no control of retention. Now, that in turn can have two variants; you can have the one such as Saccharomyces cereviseae which synthesizes glycerol and leaks it (does not control the leaking) or you can have a case such as Dunaliella. It controls the synthesis of glycerol and does not leak it. And finally, you can have regulation of synthesis and retention and this is what the Debaromyces hensinia does. So here are five different strategies for a start and if one is looking for some sort of genetic manipulative technique then I think you should know what it is you are likely to turn this organism into. Because it will have quite profound effects on the number of markers that need to be changed. Here is another interesting question one might raise. What is the physiological difference between so-called halophilic and the non-halophilic species of Dunaliella? I was talking to Dr. Ben-Amotz earlier on and he doesn't know. But the fact remains there are some which are marine species and they grow a lot faster than the halophilic species. And they do this pretty much over the entire salt concentration range. But the marine species have a lower salt optimum than the other but not withstanding the lower salt optimum they still grow faster than the halophilic species even at high salt concentration. What are the differences between them? We can identify quite a number of physiological differences between them such as response to stress and so on. But we can't say anything about why they should have these basic differences in their salt relations. This is another area to be studied. Then, of course, there is the question which has cropped up several times today and is going to be quite important. What's the signal that triggers off the osmoregulator response? I would assume it is likely to be a different signal according to which particular physiological strategy an organism has evolved. I would expect, although the expectation might not be right, that an organism which controlled osmotic balance by regulating permeation would perhaps respond to a different sort of signal from one which controlled metabolism. And another question which has already been mentioned a couple of times and which Dr. Valentine was rather keen to have me pursue is "what exactly is the role of temperature response to water stress." The relation of temperature to osmotic stress is pretty well established in the literature. There's a good physical chemical reason why this also should be true. It is generally true that the optimum growth temperature of a microbe can be increased by increasing the solute concentration of the medium in which it is grown. And as I said in answer to a question this afternoon that sort of thing can change what appears to be an obligate requirement for a certain solute concentration into a facultative one.

R. C. Valentine: Thank you for mentioning the relationship to thermal tolerance. Dr. Brown tells me that in some cases there may be a difference of 5 C in terms of pushing the maximum temperature up. This might be very important in fermentation and in industrial use of yeast and other organisms if indeed it became possible to, say, genetically manipulate the optimum temperatures of microorganisms. Yeast, for example, don't adapt to very high temperatures of above 40 C or so. Five degrees might really be a lot if you could improve fermentation yeast to this extent. That might be a start toward a thermal tolerant fermentation.

A. D. Brown: I am not quite sure at what you getting at, but if you are looking to faster reaction rates, by raising the temperature then you are likely to be disappointed because these two processes are operating in opposition, the high solute concentration (and this is true whether you are talking about electrolytes or nonelectrolytes) generally slow reaction rates down. And the temperature increases them so I think you might finish up with an optimum or maximum rate which is not very different from the one you started with. It just occurs with a higher temperature that's all.

R. C. Valentine: Do thermophiles osmoregulate and what is the relationship of osmoregulation and thermophily?

A. D. Brown: Is the mechanism known in general of how organisms living at, say, 55 C or above stand the high temperatures? Well, as far as I understand it they have heat resistant proteins. And when you get to the really extreme thermophiles like those that grow at 90 C, they certainly do. And not only do they have novel proteins but they have novel lipids as well. So that the whole chemistry becomes quite characteristic and strange. In that sense they are like the extreme halophiles. The whole protein chemistry, the whole genetic structure, is changed.

A. Ben-Amotz: I want to add a few sentences about the halotolerant blue-green algae, work mentioned at the Boyce-Thompson symposium. I'm surprised nobody spoke about these interesting organisms here. The blue-green algae are bacteria. They have several advantages over higher cells. Some are halotolerant; they can live in salinities up to about three or four molar. Some of them are even thermophilic as well growing in solar lakes at temperatures up to about 50 C to 60 C. And finally, two important points are that blue-green algae can fix nitrogen, which eukaryotes cannot, and photosynthetically evolve hydrogen using their N_2 fixation machinery. In the next few years much more work should be dedicated to blue-green algae. Another area of great interest is the membrane properties of *Dunaliella* which permits large concentrations of glycerol to accumulate in the cell, lipid biochemistry, etc. The triggering mechanism of osmoregulation in *Dunaliella* should also be explored in the future. How does the algae know when and how much glycerol to make? We know that enzyme levels change, but we don't know how much and in what direction. How much is due to enzyme regulation or new protein synthesis. What is the genetic component?

All these questions are still open for more research. We are also very interested in cellular compartmentalization. In other words, where is glycerol accumulated in cellular organelles, etc. What is the relation between glycerol concentration in one compartment and glycerol concentration in the other compartment, and if there is any difference in concentration between the two compartments, what is the level of the concentration difference? Finally, we come to the question of industrial application. Many other algae in addition to the ones of our project might have industrial application for organic compounds such as sortibol, etc. It would be a good idea to survey a variety of habitats such as salt marshes, etc., to isolate new and useful organisms for applications and also for studies of basic mechanisms of osmoregulation.

A. D. Brown: Could I make just one comment about the site of glycerol synthesis. We have looked for the distribution of the glycerol dehydrogenase in *Dunaliella* and it does seem to be located in the cytoplasm.

J. Raven: Model organisms have been noted quite a bit here. I think we have to be very careful in choosing model organisms. *Chlamydomonas*, for example, has been suggested as a useful tool for osmoregulation. It has perhaps the best categorized genetics of any eukaryotic phototroph. However, we have to be very careful about extrapolating from one organism to another. For example, cells with rigid cell walls versus wall-less types may behave differently. Translation of results from wall-less cells to walled plant cells should be done with caution.

SECTION III

OSMOREGULATORY MECHANISMS IN PLANTS

THE ROLE OF ORGANIC SOLUTES IN OSMOREGULATION IN HALOPHYTIC HIGHER PLANTS

R. L. Jefferies

Department of Botany
University of Toronto
Toronto, Ontario, Canada M5S 1A1

INTRODUCTION

Excessive salt accumulations prevent or limit the growth of crops on at least 50 million hectares of agricultural land, particularly in areas where arid or semi-arid conditions exist (Carter, 1975). If these lands are to be used to increase plant productivity, it is pertinent to examine the characteristics of plants which have evolved under the influence of natural selection in saline environments, in order to recognize characters that are likely to increase the fitness of plants in such habitats. Halophytes growing in coastal and estuarine ecosystems are of particular interest, because of readily available supplies of saline water and nutrients (with the possible exception of nitrogen). Coastal environments are nutrient sinks, and the tidal input of large quantities of water and mineral ions, represents an energy subsidy to these ecosystems (Odum, 1974). In arid regions, in contrast, the combined effects of drought and salinity present a formidable obstable to the breeder attempting to introduce varieties which will give an adequate yield.

Within coastal systems, although there is an abundance of water, the osmotic potential (Ψ_s) of the saline solution around plant roots normally ranges between -0.5 to - 6.0 MPa, which is approximately equivalent to concentrations of sodium chloride from 80 mM to 1200 mM respectively. The Ψ_s of seawater is -2.4 MPa (c. 500 mM Na^+). The average concentrations of sodium and chloride ions in seawater are approximately 100 times greater than corresponding values for fresh water (Phillips, 1972). These data indicate the wide range of osmotic potentials and salinities that coastal halophytes tolerate at all stages of their life cycle. In agricultural ecosystems crop plants may be subject to high salinities only at certain stages of their life cycle when rates of evapotranspiration are high.

CHARACTERISTICS OF COASTAL HALOPHYTES

Perhaps one of the most striking characteristics of salt marsh species, in contrast to many crop species, is the perennial habit. With the exception of members of the Chenopodiaceae, the majority of halophytes in coastal marshes of Europe and North America are perennial and frequently long-lived (Gimingham, 1964; Jefferies, 1972). The perennial habit is characteristic of many plants growing in "stress" environments. In addition, as much as 70% of the total biomass of herbaceous perennial halophytes may be below-ground. Perennating organs account for much of this biomass. Estimates of the average net annual primary production for *Spartina* marshes on the east coast of North America, which have been studied extensively, range from about 500 g m^{-2} yr^{-1} at 38 N to about 700 g m^{-2} yr^{-1} at 28 N. Values for individual marshes may be considerably higher (Turner, 1976). It must be remembered that these values are rarely corrected for the internal salt content of the tissue which may account for over 50% of the total dry weight. Nevertheless, members of the genus *Spartina* (a C_4 genus) may achieve very high rates of net productivity during the summer. C_4 plants not only have high growth rates but appear to use nitrogen efficiently (Brown, 1978). There is also evidence that nitrogen-fixing bacteria inhabit the rhizosphere in stands of *Spartina alterniflora* (Valiela and Teal, 1979). Such plant communities usually are restricted to the lower levels of salt marshes where tidal inundations are frequent. Elsewhere in salt marshes, particularly in the upper levels, hypersaline conditions, low water potentials, drought conditions and a shortage of nitrogen are prevalent. At these sites growth rates of plants are frequently low, and clonal growth predominates. Sexual reproduction is infrequent in these perennial halophytes and the amount of dry matter devoted to sexual reproduction is frequently less than 10% of the maximum above-ground standing crop. Populations appear to have undergone genetic differentiation in response to these environmental conditions (Jefferies, 1977). Within salt marsh ecosystems in general genetic differentiation between populations is characteristic of most species (Gray et al., 1979), and much of this variation can be correlated with environmental heterogeneity.

I have described briefly the characteristics of coastal halophytes, because I wish to use this information in a discussion of the role of organic compounds in osmoregulation in higher plants. This is because an important question which requires resolving is the extent to which the processes of growth, reproduction and maintenance in halophytes are competing for limited resources. The concept of allocation depends absolutely on the idea that different structures or activities are alternatives, that a gain in one as a result of selection, must be offset by a loss in another (Harper, 1977). The so-called maintenance costs of living in a saline environment may result in a reduction in growth rate and reproductive output, because a limited resource, such as nitrogen, is utilized for maintenance. These differences in the allocation of resources between individuals from different saline areas may be the result of phenotypic plasticity. However, as mentioned above, genetic variation is widespread in populations of coastal halophytes and differences in allocation of resources between populations may reflect genetic differentiation of the populations in response to environmental heterogeneity.

ORGANIC SOLUTES

There is now considerable circumstantial evidence that organic solutes may serve as compatible osmotica within the cytoplasm of cells of halophytes (Hellebust, 1976; Wyn Jones et al., 1977a; Flowers et al., 1977). Electrolyte concentrations in the cytoplasm compared to those in the vacuole are relatively low, but organic solutes contribute to the osmotic potential, thereby maintaining osmotic equilibrium between the cytoplasm and vacuole. These highly soluble organic substances are low molecular weight compounds which apparently do not interfere with the activity of soluble enzymes (Stewart and Lee, 1974; Flowers et al., 1978; Pollard and Wyn Jones, 1979). Wyn Jones et al. (1979) have proposed that there is a broad consistency in the ionic composition of the cytoplasm of eukaryotic cells, such that the cytoplasm has a relatively constant ionic environment in which metabolism can occur in a controlled and regulated manner. This property of eukaryotic cells is in marked contrast to that of prokaryotic halophytic bacteria in which enzymes are active and stable in solutions of high ionic strength (Lanyi, 1974; Brown, 1976).

The different types of organic compounds which serve as compatible osmotica are given in Table I. The list is not comprehensive; further studies undoubtedly will lead to the identification of additional organic solutes which fulfill this role. A wide range of organic solutes including polyols, amino acids, methylated quaternary compounds and reducing sugars serve as possible compatible osmotica in halophytes. Within an individual plant a number of these compounds may be present at relatively high concentrations. There is also evidence that within a genus similar compounds are produced by both halophytic and glycophytic species, although the concentrations present in the tissues may differ (Jefferies, unpublished).

The relationship between osmotic potential and the molal concentration of different solutes is shown in Figure 1. Although molal solutions of organic solutes have approximately the same osmotic potential (-2.5 to -2.8 MPa), amounts of carbon and nitrogen required to achieve these potentials are substantially different, depending on which solute is being examined. For example, amounts

Table I. Proline, methylated quaternary ammonium compounds and sorbitol in coastal halophytes (modified after Stewart et al., 1979; Rozema, 1979; Cavalieri and Huang, 1979).

Group A. Methylated quaternary ammonium accumulators:

Atriplex hastata
Atriplex patula
Beta maritima
Halimione portulacoides
Salicornia europaea
Suaeda maritima

Group B. Proline accumulators:

Cochlearia officinalis
Glaux maritima
Juncus maritimus
Puccinellia distans
Puccinellia maritima
Spergularia marina
Spergularia media
Triglochin maritima

Group C. Proline and methylated quaternary ammonium accumulators:

Agrostis stolonifera
Armeria maritima
Aster tripolium
Festuca rubra
Limonium vulgare
Spartina anglica

Group D. Proline and miscellaneous solute accumulators:

Distichlis spicata
Juncus roemerianus
Limonium carolinianum
Spartina alterniflora
Spartina patens

Group E. Sorbitol accumulators:

Plantago martima

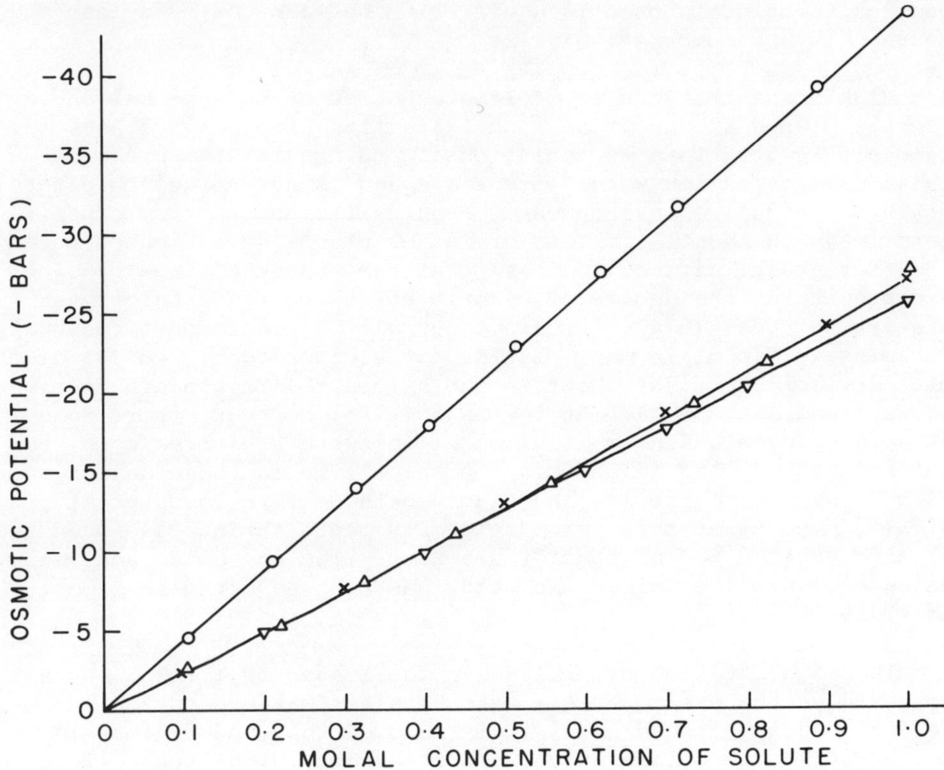

Fig. 1. Relationship between the molal concentration of a solute and the corresponding osmotic potential (20 C) (O, sodium chloride; Δ, sucrose; X, glycinebetaine; ∇, proline) (Data calculated from Smith and Smith, 1940, and Weast, 1976).

of carbon present in molal solutions of sucrose, glucose, proline and glycinebetaine are 144, 72, 60 and 60 grams respectively. In plants in which carbon or nitrogen are limiting, certain organic solutes may be selected because these solutes generate a low demand for a limited resource. It is of interest that halophytes, such as _Salicornia europaea_, which grow in highly saline areas of salt marshes do not appear to accumulate large amounts of mono and disaccharides in their tissue; instead highly soluble nitrogeneous compounds act as compatible organic solutes. Not only would large amounts of sugars (and carbon) be required in order to lower the osmotic potential to values below -2.4 MPa (seawater), but also high concentrations of sugars appear to decrease rates of enzymic activity (Rozema et al., 1978). Plants which accumulate mono- and disaccharides in relatively large amounts appear to grow in brackish marsh or strand-line habitats, in which hyposaline conditions prevail much of the time. Under these conditions amounts of sugar required to maintain osmotic equilibrium between the cytoplasm and the vacuole are comparatively low.

High concentrations of proline are present in some halophytes when the plants are grown at high salinities (Stewart and Lee, 1974; Treichel, 1975). This accumulation of proline is widespread in plants subject to low water potentials and is not unique to plants growing in a saline environment. Results from short-term experiments in which shoots, tillers or intact plants have been cultured at different salinities, indicate that the amount of free proline is dependent on the degree of osmotic stress in these halophytes (Stewart and Lee, 1974). Proline, normally a minor constituent of the amino acid pool in many plants, may account for 10 to 20% of the total dry weight of the shoot in the halophyte, _Triglochin martima_ (Stewart and Lee, 1974). At low salinities, however, there does not appear to be a linear relationship between an increase in proline and an increase in salinity, particularly in those halophytes which produce both proline and glycinebetaine (Wyn Jones et al., 1977a). Amounts of this amino acid only begin to increase when the salinity exceeds 50% of that of seawater. The non amino acid analogue of proline, pipecolic acid, has been reported in _Limonium_ (Goas, 1965).

The concentration of proline in the leaves of _Triglochin martima_ expressed on the basis of cell water content may exceed 200 mM (Jefferies, unpublished). If it is assumed that the proline is located in the cytoplasm which occupies about 10% of the cell volume, the calculated osmotic potential for this compartment is approximately -5.6 MPa. This value is considerably lower than the osmolarity of the cell sap (-2.9 MPa), hence some of the proline must be located in the vacuole as well as the cytoplasm, in order for osmotic equilibrium to be maintained. Such a distribution would provide a readily available storage pool of nitrogen in the vacuole for growth and development. In addition, a rapid fall in internal osmotic potential could be achieved by a movement of proline into the cytoplasm from the vacuole and an increase in the ionic concentration of the vacuolar sap.

Of the methylated quaternary ammonium compounds, glycinebetaine appears to be the most widespread (Storey and Wyn Jones, 1975; Storey et al., 1977), although other betaines such as alaninebetaine are also present in some halophytes (Stewart et al., 1979). Glycinebetaine is produced in a number of higher plants, particularly in the Gramineae and Chenopodiaceae. Evidence for the preferential localization of glycinebetaine in the cytoplasm has been obtained from whole vacuole isolations (Wyn Jones et al., 1977a). Pollard and Wyn Jones (1979) have reported that concentrations of glycinebetaine up to 500 mM did not inhibit the activities of malate dehydrogenase and pyruvate kinase obtained from barley plants. In some plants, such as Spartina townsendii a correlation coefficient of 0.99 was obtained between the concentration of glycinebetaine in the tissues and the osmotic pressure of the sap (Storey and Wyn Jones, 1978). Atriplex spongiosa maintained a fairly constant osmotic potential difference between the leaf and the external solution as the potential of the latter dropped. The glycinebetaine concentrations in the shoot correlated closely with the sap osmotic pressure (Wyn Jones et al., 1977b). However, in other halophytes which produced substantial quantities of glycinebetaine the close correlation was not always apparent. The osmotic pressure of Suaeda monoica was high, even when plants were grown in a non saline solution (Storey and Wyn Jones, 1979). An addition of 50 mM NaCl to the medium resulted in a doubling of the shoot osmotic pressure, which thereafter remained constant over a ten-fold increase in external salt concentrations. However, glycinebetaine concentrations, expressed as µmoles gfw^{-1}, were not correlated closely with the osmotic pressure of the cell sap of the leaves. While glycinebetaine synthesis may be induced by salinity, the accumulation of this solute is a constitutive property of Suaeda monoica, as the solute is present in plants grown in the absence of salt. In a closely related halophytic species, Suaeda maritima, glycinebetaine likewise was present in the leaves when the plants were grown in the absence of sodium chloride (Flowers and Hall, 1978). However, the level of this quaternary ammonium compound increased in plants from 1.8 mg $gfwt^{-1}$ in the control solution to 4.40 mg $gfwt^{-1}$ when plants were grown in a saline solution. Flowers and Hall estimate that the latter concentration is equivalent to a concentration of some 41 mM when expressed on the basis of plant water. They further estimate that if this were confined to the cytoplasm it would be equivalent to a concentration of 830 mM and generate an osmotic potential of around -2.5 MPa.

Besides the presence of glycinebetaine, leaves of Spartina townsendii also contain choline and dimethylsulphoniumpropionate (Storey and Wyn Jones, 1979). This and other substituted sulphonium compounds are also found in a number of marine algae and higher plants and are structurally similar to glycinebetaine (cf. Larher and Hamelin, 1975a,b).

A number of halophytes accumulate neither imino acids nor methylated quaternary ammonium compounds. Plantago maritima accumulates high levels of the polyhydric alcohol, sorbitol, under saline conditions (Ahmad et al., 1979; Jefferies et al., 1979). This polyol has also been shown to act as a compatible osmotic solute in a small marine green alga, Stichococcus bacillaris (Brown and Hellebust, 1978). Although a number of polyols such as glycerol, arabitol, mannitol, volemitol, cyclohexanetetrol and the polyol derivative isofloridiside have been shown to have osmoregulatory functions in fungi and algae (Brown, 1976; Hellebust, 1976; Kauss,

1978), there is little evidence of their role as compatible osmotica in halophytic higher plants at present.

As Flowers et al. (1977) indicate, the role of carbohydrates in the osmotic adjustment of halophytes to high salinity is not resolved. In Suaeda maritima, for example, sucrose concentrations are maintained at about 1 mM in tissues of plants subject to different salinities (Flowers et al., 1977). The amounts of sucrose were three times those of glucose and fructose, but the total sugar content was only 8 mM expressed on an overall water content basis. It appears unlikely that di- or monosaccharides make a substantial contribution to the osmotic pressure in most halophytes. For example, Greenway (1973) has calculated that 30 g l^{-1} of hexose would be required in the cytoplasm to maintain osmotic equilibrium, if the concentration of sodium chloride in the vacuole was 100 mM. For many species this would represent approximately 20% of the dry weight. In addition, the activities of some enzymes are inhibited in the presence of moderate concentrations of sucrose (Rozema et al., 1978). The accumulation of soluble sugars appears to be characteristic of halophytes and brackish water species which are members of the Poaceae, Cyperaceae and Juncaceae (Albert and Popp, 1977).

Aside from CAM plants, organic acid contents do not rise appreciably in plants grown under saline conditions; in fact in Suaeda maritima the highest organic acid levels were recorded in plants grown in tap water (Flowers and Hall, 1978). In many halophytes sodium ions are largely balanced by chloride ions, although high oxalate concentrations are found in some species of Atriplex (Osmond, 1968).

Much of the physiological and ecological evidence for the involvement of organic solutes in osmoregulation in higher plants is circumstantial and indirect. It is evident that the presence of these compounds in the tissues of higher plants is not always associated with saline conditions. An assumption, for which there is little evidence at present, is that these solutes are restricted to the cytoplasm of the cells of higher plants. Some physiological and histochemical evidence has been produced by Wyn Jones and Storey (1977a) for the preferential localization of glycinebetaine in the cytoplasm of beet cells but overall, a lot more evidence is required. The small size of higher plant cells, and the difficulty of determining accurately cytoplasmic and vacuolar volumes, has meant that progress has been slow in determining both the location of these solutes, and the magnitude of their fluxes between different cell compartments. The exacting criteria laid down by Kauss (1977) for determining whether soluble cellular constituents play a role in osmotic regulation can only be applied at present to single cells, and not to the tissues of higher plants.

Most of the studies which show a close correlation between an increase in the amount of an organic solute and an increase in external salinity are based on short term experiments using tillers, shoot systems or else intact plants. In natural environments, the correlation mentioned above does not always hold. Seasonal changes in the concentration of solutes, such as proline and glycinebetaine, in the tissues of halophytes are not correlated directly with changes in external salinity, or with the water potential of the tissue, or with the osmotic potential of the cell sap (Stewart and Lee, 1974; Jefferies et al., 1979). Very high concentrations of organic solutes may be present in the tissues of halophytes during periods when relative growth rates are high (i.e., 0.8-1.0 g g^{-1} wk^{-1}). This contrasts with glycophytes in which high concentrations of solutes are associated with a near cessation of growth (Wyn Jones and Storey, 1978). Amounts of organic solutes in the tissues frequently appear to far exceed those required to maintain osmotic equilibrium between the cytoplasm and vacuole, assuming the cytoplasm to be 10% of the cell volume. In other words, these organic solutes have additional roles, other than osmoregulation, and the status and fate of these compounds in different tissues cannot be divorced from the overall economy of nitrogen and carbon compounds in whole plants during growth and development.

Plants use a limited number of compounds for the transport and storage of the majority of their nitrogen (Miflin and Lea, 1977). For example, proline appears to play a role in nitrogen storage; it is a major component of the storage protein of cereal seeds, and also is important in the development of seedlings of maize (Barnard and Oaks, 1970). Miflin and Lea (1977) suggest that the evolution of a limited range of amino acids as general nitrogen-transport compounds is probably related to their specific properties and their central position in pathways of nitrogen metabolism. Whether the movement of proline accounts for most of the redistribution of nitrogen in those halophytic plants, which contain large quantities of proline in their leaves, is uncertain.

If ionic concentrations in the cytoplasm are low then enzymes of halophytes are likely to be inhibited by high concentrations of electrolytes. This indeed appears to be the case, as enzymes from salt tolerant plants are neither salt resistant nor salt requiring (Flowers et al., 1977). A number of enzymes are stimulated by solutions of monovalent ions up to 150 mM but the stimulation appears to be a non specific electrolyte effect. Franks and Eagland (1975) have argued that these concentrations suppress the Donnan diffuse double layer of proteins and maximize folding without causing conformational instability. However, some qualifying statements are needed as enzymes do not show identical responses to an increase in the ionic concentration of the media. In *Suaeda maritima*, for example, malate dehydrogenase exists in both low and high molecular weight forms which differ in their sensitivity to sodium chloride

(Flowers et al., 1976). The high molecular weight form is activated by a higher salt concentration (100 mM) than the low molecular weight form (50 mM), and an increase in the amount of the more tolerant form occurs when plants are grown under highly saline conditions. A number of studies (cf. Flowers et al., 1977) have shown that the inhibitory effects of high concentrations of ions may be alleviated partially by an increase in the substrate concentration. It is likely that further differences in the responses of enzymes to inorganic ions will be found; enzymes likely to be particularly sensitive to inorganic ions are those associated with protein synthesis and ribosome function (Wyn Jones et al., 1979).

The general situation is eukaryotic cells, therefore, is that the evolution of enzymes which are stable in solutions of high ionic strength has not occurred, as in prokaryotic bacteria (Lanyi, 1974; Brown, 1976).

POPULATION DIFFERENCES IN THE ACCUMULATION OF ORGANIC SOLUTES IN A HALOPHYTE

Seasonal changes in the concentrations of soluble nitrogen and proline in different organs, expressed as a percentage of the total nitrogen in plants of *Triglochin maritima* from two populations, are shown in Figure 2. The percentage contribution soluble nitrogen makes to the total nitrogen in the plant is very high, and much of the soluble nitrogen is proline in this species. Early in the season in excess of 60% of the total nitrogen is present as soluble nitrogen. The swollen leaf bases are an important source and sink for nitrogen during the annual growth cycle. Early in the season there is a mobilization of insoluble nitrogen in the root/rhizomes and an increase in amounts of soluble nitrogen in the swollen leaf bases. Further redistribution of nitrogen occurs during the summer, followed in autumn by an accumulation of insoluble nitrogen in the leaf bases, rhizomes and roots, and a corresponding fall in soluble nitrogen in the tissues. Amounts of proline as a percentage of soluble nitrogen in different organs do not change appreciably during the season. The population grows in a marsh where hypersaline conditions frequently prevail in the summer months, but, as mentioned previously, the overall correlation between the osmotic potential of the cell sap of leaves and the concentration of proline is poor. However, in individuals from an adjacent population which grow at the edge of an old dune where conditions are hypersaline during summer, amounts of soluble nitrogen and proline in the tissues are less, and much of the nitrogen is present as insoluble nitrogen. Under greenhouse conditions the growth rate of individuals from this population is faster than the corresponding rate for individuals from the other population. This suggests that there is genetic differentiation with respect to growth rate between the two populations. When plants are grown from seed from the two populations at different salinities, not only are there differences in growth rate, but the plants from the populations which grows in the hypersaline site accumulate higher amounts of proline in both the root and shoot than plants from the other population (Dillon, 1978). In summary therefore, adaptation to hypersaline conditions in this population of Triglochin maritima is associated with a low growth rate, the presence of large amounts of soluble nitrogen, especially proline, in the tissues, and an efficient recycling of nitrogen between above- and below-ground tissues. The presence of large amounts of proline in excess of cytoplasmic osmotic requirements has considerable selective value in this perennial population. Proline may act as a nitrogen storage compound, the main form of transported nitrogen, a compatible osmotic solute, and in addition, because it is closely related to glutamate, it can easily be converted to other amino acids and acids of the tricarboxylic acid cycle. The occurrence of large amounts of proline also acts as a safeguard against environmental unpredictability. A rapid rise in external salinity may be associated with a movement of proline from the vacuole to the cytoplasm, and a commensurate uptake of inorganic ions into the vacuole, as the plant lowers its osmotic potential.

What of other species which accumulate differenc organic compounds that act as compatible osmotic solutes? Although both proline and quaternary ammonium compounds may act as compatible osmotic solutes in Limonium vulgare, amounts of proline are low and the turnover rate of quaternary ammonium compounds in plants appears to be slow (Hanson and Nelsen, 1978), hence it is unlikely that these solutes play such an important role in nitrogen metabolism as in Triglochin maritima.

PATTERNS IN THE DEVELOPMENT OF OSMOTIC POTENTIAL IN HALOPHYTES

Two distinct patterns of the development of osmotic potential have been recognized in animals and plants (Prosser, 1970; Wyn Jones, 1977a). Some species appear to maintain a fairly constant osmotic potential gradient between the leaf and the external solution as the potential of the latter falls (osmoconformers). In other species there is an initial rapid fall in the osmotic potential of leaves as the external salinity increases, thereafter the potential remains steady over a wide range of external salinities (osmoregulators). An explanation of the significance of these two types of strategy is missing. In the few examples so far studied in plants, the majority of osmoregulators appear to be highly succulent species. They include the perennia, Halimione portulacoides, (and possible Aster tripolium) and the annuals Salicornia europaea (Figure 3b) and Suaeda maritima. These plants grow in sediments in which large diurnal changes in salinity and water potential may occur in the upper layers, as the exposed sediment dries out between tidal cycles. The maintenance of an osmotic potential in the leaves considerably below that of external saline solution avoids continual osmotic adjustment as external conditions change. Frequent alterations to the osmotic potential of plant tissues involving the synthesis of

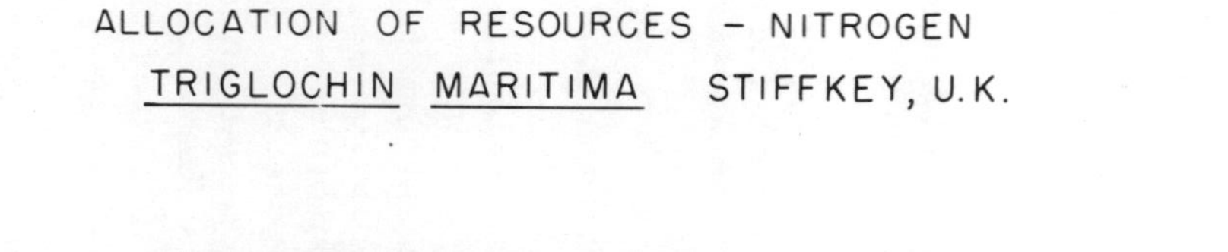

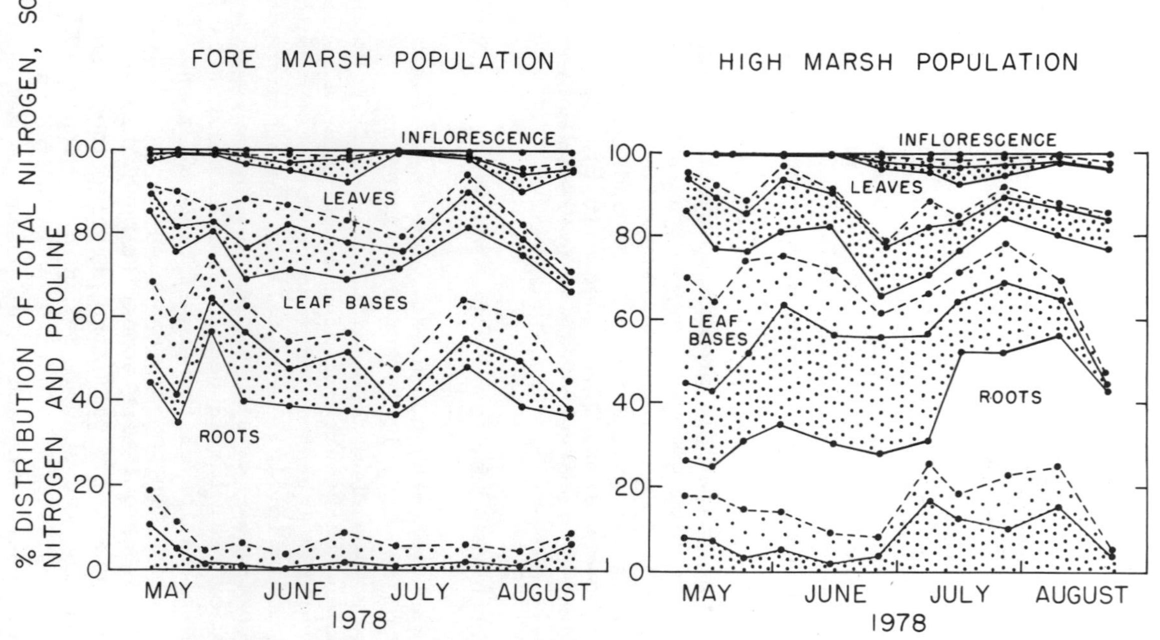

Fig. 2. The distribution, in different plant organs, of insoluble nitrogen (clear areas) and soluble nitrogen (stippled areas) expressed as a percentage of the total nitrogen in plants of two populations of <u>Triglochin maritima</u> during the growing season of 1978. The heavily stippled areas indicate the amounts of the soluble nitrogen which are present as proline in each plant organ.

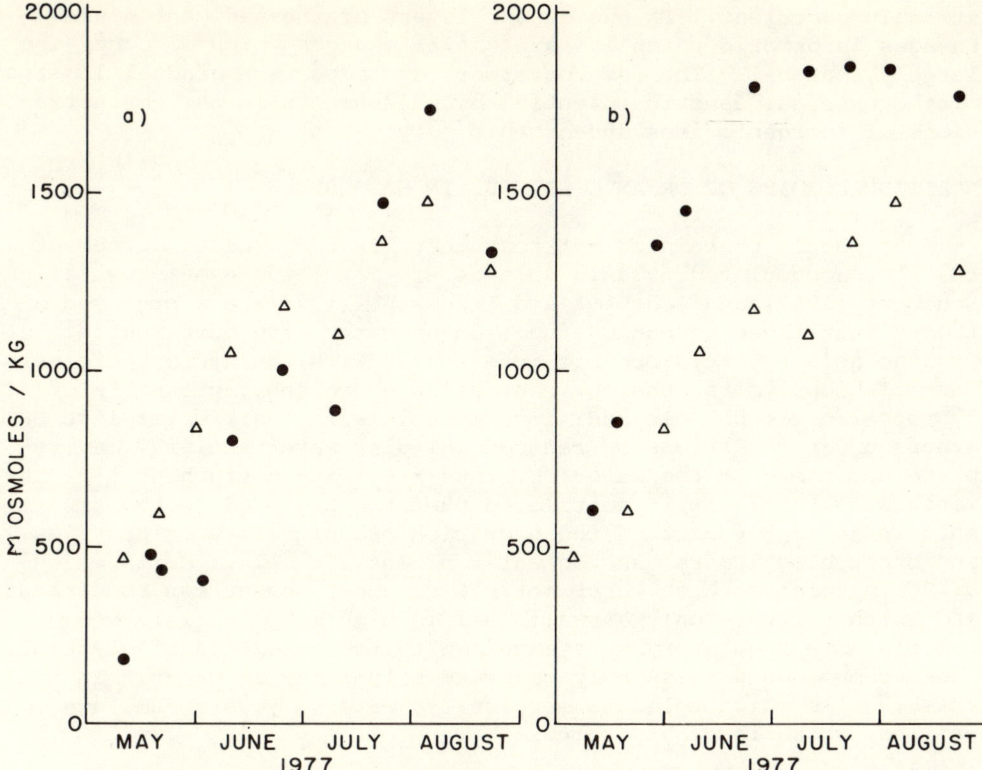

Fig. 3. Osmolarity of cell sap expressed from the shoots of plants of (a), Plantago maritima (o) and (b), Salicornia europaea (o), growing in Stiffkey salt marsh, Norfolk, U.K. (Sap expressed with a hydraulic press and a Wescor osmometer used to measure osmolarity.) Δ, osmolarity of the bulk soil solution in the rooting zone.

organic solutes and the movements of these organic solutes and inorganic ions within tissues may be inefficient in relation to energy costs, growth, and the short time scale over which these changes need to occur. It is of interest that in halophytes which are osmoregulators, glycinebetaine appears to be a compatible osmotic solute. This compound, which has a low turnover rate, and is regarded as an end product of metabolism, is present in plants that maintain a steady low internal osmotic potential well below the fluctuating values characteristic of the external osmotic potential.

In contrast, the osmoconformers (e.g., Plantago maritima, Figure 3a) include deep-rooted perennial plants, which are not strongly succulent. In the deeper layers of the sediment diurnal changes in osmotic potential are small; changes which do occur are largely seasonal. The osmoconformers accommodate a gradual lowering of the external osmotic potential by adjustments in the concentrations of inorganic ions and organic solutes.

ADDITIONAL ROLES OF ORGANIC SOLUTES IN HALOPHYTES

It should be pointed out that there is not general agreement that the accumulated organic solutes are involved in osmoregulation. Schobert (1977) and Schobert and Tzchesche (1978) have proposed a theory that these compounds function as water structure regulators. Proline acts by a hydrophobic interaction with the hydrophobic surface of biopolymers, thereby converting hydrophobic groups into hydrophilic groups. In addition, polyols with their water-like OH-groups under conditions of reduced cellular water activity may replace positions of the water-OH, thus repairing disturbances of the water structure. It is postulated that the proposed mechanisms result in an improvement of the hydration or solubility of proteins and other biopolymers, as the water potential falls. Cells which maintain relative high water activities under non stress conditions and which tolerate only a small fall in osmotic potential, use polyols as a regulatory substance. With the exception of glycerol, these compounds possess only moderate solubility in water. In contrast, cells which tolerate more severe water stress accumulate proline, which is highly soluble.

There is some evidence that these organic solutes may act as colligative protecting agents in enabling halophytes to tolerate freezing. In halophytes such as *Cochlearia anglica*, *Limonium vulgare* and *Halimione portulacoides* concentration of sugars and organic acids increased in winter (Kappen, 1969; Kappen and Maier, 1973; Maier and Kappen, 1979). Such evidence indicates the considerable versatility of these organic solutes.

METABOLIC PATHWAYS

As mentioned earlier, the significance of these organic solutes is related both to their specific properties and to their position in nitrogen and carbon metabolism. Proline is an excellent example of a molecule that can easily be derived from, or give rise to, other molecules (Miflin and Lea, 1977). The pathway involved in the biosynthesis of proline from glutamate is reasonably well established, although the control and regulatory mechanisms determining the rates of enzymic activities are poorly understood. As far as I am aware this pathway has never been examined in halophytic higher plants, although at present, there is no reason for believing that novel pathways are involved in the biosynthesis of this inorganic solute in halophytes.

Much less is known about the biosynthesis of the quaternary ammonium compounds in plants. Glycinebetaine is thought to arise from the exhaustive methylation of glycine (Bowman and Rohringer, 1970; Cromwell and Rennie, 1954; Delwiche and Bregoff, 1958; Hansen and Nelsen, 1978). There is a marked difference in the metabolic fates reported for proline and betaine. Proline can be incorporated into protein or oxidized via Δ-pyrroline-5-carboxylate to glutamate and thence to α-ketoglutarate. Proline accumulated during water stress is rapidly metabolized via these pathways when the stress is alleviated (Stewart, 1972; Stewart et al., 1977). Betaine is apparently a stable metabolic end product (Bowman and Rohringer, 1970), although in some cases it may act as a methyl donor (Byerrum et al., 1956). Hanson and Nelson (1978) have shown that when barley plants are subject to water stress, betaine accumulates, but on addition of water to the cultures betaine levels decline slowly, suggesting that the compound is an end product with a slow turnover rate.

The biosynthesis of sugar alcohols involves the reduction of the aldehyde or ketone oxygen in the C_1 position of sugars to give the corresponding sugar alcohol. Thus D-glucose can be reduced to give sorbitol (Bidwell, 1974).

BREEDING FOR SALINITY TOLERANCE

The use of selective conditions to recover defined biochemical mutant types in higher plants is poorly developed. As pointed out by Rice and Carlson (1975) plants are large organisms, they have long generation times and large nutrient pools. In addition, they are composed of cells that represent many distinct differentiated states, and genetically they are frequently polyploid. These characteristics of the whole plant which permit adaptation to a changing environment do not allow the use of biochemical selection techniques to isolate mutants. Most of the known mutants in higher plants are morphological variants and an observed phenotypic change is separated from the underlying molecular change(s) by several levels of organization, hence it is very difficult to describe the molecular changes which have taken place in mutants. An additional complication is that many important characters of agronomic interest, such as yield and stature, are under polygenic control so that the final phenotype is a product of a large number of biological processes. These characteristics of plants indicate the formidable genetic complexity that will need to be manipulated in plant breeding programs for salinity tolerance.

The growth and reproductive strategies of coastal halophytes are very different from those of most crop plants. The well-developed perennial habit, a low rate of sexual reproduction, the maintenance of large pools of inorganic and organic solutes that fulfill a variety of roles, and the efficient recycling of essential elements are characteristics that are common in plants from "stress" environments. Besides these, the energetic costs of growing in a saline environment, which is considered elsewhere in this volume, appear to be substantial. The crop plants of coastal saline areas in the future may be very different from varieties available today. It is significant that in certain countries, such as Israel, some halophytes, such as species of Atriplex, are irrigated with sea-water and cultivated for fodder. Above all else, there is an urgent need for information on the growth physiology of halophytic higher plants -- a need which is widespread in many areas of whole plant physiology.

ACKNOWLEDGMENTS

I wish to thank Drs. S. C. H. Barrett, J. Dainty, and J. A. Hellebust for constructive criticism of the manuscript. Ms. G. Richardson kindly typed the manuscript.

REFERENCES

Ahmad, I., Larher, F., and Stewart, G. R., 1979, Sorbitol, a compatible osmotic solute in Plantago maritima, New Phytol., 82:671.

Albert, R. and Popp, M., 1977, Chemical composition of halophytes from the Neusiedler Lake Region in Austria, Oecologia, 27:157.

Barnard, R. A. and Oaks, A., 1970, Metabolism of proline in maize root tips, Can. J. Bot., 48:1155.

Bidwell, R. G. S., 1974, "Plant Physiology," Macmillan, New York.

Bowman, M. S. and Rohringer, R., 1970, Formate metabolism in healthy and rust-infected wheat, Can. J. Bot., 48:803.

Brown, A. D., 1976, Microbial water stress, Bact. Rev., 40:803.

Brown, L. M. and Hellebust, J. A., 1978, Sorbitol and proline as intracellular osmotic solutes in the green alga Stichococcus bacillaris, Can. J. Bot., 56:676.

Brown, R. H., 1978, A difference in N use efficiency in C_3 and C_4 plants and its implication in adaptation and evolution, Crop Sci., 18:93.

Byerrum, R. U., Sato, C. S., and Ball, C. D., 1956, Utilization of betaine as a methyl group donor in tobacco, Plant Physiol., 31:374.

Carter, D. L., 1975, Problems of salinity in agriculture, in: "Plants in Saline Environments," A. Poljakoff-Mayber and J. Gale, eds., Springer-Verlag, Berlin.

Cavalieri, A. J. and Huang, A. H. C., 1979, Evaluation of proline accumulation in the adaptation of diverse species of marsh halophytes to the saline environment, Amer. J. Bot., 66:307.

Cromwell, B. T. and Rennie, S. D., 1954, The biosynthesis and metabolism of betaines in plants. 3. Studies on the biosynthesis of glycine-betaine in seedlings of wheat (Triticum vulgare VIII.), Biochem. J., 58:322.

Delwiche, C. C. and Bregoff, H. M., 1958, Pathway of betaine and choline synthesis in Beta vulgaris, J. Biol. Chem., 223:430.

Dillon, E. M., 1978, Growth and metabolic responses of salt marsh halophytes in response to salinity, MSc. Thesis, University of Toronto.

Flowers, T. J., Troke, P. F., and Yeo, A. R., 1977, The mechanism of salt tolerance in halophytes, Ann. Rev. Plant Physiol., 28:89.

Flowers, T. J. and Hall, J. L., 1978, Salt tolerance in the halophyte, Suaeda maritima (L.) Dum.: The influence of the salinity of the culture solution on the content of various organic compounds, Ann. Bot., 42:1057.

Flowers, T. J., Hall, J. L., and Ward, M. E., 1978, Salt tolerance in the halophyte, Suaeda maritima (L.) Dum: Properties of malic enzyme and PEP carboxylase, Ann. Bot., 42:1065.

Franks, F. and Eagland, D., 1975, The role of solvent interactions in protein conformation, C.R.C. Crit. Rev. Biochem., 3:165.

Gimingham, C. H., 1965, Maritime and sub-maritime communities, in: "The Vegetation of Scotland," J. H. Burnett, ed., Oliver and Boyd, Edinburgh.

Goas, M., 1965, Sur le metabolisme azote des halophytes: etude des acides amines et amides libres, Bull. Soc. Fr. Physiol. Veg., 11:309.

Gray, A. J., Parsell, R. J., and Scott, R., 1979, The genetic structure of plant populations in relation to the development of salt marshes, in: "Ecological Processes in Coastal Environments," R. L. Jefferies and A. J. Davy, eds., Blackwell Scientific Publications, Oxford.

Greenway, H., 1973, Salinity, plant growth and metabolism, J. Aust. Inst. Agric. Sci., 39:24.

Hanson, A. D. and Nelsen, C. E., 1978, Betaine accumulation and ^{14}C formate metabolism in water-stressed barley leaves, Plant Physiol., 62:305.

Harper, J. L., 1977, "The Population Biology of Plants," Academic Press, London.

Hellebust, J. A., 1976, Osmoregulation, Ann. Rev. Plant Physiol., 27:485.

Jefferies, R. L., 1972, Aspects of salt-marsh ecology with particular reference to inorganic plant nutrition, in: "The Estuarine Environment," R. S. K. Barnes and J. Green, eds., Elsevier, Amsterdam.

Jefferies, R. L., 1977, Growth responses of coastal halophytes to inorganic nitrogen, J. Ecol., 65:847.

Jefferies, R. L., Davy, A. J., and Rudmik, T., 1979, The growth strategies of coastal halophytes, in: "Ecological Processes in Coastal Environments," R. L. Jefferies and A. J. Davy, eds., Blackwell Scientific Publications, Oxford.

Jefferies, R. L., Rudmik, T., and Dillon, E. M., 1979, The responese of halophytes to high salinities and low water potentials, Plant Physiol., 64:989.

Kappen, L., 1969, Frostresistenz einheimischer Halophyten in

Beziehung zu ihrem Salz-, Zucker- und Wassergehalt im Sommer und im Winter, Flora Abt. B., 158:232.

Kappen, L. and Maier, M., 1979, Bedeutung einiger nichtflüchtiger Carbonsäuren für die Frostresistenz des Halophyten Halimione portulacoides unter dem Einfluss verschieden hoher Kochsalzbelastung, Oecologia, 12:241.

Kappen, L. and Maier, M., 1979, Cellular compartmentalization of salt ions and protective agents with respect to freezing tolerance of leaves. Investigations with the halophyte Halimione portulacoides (L.) Aellen, Oecologia, 38:303.

Kauss, H., 1977, Biochemistry of osmotic regulation, in: International Review of Biochemistry, Plant Biochemistry II, Vol. 13, D. H. Northcote, ed., University Park Press, Baltimore.

Kauss, H., 1978, Osmotic regulation in algae, Prog. in Phytochem., 5:1.

Lanyi, J. K., 1974, Salt-dependent properties of proteins from extremely halophytic bacteria, Bact. Rev., 38:272.

Larher, F. and Haemlin, J., 1975a, L'acide dimethylsulfonium-3 propanoique de Spartina anglica, Phytochem., 16:2019.

Larher, F. and Haemlin, J., 1975b, L'acide-trimethylaminoproionique des rameau de Limonium vulgare Mill., Phytochem., 14:205.

Miflin, B. J. and Lea, P. J., 1977, Amino acid metabolism, Ann. Rev. Plant Physiol., 28:299.

Odum, E. P., 1974, Halophytes, energetics and ecosystems, in: "Ecology of Halophytes," R. J. Reinold and W. H. Queen, eds., Academic Press, New York.

Osmond, C. B., 1968, Acid metabolism in Atriplex, Aust. J. Biol. Sci., 21:1119.

Phillips, J., 1972, Chemical processes in estuaries, in: "The Estuarine Environment," R. S. K. Barnes and J. Green, eds., Elsevier, Amsterdam.

Pollard, A. and Wyn Jones, R. G., 1979, Enzyme activities in concentrated solutions of glycinebetaine and other solutes, Planta, 144:291.

Prosser, C. L., 1973, "Comparative Animal Physiology," 3rd Ed., W. B. Saunders, Philadelphia.

Rice, T. B. and Carlson, P. S., 1975, Genetic analysis and plant improvement, Ann. Rev. Plant Physiol., 26:279.

Rozema, J., 1979, Population dynamics and ecophysiological adaptations of some coastal members of the Juncaceae and Gramineae, in: "Ecological Processes in Coastal Environments," R. L. Jefferies and A. J. Davy, eds., Blackwell Scientific Publications, Oxford.

Rozema, J., Buizer, D. A. G., and Fabritius, H. E., 1978, Population dynamics of Glaux maritima and ecophysiological adaptations to salinity and inundation, Oikos, 30:539.

Schobert, B., 1977, Is there an osmotic regulatory mechanism in algae and higher plants?, J. Theor. Biol., 68:17.

Schobert, B. and Tzchesche, H., 1978, Unusual solution properties of proline and its interaction with proteins, Biochim. Biophys. Acta, 541:270.

Smith, P. K. and Smith, E. R. B., 1940, Thermodynamic properties of solutions of amino acids and related substances: V. The activities of some hydroxy- and n-methylamino acids and proline in aqueous solution at twenty-five degrees, J. Biol. Chem., 132:57.

Storey, R. and Wyn Jones, R. G., 1975, Betaine and choline levels in plants and their relationship to NaCl stress, Plant Science Letters, 4:161.

Storey, R., Ahmad, A., and Wyn Jones, R. G., 1977, Taxonomic and ecological aspects of the distribution of glycinebetaine and related compounds in plants, Oecologia, 27:319.

Storey, R. and Wyn Jones, R. G., 1978, Salt stress and comparative physiology in the Gramineae. III. Effect of salinity upon ion relations and glycinebetaine and proline levels in Spartina x townsendii, Aust. J. Plant Physiol., 5:831.

Storey, R. and Wyn Jones, R. G., 1979, Responses of Atriplex spongiosa and Suaeda monoica to salinity, Plant Physiol., 65:156.

Stewart, C. R., 1972, Proline content and metabolism during rehydration of wilted excised leaves in the dark, Plant Physiol., 50:679.

Stewart, C. R., Boggess, S. F., Aspinall, D., and Paleg, L. G., 1977, Inhibition of proline oxidation by water stress, Plant Physiol., 59:930.

Stewart, G. R. and Lee, J. A., 1974, The role of proline accumulation in halophytes, Planta, 120:279.

Stewart, G. R., Larher, F., Ahmad, I., and Lee, J. A., 1979, Nitrogen metabolism and salt tolerance in halophytes, in: "Ecological Processes in Coastal Enviornments," R. L. Jefferies and A. J. Davy, eds., Blackwell Scientific Publications, Oxford.

Treichel, S., 1975, Der Einfluss von NaCl auf die Prolinkonzentration verschiedener Halophyten, Zeit. für Pflanzenphysiol., 76:56.

Turner, R. E., 1976, Geographic variations in salt marsh macrophyte production: a review, Contrib. Mar. Sci., 20:47.

Valiela, I. and Teal, J. M., 1979, The nitrogen budget of a salt marsh ecosystem, Nature, 280:652.

Weast, R. C., ed., 1976, Handbook of Chemistry and Physics, 57th Edition, C.R.C. Press, Cleveland, Ohio.

Wyn Jones, R. G., Storey, R., Leigh, R. A., Ahmad, N., and Pollard, A., 1977a, A hypothesis on cytoplasmic osmoregulation, in: "Regulation of Cell Membrane Activities in Plants," E. Marre and O. Ciferri, eds., North Holland, Amsterdam.

Wyn Jones, R. G., Storey, R., and Pollard, A., 1977b, Ionic and osmotic regulation in plants, particularly halophytes, in: "Transmembrane ionic exchanges in plants," M. Thellier, A. Monnier, M. Demarty, and J. Dainty, eds., CNRS, Paris.

Wyn Jones, R. G. and Storey, R., 1978, Salt stress and comparative physiology in the Gramineae. IV. Comparison of salt stress in _Spartina_ x _townsendii_ and three barley cultivars, _Aust. J. Plant Physiol._, 5:839.

Wyn Jones, R. G., Brady, C. J., and Speirs, J., 1979, Ionic and osmotic relations in plant cells, _in_: "Recent Advances in the Biochemistry of Cereals," D. L. Laidman and R. G. Wyn Jones, eds., Academic Press, New York.

AN ASSESSMENT OF QUATERNARY AMMONIUM AND RELATED COMPOUNDS AS OSMOTIC EFFECTORS IN CROP PLANTS

R. Gareth Wyn Jones

Department of Biochemistry & Soil Science
University College of North Wales,
Bangor Gwynedd LL57 2UW, Wales

INTRODUCTION

In this paper I shall attempt to assess whether or not the accumulation of certain amino acids and their N-methylated derivatives, particularly glycinebetaine, is a specific adaption related to enhanced salt tolerance in, at least, some halophytic species. As a result I shall neglect other extremely important aspects of salt tolerance and toxicity but these have been considered elsewhere (Wyn Jones, 1980; Wyn Jones et al., 1979).

A number of pertinent questions may be asked regarding the relationship of these compounds, especially glycinebetaine and proline, to salt tolerance. 1) Is their accumulation specially related to tolerance or a manifestation of a stress-induced metabolic lesion? 2) Are glycinebetaine, proline and related compounds involved in osmotic adaptation? 3) Do these compounds have roles in addition to or instead of those of osmotic effectors? 4) How is their accumulation related to other changes in C and N metabolism and ion relations? 5) Have these compounds any value as selection criteria in the breeding of salt tolerance agricultural crops?

Our answers even to this limited range of questions are hedged by uncertainty. However sufficient evidence has emerged in the last few years for some tentative assessments to be made.

ACCUMULATION: ADAPTIVE CHANGE OR STRESS-INDUCED LESION

A wide range of organic compounds has long been known to be accumulated in higher plants subject to salt stress. It is improbable that they all have a positive adaptive significance.

Indeed Stroganov (1964) considered that the build-up of certain amines in stressed tissues was a major contribution to salt toxicity. If the adaptive significance of a compound is to be established, I suggest that the following criteria should be considered.

1) The quantitative distribution of the compound revealed by field surveys should be characteristic of certain habitats and, possibly, of families usually associated with such habitats.

2) The compound should be constitutively accumulated in tolerant but not in sensitive species.

3) Accumulation should be enhanced by a moderate stress, i.e., one which either enhances or does not impair the growth of a tolerant species. (Since low to moderate salinities, 50-200 mM NaCl, promote the growth of some halophytes, this concept is readily applicable to salinity but may be less relevant to other stresses.)

4) Exogenous application of the compound might promote *in vivo* the tolerance of the whole organism or *in vitro* enhance the tolerance of specific metabolic processes.

Glycinebetaine

Analysis of vegetative tissue from a great number of collected species have shown glycinebetaine to be present in substantial concentrations (10 to 100 μmoles g^{-1} fresh weight) in many, but by no means all, halophytes. There is a clear taxonomic pattern underlying its distribution as it is universally found in members of the Chenopodiaceae, a family typified by a halophytic distribution, but is barely detectable in members of other families, for example, the Cyperaceae or the Plantaginaceae. Amongst the Gramineae, glycinebetaine accumulation is common in some tribes, e.g., the Hordeae and the Chlorideae but not in others, e.g, Festuceae and the Maydeae. A compliation of existing ecological and taxonomic data has recently been made (Wyn Jones and Storey, 1979) and the general pattern confirmed by more recent work (Gorham et al., 1980). The compound is also concentrated in partially dehydrated tissues of unstressed cereals; seed embryo and aleurone cells (Chittenden et al., 1978), anthers and paleas (Pearce et al., 1976). Accumulation has also recently been observed in *Aster tripolium* petals. A broader comparative biochemical survey also reveals an association of glycinebetaine with saline habitats, it is a compound widely distributed in marine invertebrates (Schoffeniels and Gilles, 1972) and it promotes the tolerance of a halophytic bacterium (Rafaeli-Eshkol and Avi-Dor, 1968).

In relation to the second criterion, glycinebetaine is constitutively accumulated in the leaves of halophytes such as *Suaeda*, *Salicornia* and, to a lesser extent, *Spartina* species (Storey and

Wyn Jones, 1978, 1979). In the two chenopods these high constitutive levels appear associated with an osmo-regulatory adaptation, i.e., one involving a high and relatively constant sap osmotic pressure over a wide range of external salinities. Conversely I know of no salt sensitive glycophyte in which significant constitutive synthesis of glycinebetaine has been observed in green tissue. A rather high level has however been reported in etiolated wheat coleoptiles (Cromwell and Rennie, 1954).

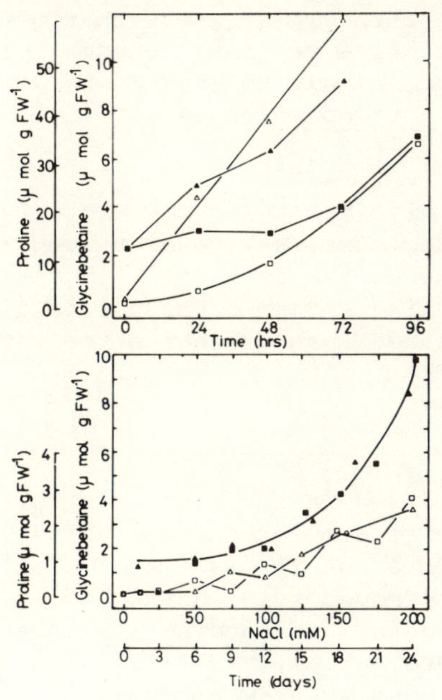

Fig. 1. (a) Atriplex spongiosa, (b) Spinacea oleracea grown at different salinities (Storey and Wyn Jones, 1979). Effect on leaf osmolarity (Δ,▲) and glycinebetaine (□,■) and proline (O,●) contents.

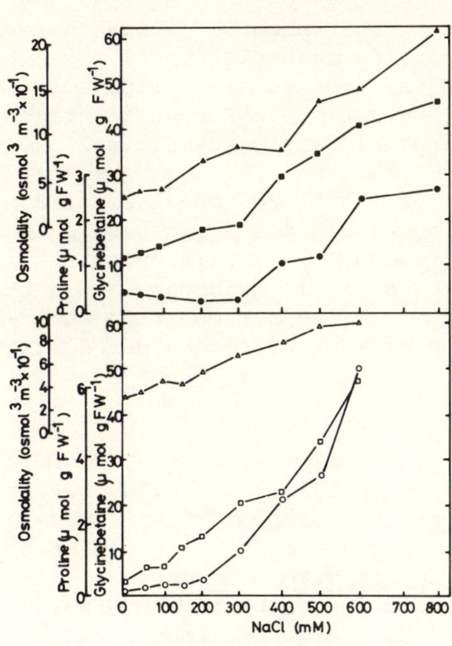

Fig. 2. (a) H. vulgare subjected to shock (≡ 450 mosomol), (b) H. vulgare subjected to incremental stress (Wyn Jones and Storey, 1978). Symbols: open-proline; closed-glycinebetaine; Δ,▲ salt; □,■ osmotic stress.

The third consideration is induction under a stress which does not impair or may even promote the growth of the tolerant species. Accumulation of glycinebetaine under such conditions has been observed in the osmoconformer Atriplex spongiosa (Figure 1a). In the more salt-sensitive species, Spinacia oleracea, a significant increase in leaf glycinebetaine was observed at salinities (50 to 100 mM NaCl) which had little influence on growth (Coughlan and

Wyn Jones, 1980) (Figure 1b). Similar evidence is available for *Spartina townsendii* (Storey and Wyn Jones, 1978).

The influence of exogenous glycinebetaine both on whole seedlings and on enzymes and organelles *in vitro* will be described later. Nevertheless it must be noted here that added glycinebetaine ameliorates salt stress in seedlings and partially protects *some* enzymes against salt inhibition (Pollard and Wyn Jones, 1979).

The weight of evidence therefore strongly suggests that glycinebetaine accumulation is a positive adaptive response in some species. Other factors are undoubtedly involved in salt tolerance and it is abundantly clear that tolerance does *not depend* on an ability to accumulate glycinebetaine.

There are some indications that glycinebetaine may protect some organisms against cold and osmotic stresses (Bokarev and Ivanova, 1971; Dulaney et al., 1969), but the data are very sparse. The analytic evidence in plants and animals seem to associate glycinebetaine accumulation with salt rather than drought stress; this point will be reconsidered later.

Proline

If proline accumulation is considered in the light of the same five criteria, a similar but rather more complex and confusing picture emerges. Field surveys indicate that proline accumulation is perhaps most characteristic of some monocotyledonous species such as *Triglochin* species (Juncaginaceae) and *Puccinellia* in the Festuceae tribe of the Gramineae (Storey et al., 1977; Stewart, 1978). High proline levels are also consistently reported in the dicot-*Spergularia* (Caryophyllaceae). However, when growth culture experiments have been undertaken, it has been frequently observed that initial proline levels are low and major accumulation only occurs at salt or water stresses which severely inhibit growth (Figure 1) (Treichel, 1975; Cavalieri and Huang, 1979). In barley proline accumulation is produced by a highly damaging, possibly lethal salt or osmotic shock (Hanson and Nelson, 1978; Wyn Jones and Storey, 1978), whereas a gradual incremental salt stress leads to higher glycinebetaine than proline levels (Figure 2). Finally some sensitive glycophytes have fairly high constitutive proline levels and accumulate quite high levels when stressed despite their salt sensitivity (Wyn Jones and Storey, 1979).

This evidence would tend to associate proline accumulation with salt toxicity rather than tolerance but this interpretation is clearly inadequate. In *Puccinellia maritima* (Figure 3) the low-salt proline level is fairly high and increases linearly with external salinity (Stewart et al., 1978). The ecophysiological studies of Jefferies in the accompanying paper also provide

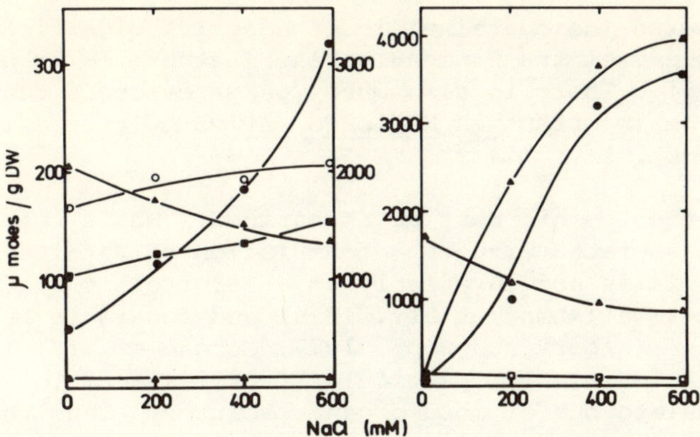

Fig. 3. Effect of salinity of Puccinellia maritima. (a) right -- total organic nitrogen (▲), soluble organic nitrogen (■), α-NH_2-N (O), proline (●), quaternary ammonium compound (Δ). (b) left -- K (Δ), Na (▲), Cl (O), Mg (□).

compelling evidence for the importance of proline in the salt tolerance and general metabolism of Triglochin. Bar-Nun and Poljakoff-Mayber have observed a positive effect of endogenous proline on salt stressed roots (Bar-Nun and Poljakoff-Mayber, 1977). It is also difficult to escape the obvious deduction from the wealth of data on algae, bacteria and various invertebrates which accumulate proline following hyperosmotic stress (Measures, 1975; Schobert, 1974; Schoffeniels and Gilles, 1970). Furthermore high concentrations of the compound occur in dehydrated plant tissues such as pollen (Virtanen and Kan, 1955). Thus we can distinguish both a general proline accumulation in response to water stress in almost all plant tissues and a rather specific accumulation of proline in certain halophytes. The relationship of these observations is not precisely clear but it may not be too fanciful to suggest that the latter is a special adaptation based on the former. It is however not proven that proline accumulation under intense water stress in agronomic species (Hansen and Nelson, 1978) has a positive adaptive value rather than reflecting the internal water stress (Hanson et al., 1977). Nevertheless the criteria previously advocated to assess adaptive significance may be too restrictive. A compound accumulated at inhibitory stresses, in extremis could have a role in tiding the plant over a temporary but highly damaging stress; such an effect could be difficult to substantiate. The question remains: is what is good for Triglochin maritima good for Hordeum vulgare?

Other Betaines

While it is impossible to discuss the full range of accumulated nitrogenous compounds in this paper, the work of Larher must be mentioned as he has obtained good evidence that the accumulation of β-alaninebetaine (homobetaine) is of adaptive significance in salt tolerant members of the Plumbaginaceae (Larher, 1976; Larher and Hamelin, 1975). There is also much weaker evidence that proline-betaine may be important in *Medicago sativa* cultivars (Wyn Jones and Owen, 1980).

Two further points must be emphasized. While tolerance is frequently associated with the accumulation of betaines and/or amino acids, it is not invariably so. Sorbitol is accumulated in *Plantago maritima* (Ahmad et al., 1979) and sugars in many Juncaceae and Cyperaceae (Albert and Popp, 1977; Gorham et al., 1980) often, but again not invariably, particular families seem to preferentially accumulate one or more organic solutes. Thus there is considerable evolutionary variation, even latitude, in organic solute accumulation. However, as has been discussed elsewhere, ion relations in the cytoplasm are probably more stringently controlled and more crucial to stress resistance (Wyn Jones et al., 1979).

GLYCINEBETAINE AND PROLINE AS OSMOTIC EFFECTORS

In culture experiments glycinebetaine tissue contents are strongly correlated with sap osmotic pressure in a number of species (Storey and Wyn Jones, 1978, 1979). The proline accumulation in *Puccinellia* (Figure 3) suggests a similar relationship although in some tissues, proline levels only increase at very high tissue osmotic pressures (Figure 1). However in almost all cases in vacuolated cells, the absolute concentrations of these solutes are too low to contribute significantly to the total sap osmotic pressure. There is both direct and indirect evidence to suggest that the inorganic ions are compartmented in higher plant cells so that K^+ is preferentially located in the cytoplasm while Na^+ and Cl^- are partially occluded in the vacuole (Wyn Jones et al., 1979; Jeschke, 1979). Furthermore high electrolyte concentrations (including K^+ salts) are probably not tolerated in cytoplasm. Thus there is a requirement for a non-toxic cytoplasmic osmoticum in cells exposed to hyper-saline stress or for a decrease in the cytoplasmic matrix potential.

It has been suggested the proline and glycinebetaine act as cytoplasmic osmotic effectors in particular species (Stewart and Lee, 1974; Storey and Wyn Jones, 1975). For these hypotheses to be sustained, those solutes must be shown to be preferentially concentrated in the cytoplasm (cytosol) and to be nontoxic to cytoplasmic activities and processes. Their compatibility at concentrations of 500 mM and more with enzyme activities, mitochondrial and chloroplast function and polysomes and protein synthesis have been fairly well established (Wyn Jones et al., 1979; Wyn Jones and Storey, 1979; Pollard and Wyn Jones, 1979). However irrefutable evidence for preferential cytoplasmic localization has proved more elusive. Some cytochemical evidence for glycinebetaine being exclusively cytoplasmic has been obtained (Hall et al., 1978). We have calculated approximate tonoplast gradients by using the whole vacuole isolation technique and making use of the pigment betanin as an intravacuolar marker (Table I). A tonoplast gradient of about 15:1 in favor of the cytoplasm was observed for glycinebetaine which, assuming a conservative cytoplasm-vacuole ratio of 1:10 and finding a sap glycinebetaine content of 20-25 mM, gave a cytoplasmic glycinebetaine concentration of around 120 mM. This is of course only a rough estimate but indicates that this solute would make a significant contribution to the cytoplasmic osmotic pressure when the total sap osmotic pressure is around 500 milliosmolal (cf. Wyn Jones et al., 1977). The same work indicates a comparable tonoplast proline gradient; Goring et al. (1978) also have evidence for the cytoplasmic accumulation of the latter.

There is therefore, quite strong evidence to support the hypothesis that these compounds act as nontoxic cytosolic osmotica. An alternative view that proline, in particular, acts by maintaining the hydration sphere of protein rather than contributing to the colligative properties of the cytosolic fluid has been advocated by Schobert (Schobert, 1977; 1979).

It does not necessarily follow that other betaines necessarily have equivalent functions in other tissues. Larher (1976) observed that β-alaninebetaine was largely present as the choline ester in *Armeria maritima* -- a polycation and not a zwitterion. His cytochemical evidence suggested that the compound might be largely vacuolar. While the data are not decisive, they clearly preclude sweeping generalization on the role of betaines.

NON-OSMOTIC ROLES OF GLYCINEBETAINE

Enzyme and Membrane 'Protection'

In their early work Avi-Dor and his colleagues (Rafaeli-Eshkol and Avi-Dor, 1968) found that choline on conversion to glycinebetaine, and the betaine itself, enhanced the salt tolerance of a moderately salt tolerant bacterium. In later work they showed that glycinebetaine could be actively accumulated in this bacterium up to 800 mM but concluded that the effect of extracellular glycinebetaine on the respiratory membranes was the dominant cause of salt stress protection (Shkedy-Vinkler and Avi-Dor, 1975). It may be noted that protection was increased by the sequential N-methylation from glycine to glycinebetaine (Table II). We have observed some stabilization of spinach chloroplasts in glycinebetaine (Larkum and Wyn Jones, 1979) but it is far from clear whether the compound has a general effect on membrane stability. More specific effects on ion fluxes will be discussed later.

Table I. Distribution of glycinebetaine and proline between cytoplasm and vacuole of red beet storage tissue.

	Glycinebetaine	Proline
Solute: betanin ratio in homogenate	2616	144
Solute: betanin ratio in isolated vacuoles	1145	60
% solute in vacuole	45	43
Relative cytoplasmic concentration	14±4 (SE)	18±8

Values given are means of 5 separate experiments.

Table II. Stimulation of a metabolic event by glycinebetaine and related compounds.

	Stimulation of respiration of bacterium Ba_1 1.8 mM KCl*	Stimulation of barley MDH inhibited by 300 mM NaCl**
Glycinebetaine	200	167
Dimethylglycine	170	125
Sarcosine	150	84
Glycine	110	80
Tetramethylammonium Cl	100	--

* Solutes at 60 mM (Shkedy-Vinkler and Avi-Dor, 1975).
** Solutes at 500 mM (Pollard and Wyn Jones, 1979).

In studies on the comparative toxicity of glycinebetaine and inorganic ions to enzymes, it was found that glycinebetaine was able to partially protect some enzymes against salt toxicity (Table II); an effect again related to the degree of methylation of the N group. A detailed analysis of this phenomenon suggested that 'protection' occurred via solvent-solute interactions and not by a direct binding of glycinebetaine to the enzymes (Pollard and Wyn Jones, 1979). Further physical chemical studies also strongly indicate that, at least up to 500 mM, a reasonable physiological range for most vegetative cells, glycinebetaine does not discernably 'bind' to proteins (Pollard, 1979). These data conflict somewhat with the model advanced by Schobert which seeks to explain the role of proline and glycinebetaine by postulating a solute-protein 'binding' and a solute-water 'binding' which collectively lead to a maintenance of the protein hydration sphere. Schobert (1979) has obtained evidence of specific proline water interactions under conditions of low water activity. However, in attempting to assess these rival interpretations a crucial factor must be the concentration range being considered. In relatively dilute solution up to 1 molar the conventional colligative properties of solutes are probably dominant and this, in my opinion, is the situation in most vegetative tissues. However in very concentrated solutions, molecular interactions between solute and water and anomalous water binding may well become significant.

Effects on Ion Transport

It has already been noted that exogenous choline and glycinebetaine partially alleviate salt stress in axenic maize and barley seedlings (Table III). Inorganic analyses showed a significant difference in the tissue Na^+ contents and this problem has been pursued further. The time course of the effect of glycinebetaine on Na^+ uptake into barley seedlings revealed an unexpected phenomenon as the initial rate of Na^+ uptake was increased by glycinebetaine but the effect was reversed by 24 hours (Figure 4). More detailed experiments have been carried out with excised roots (Ahmad, 1978) and, in summary, it has been found that the initial effect on Na^+ fluxes depends on the intracellular loading of glycinebetaine, or a consequence thereof, and not the extracellular solute (Figure 5). Analysis of influx kinetics further suggests that the quasi-steady state influx into the vacuole rather than the plasma membrane influx is influenced by glycinebetaine. Using the efflux technique perfected by Jeschke (1979), it has been possible to show that the vacuolar Na content is increased, either directly or indirectly, by glycinebetaine loading (Table IV).

RELATIONSHIP TO GENERAL CARBON AND NITROGEN METABOLISM

The general role of proline in the metabolism of *Triglochin* is discussed by Jefferies (in this volume) and it is clear that the

Table III. Effect of exogenous glycinebetaine (gb) on growth and ion contents of seedlings.

Maize (10 seedlings)	Control	Salt Treatment no gb	(100 mM NaCl) gb (1 mM)
Yield (g FW^{-1} plant)			
shoots	3.5	1.35	2.0
roots	2.0	0.8	1.0
Ion contents (mmol gDW^{-1})			
shoot K^+	1.15	0.73	0.77
Na^+	1.08	2.4	2.0
Cl^-	0.15	2.0	1.8
root K^+	0.7	0.28	0.26
Na^+	0.1	1.9	1.5
Cl^-	0.1	0.7	0.6

Barley (16 replicates)	Salt Treatment no gb	(25 mM NaCl) gb
Ion contents (mM)		
shoot K^+	176	183
Na^+	18.5*	15.7*
root K^+	102	103
Na^+	25.7**	19.6**

*P <0.5; **P <0.01; Data from Ahmad (1978)and Storey (1976).

compound has a broad significance in the nitrogen economy of that plant. Glycinebetaine may also account for a large percentage of the leaf nitrogen in Suaeda and other species (Storey et al., 1977) but there are other marked differences between the behavior of the two compounds. While proline turns over very quickly and is rapidly degraded following the removal of stress, glycinebetaine is remarkably constant (Ahmad and Wyn Jones, 1979). Hanson and Nelson (1978) even suggest that the level of glycinebetaine could be used as a measure of accumulated stress. Thus, unlike proline, glycinebetaine is unlikely to represent a potential store of energy and nitrogen.

However there may be an association between the stability of glycinebetaine and a high retention of Na in vacuoles (Jeschke, 1979), since this could provide a system of stable osmotic adaptation. Both the ecological relationship between glycinebetaine and salt tolerance rather than drought tolerance, and the effect of

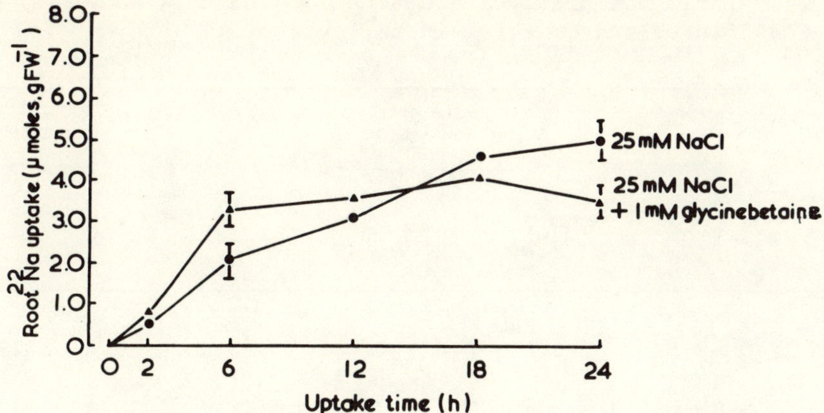

Fig. 4. Effect of glycinebetaine (1 mM) on the time course of ^{22}Na from NaCl (25 mM) into the roots of intact seedlings growing in the Hoagland's medium 47.

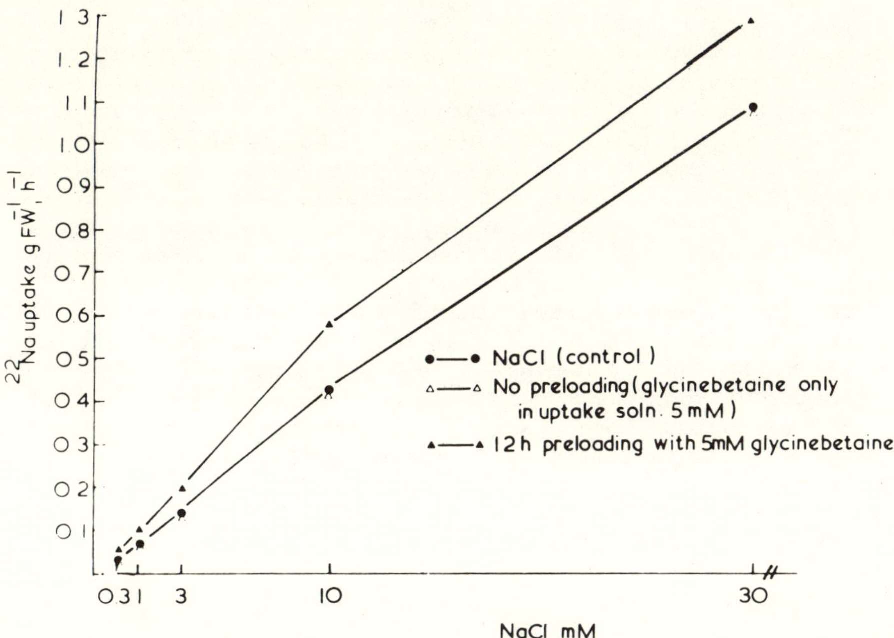

Fig. 5. Concentration-dependent influx of ^{22}Na (40/30 uptake:wash regime) into excised barley roots either preloaded with glycinebetaine (5 mM, 12 h, ▲) or with glycinebetaine (Δ) in the uptake solution. Control with glycinebetaine excluded from both preolading and uptake solutions (●) (Ahmad, 1978).

Table IV. Influence of glycinebetaine loading on steady state Na fluxes and contents in barley roots.*

Efflux conditions	Fluxes (μmole $gFW^{-1}h^{-1}$) ϕ_{oc}	ϕ_{co}	ϕ_{cv}	ϕ_{cx}(R)	Contents (μmoles gFW^{-1}) Q_c	Q_v
0.2 mM K, 5 mM Na	0.65	0.41	0.047	0.25	0.38	23.0
0.2 mM K, 5 mM Na**	0.70	0.47	0.082	0.22	0.49	30.8

* Unpublished data of Wyn Jones, Ahmad and Jeschke.
** Preincubated with glycinebetaine, 1 mM 18 h.

intracellular glycinebetaine on Na fluxes appear to point to this type of integrated adaptation. These observations do not account for all the phenomena observed in Table III and Figure 4 but will explain the short-term promotion of root Na^+ uptake by glycinebetaine. More important perhaps, they provide evidence that the cytosolute, glycinebetaine, has more than a passive osmotic role and that important interactions can take place between organic and inorganic solutes which in themselves influence compartmentation.

The biosynthetic pathway to glycinebetaine and in particular its control remains to be fully elucidated. There is no doubt that choline can be rapidly converted to glycinebetaine in tissues such as spinach but possibly not in non-accumulators such as maize. The putative primary precursor of glycinebetaine, ethanolamine is closely related biochemically to photorespiration and this interrelationship also requires further consideration.

GLYCINEBETAINE AND PROLINE IN BREEDING

As has been discussed in other contributions to this volume by Epstein and Ramage, salt tolerant cultivars of major crops must be bred and simple selection criteria are required in such programs. The question must be asked whether glycinebetaine or proline provide such criteria. Despite initial enthusiasm the use of proline as a criterion for drought tolerance in barley has been shown to be ill-advised (Hanson et al., 1977). Equally there is no case in barley or other cereals to suppose that either low or high salt proline levels would be an adequate criterion for salt tolerance, indeed, probably the reverse (Wyn Jones and Storey, 1978).

In comparison of three barley cultivars Wyn Jones and Storey (1978) did not find a positive correlation between tolerance and glycinebetaine content. However if a wide range of species, particularly in the Chenopodiaceae, are considered, a rough correlation between constitutive glycinebetaine content and tolerance is observed. There is insufficient data to show whether or not this is the case in the Gramineae but it would, at best, only be expected to apply in certain tribes.

However, such data are only meaningful if considered in the perspective of the overall ionic and osmotic regulation of particular species. In barley tolerance is closely correlated with the ability to exclude Cl^- and probably Na^+ from the shoots, thus maintaining a rather constant total shoot cation content on a dry weight basis. Some osmotic adaptation is brought about by a drop in shoot water content. Thus there is no reason to expect a compound postulated to act as a cytosolute at osmolalities in excess of about 400 milliosmolal to be correlated with tolerance by exclusion.

Tolerance by exclusion may have limitations as the capability of osmotic adjustment may be limited or involve a significant proportion of fixed carbon (Wyn Jones, 1980). Tolerance by control of K^+, Na^+ and Cl^- fluxes and the efficient sub-cellular compartmentation of the ions has the advantage of using salt as a major, cheap, cellular osmoticum. If selection can be made for this type of adaptation then a positive correlation between tolerance and compatible solute content could be expected. However from our existing data, there is no evidence that the ability to accumulate the compatible solute is the primary limitation on tolerance and prime facie it would appear more likely that some measure of either passive ion permeability or active ion transport could provide a more relevant criterion.

ACKNOWLEDGMENTS

I would like to gratefully acknowledge the support and cooperation of my colleagues Prof. W. D. Jeschke and Drs. Storey, Leigh, Ahmad, Pollard, Gorham, and Coughlan in the work and developing the ideas described in this paper. We are also grateful to the SRC and NERC for financial support.

REFERENCES

Ahmad, N., 1978, Aspects of glycinebetaine phytochemistry and metabolic functions in plants, Ph.D. Thesis, Univ. of Wales, Cardiff.

Ahmad, N. and Wyn Jones, R. G., 1979, Comparison of glycinebetaine and proline turnover in barley seedlings released from salt and water stress, Plant Sci. Lett., 15:231.

Ahmad, I., Larher, F., and Stewart, G. R., 1979, Sorbitol, a compatible osmotic solute in Plantago maritima, New Phytol., 82: 671.

Albert, R. and Popp, M., 1977, Chemical composition of halophytes from the Neusiedler lake region of Austria, Oecologia (Berl.), 27:1157.

Bar-Nun, N. and Poljakoff-Mayber, A., 1977, Salinity stress and the content of proline in roots of Pisum sativum and Tamarix tetragyna, Ann. Bot., 41:173.

Bokarev, K. S. and Ivanova, R. P., 1971, The effect of certain derivatives and analogs of choline and betaine on content of free amino acids in leaves of two species of potato differing with respect to frost resistance, Soviet Plant Physiol., 18: 302.

Cavalieri, A. J. and Huang, A. H. C., 1979, Evaluation of proline accumulation in the adaptation of diverse species of marsh halophytes to the saline environment, Amer. J. Bot., 66:307.

Chittenden, C. G., Laidman, D. L., Ahmad, N., and Wyn Jones, R. G., 1978, Amino acid and quaternary nitrogen compounds in the germinating wheat grain, Phytochem., 17:1209.

Coughlan, S. and Wyn Jones, R. G., 1980, Some responses of Spinacea oleracea by salt stress, J. Exp. Bot., (in press).

Cromwell, B. T. and Rennie, S. D., 1954, The biosynthesis and metabolism of betaines in plants. III. Studies on the biosynthesis of precursors of glycinebetaine in seedlings of wheat (Triticum vulgare), Biochem. J., 58:322.

Dulaney, E. L., Dulaney, D. D., and Rickes, E. L., 1969, Factors in yeast extract which relieve growth inhibition of bacteria in defined medium of high osmolarity, Develop. Ind. Microbiol., 9:260.

Gorham, J., Hughes, L., and Wyn Jones, R. G., 1980, Strategies of solute accumulation in salt marsh plants and their relatives, Oecologia, (in preparation).

Goring, M., Dreier, W., and Heinke, F., 1978, Zytoplasmatische Osmoregulation durch Prolin bei Wurzeln von Zea mays L., Biologische Rundschau, 15:377.

Hall, J. L., Harvey, D. M. R., and Flowers, T. J., 1978, Evidence for the cytoplasmic localization of betaine in leaf cells of Suaeda maritima, Planta, 140:59.

Hanson, A. H. and Nelson, C. E., 1978, Betaine accumulation and ^{14}C formate metabolism in water stressed barley leaves, Plant Physiol., 62:305.

Hanson, A. D., Nelsen, C. E., and Everson, E. H., 1977, Evaluation of free proline accumulation as an index of drought resistance using two contrasting barley cultivars, Crop Sci., 17:720.

Jeschke, W. D., 1979, Univalent cation selectivity and compartmentation in cereals, in: "Recent Advances in the Biochemistry of

Cereals," D. L. Laidman and R. G. Wyn Jones, eds., Academic Press, London and New York.

Larher, F., 1976, Sur quelques particularities due metabolisme azote d'une halophyte: Limonium vulgare, Thèse Doct. Sc. Nat., Rennes.

Larher, F. and Hamelin, J., 1975, L'acide β-trimethylamino propionique des rameaux de Limonium vulgare, Mill., Phytochem., 14:205.

Larkum, A. W. D. and Wyn Jones, R. G., 1979, Carbon dioxide fixation by chloroplasts isolated in glycinebetaine: a putative cytoplasmic osmoticum, Planta, 145:393.

Measures, J. C., 1975, Role of amino acids in osmoregulation of non-halophilic bacteria, Nature, 257:398.

Pearce, R. B., Strange, R. N., and Smith, H., 1976, Glycinebetaine and choline in wheat: distribution and relation to infection by Fusarium graminearum, Phytochem., 15:953.

Pollard, A., 1979, Glycinebetaine and enzyme activity, Ph.D. Thesis, University of Wales, Cardiff.

Pollard, A. and Wyn Jones, R. G., 1979, Enzyme activities in concentrated solutions of glycinebetaine and other solutes, Planta, 144:291.

Rafaeli-Eshkol, D. and Avi-Dor, Y., 1968, Studies in halotolerance in a moderately halophilic bacterium. Effect of betaine on salt resistance of the respiratory system, Biochem. J., 109: 687.

Schobert, B., 1974, The influence of water stress on the metabolism of diatoms. I. Osmotic resistance and proline accumulation in Cyclotella meneghiniana, Z. Pflanzenphysiol., 74:106.

Schobert, B., 1977, Is there an osmotic regulatory mechanism in algae and higher plants?, J. Theor. Biol., 68:17.

Schobert, B., 1979, Die Akkumulierung von Prolin in Phaeodactylum tricornutum und die Funktion der "compatible solutes" in Pflanzenzellen unter Wasserstress, Ber. Deutsch. Bot. Ges. Bd., 92:23.

Schoffeniels, E. and Gilles, R., 1970, Osmoregulation in aquatic anthropods, in: "Chemical Zoology," Vol. V, M. Florkin and B. T. Scheer, eds., Academic Press, New York and London.

Schoffeniels, E. and Gilles, R., 1972, Ionregulation and osmoregulation in Mollusca, in: "Chemical Zoology," Vol. II, M. Florkin and B. T. Scheer, eds., Academic Press, New York and London.

Shkedy-Vinkler, C. and Avi-Dor, Y., 1975, Betaine-induced stimulation of respiration at high osmolarities in a halotolerant bacterium, Biochem. J., 150:219.

Stewart, G. R., Larher, F., Ahmad, I., and Lee, J. A., 1978, Nitrogen metabolism and salt tolerance in higher plant halophytes, in: "Ecological Processes in Coastal Environments," R. L. Jefferies and A. J. Davy, eds., Blackwell Scientific Publications, Oxford.

Stewart, G. R. and Lee, J. A., 1974, The role of proline accumulation in halophytes, Planta, 120:279.

Storey, R., 1976, Salt resistance and quaternary ammonium compounds in plants, Ph.D. Thesis, University of Wales, Cardiff.

Storey, R. and Wyn Jones, R. G., 1975, Betaine and choline levels in plants and their relationship to NaCl stress, Plant Sci. Lett., 4:161.

Storey, R. and Wyn Jones, R. G., 1978, Salt stress and comparative physiology in the Gramineae. III. The effect of salinity upon the ion relations and glycinebetaine and proline levels in S. townsendii, Aust. J. Plant Physiol., 5:831.

Storey, R. and Wyn Jones, R. G., 1979, Responses of Atriplex spongiosa and Suaeda monoica to salinity, Plant Physiol., 63: 156.

Storey, R., Ahmad, N., and Wyn Jones, R. G., 1977, Taxonomic and ecological aspects of the distribution of glycinebetaine and related compounds in plants, Oecologia, 27:319.

Strogonov, B. P., 1964, Physiological basis of salt tolerance of plants (as affected by various types of salinity), Adak. Nauk. SSSR., Translated from Russian, Israel Progr. Sci. Transl., Jerusalem.

Treichel, S., 1975, Der Einfluss von NaCl auf die Prolinkonzentration verschiedener Halophyten, Z. Pflanzenphysiol., 76:56.

Virtanen, A. J. and Kan, S., 1955, Free amino acids in pollen, Acta Chem. Scand., 9:1548.

Wyn Jones, R. G., 1980, Salt tolerance, in: "Physiological Processes Limiting Plant Productivity," C. B. Johnson, ed., Butterworth Press, London, (in press).

Wyn Jones, R. G. and Owen, E. D., 1980, Accumulation of prolinebetaine in salt stressed Medicago sativa, Plant Cell Environment, (in preparation).

Wyn Jones, R. G. and Storey, R., 1978, Salt stress and comparative physiology in the Gramineae. II. Glycinebetaine and proline accumulation in two salt and water-stressed barley cultivars, Aust. J. Plant Physiol., 5:817.

Wyn Jones, R. G. and Storey, R., 1978, Salt stress and comparative physiology in the Gramineae. IV. Comparison of salt stress in Spartina x townsendii and three barley cultivars, Aust. J. Plant Physiol., 5:839.

Wyn Jones, R. G. and Storey, R., 1979, Betaines, in: "The Physiology and Biochemistry of Drought Tolerance," L. G. Paleg and D. Aspinau, eds., Academic Press, Sydney.

Wyn Jones, R. G., Brady, C. J., and Speirs, J., 1979, Ionic and osmotic regulation in plants, in: "Recent Advances in the Biochemistry of Cereals," Academic Press, London.

Wyn Jones, R. G., Storey, R., Leigh, R. A., Ahmad, N., and Pollard, A., 1977, A hypothesis on cytoplasmic osmoregulation, in: "Regulation of Cell Membrane Activities in Plants," E. Marrè and O. Ciferri, eds., North Holland, Amsterdam.

INTEGRATION OF PHOTOSYNTHETIC CARBON METABOLISM DURING STRESS

C. B. Osmond

Department of Environmental Biology
Research School of Biological Sciences
Australian National University
Canberra City, A.C.T. 2601, Australia

INTRODUCTION

By and large, studies of photosynthesis in response to stress have pursued the reductionist philosophies of biochemistry and biophysics, and have not sought to provide integrated accounts of photosynthetic processes in successful wild plants or among the successful products of pragmatic plant breeding. Integration of biochemical, physiological and ecological aspects of photosynthesis is essential if we are to further improve upon these successful experiments. Natural selection, which is responsible for the successful plants now found in what we regard as stress environments, acts not on isolated components (such as osmoregulation) which may respond to random mutations, but upon whole integrated physiological processes which are products of the total genetic potential of the organism. Waddington (1972) reminds us that "it remains true enough to say the ultimate units, the pebbles in the concrete or genes in the organism, have been produced by random processes, (but) this is almost irrelevant to the engineering of a bridge and, in many cases, not much more relevant to the anatomical or physiological construction of the organism."

In this chapter I wish to draw on the research of colleagues to outline an integrated account of photosynthetic responses to stress at several levels in an effort to discover the consequences of osmoregulation or turgor maintenance. Turgor may be maintained in environments of low water potential (arid or saline) by means of solute uptake or solute synthesis, and by means of high tissue elasticity. Failure to maintain turgor leads to direct disruption of cell metabolism and often to closure of stomata. Both of these responses may reduce photosynthesis and growth capacity of plants in stress environments. If osmoregulation and turgor maintenance can be shown to be significant in terms of integrated photosynthetic metabolism then presumably they are pebbles of some consequence in Waddington's bridge.

INTEGRATED PHOTOSYNTHETIC METABOLISM

There have been remarkable advances in our understanding of integrated photosynthetic metabolism in the last decade. First, integration of reactions in the photosynthetic carbon reduction (PCR) cycle (Calvin cycle) with those of the photorespiratory carbon oxidation cycle (PCO) cycle (Glycolate pathway) provides a molecular basis for the interaction between O_2 and CO_2 during photosynthesis in air (Lorimer et al., 1978). For example, should stomata close in response to stress, intercellular CO_2 concentration may decline and the kinetic properties of ribulose bisphosphate carboxylase (RuP_2) oxygenase dictate that a greater proportion of carbon will then be diverted through the PCO cycle (Lawlor and Fock, 1978). Evidence that the internal source of photorespiratory CO_2 provided by these cycles can prevent photoinhibition has been presented by Cornic (1978) and Powles and Osmond (1978). Speculations as to the significance of these processes in extreme circumstances of tight stomatal closure in the light in response to stress (Osmond and Björkman, 1972; Osmond et al., 1980) remain to be tested. The integration of carbon metabolism and photochemical reactions during C_3 photosynthesis can now be modelled on a mechanistic basis (Laing et al., 1974; Peisker, 1974; Hall and Björkman, 1975; Farquhar et al., 1980).

Second, the complex biochemistry of C_4 photosynthesis is now accepted as a CO_2 concentrating mechanism which allows RuP_2 carboxylase-oxygenase to function at near to CO_2 saturation in the bundle sheath cells of these plants (Björkman, 1971; Hatch and Osmond, 1976). Consequently, the oxygenase activity of RuP_2 carboxylase in bundle sheath cells is largely suppressed, little carbon is diverted to the PCO cycle and CO_2 fixation may proceed more rapidly in C_4 plants at equal stomatal conductance. Mechanistic models of C_4 photosynthesis are not yet as advanced as those for C_3 photosynthesis (Berry and Farquhar, 1978).

Integration of these processes provides explanations of physiological phenomena, such as the higher effeciency of water use in C_4 plants, and allows us to predict that C_4 photosynthesis might be an advantage in terms of CO_2 fixation rate in any "conditions where photosynthesis would be markedly limited by the CO_2 concentration in intercellular spaces" (Björkman, 1975). As discussed below, C_4 plants tend to maintain lower intercellular CO_2 concentrations than C_3 plants under normal conditions, and these may be further reduced under water stress. Schulze et al. (1980) analyzed a large number of observations on photosynthetic CO_2 fixation in C_3 and C_4

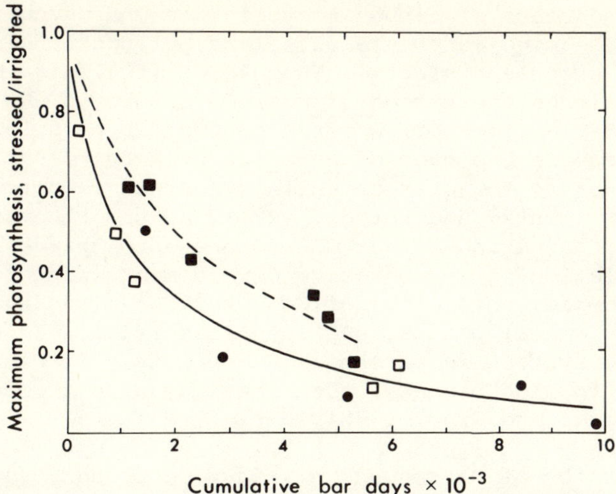

Fig. 1. Ratio of maximum net photosynthesis rate in air of droughted and irrigated plants with advancing water stress. Hammada scoparia (C_4), ■; Reaumaria regevensis (C_3), □; Zygophyllum dumosum (C_3), ●. (Data of Schulze et al., 1980, redrawn with permission).

plants in the Negev desert from spring through summer. Their data show that with progressive water stress the C_4 plant Hammada scoparia retained higher maximum CO_2 fixation rate, relative to irrigated controls, than several C_3 plants (Figure 1).

Third, the biochemical and physiological events in succulent plants with crassulacean acid metabolism (CAM) can be integrated to describe this remarkably flexible photosynthetic process (Kluge and Ting, 1978; Osmond, 1978). The CO_2 fixation capacity of CAM plants is limited by the amount of malic acid which may be stored in the vacuoles in the dark, and by their ability to engage in net CO_2 fixation in the late afternoon. In stress environments both of these net CO_2 fixation options are surrendered and CAM plants recycle respiratory CO_2 to malic acid at night, recovering it as carbohydrate in the light. Although stomata may retain totally closed throughout the light period, the internal generation of CO_2, from malic acid at first, and later from photorespiration, allows the orderly dissipation of photochemical energy in CAM plants exposed to high light intensities, thus preventing photoinhibition (Osmond, 1976; Osmond et al., 1980).

This consequence of carbon recycling in CAM may be at least as important as the high water use efficiency of nocturnal CO_2 fixation in these succulent plants. Succulence itself is important because it prolongs the period of net CO_2 uptake in the dark, up to 40 days beyond the time soil water potential drops below that of the plant tissue in the case of Ferocactus (Nobel, 1977). In some C_3 plants, exposure to water or salinity stress results in the induction of CAM (Winter and Lüttge, 1976) and under natural conditions this prolongs the period of net CO_2 uptake into the dry season, allowing completion of reproductive development (Winter et al., 1978). The complex physiology of CAM provides a mechanism for survival in these annual plants, and ensures survival of functional photosynthetic systems in perennial CAM plants. If CAM plants can be cultivated to high biomass, then they may also become highly productive (Kluge and Ting, 1978).

The diversion of carbon and nitrogen compounds from these integrated metabolic pathways for purposes of osmoregulation may be significant. Osmoregulation in leaves of higher plants appears to involve a wide range of compounds, all of which are involved in basic metabolic cycles, in contrast to osmoregulation in lower organisms which often involves specific metabolites. Jones et al. (1980) compared the compounds contributing to osmotic adjustment in Sorghum (C_4) and Helianthus (C_3). In fully expanded Sorghum leaves osmoregulation of -5 bar in response to slowly applied water stress was largely due to an increase in sugars (glucose and sucrose) and salts of malic and aconitic acids. In expanding and fully expanded Helianthus leaves osmoregulation to the extent of -4 bar was largely due to increases in cations, nitrate and carboxylic acids: sugars did not contribute to osmotic adjustment and, in fact, declined in response to stress. Evidently nitrate reduction was reduced in the stressed leaves of Helianthus. Although levels of several amino acids, including proline, increased in response to stress, these made only a minor contribution to osmoregulation. In CAM plants the nocturnal synthesis of malic acid from starch may contribute 1 to 3 bar osmoticum and restore turgor at dawn (Lüttge and Ball, 1977).

The diversion of carbon assimilated in photosynthesis for osmoregulation is only a small drain on total photosynthetic assimilation. In the Sorghum leaves studied by Jones, sucrose used in osmoregulation in response to slowly applied water stress could be provided by about 2 hours photosynthesis at the rates observed in the leaves after osmotic adjustment. However, if osmotic adjustment in nonphotosynthetic tissues is taken into account, this estimate should perhaps be increased by a factor of 10.

GAS EXCHANGE AND LEAF PHOTOSYNTHESIS

Analysis of the diffusion and metabolic components of whole

leaf photosynthetic metabolism in terms of the resistance/conductance analogy has over-emphasized the independence of these two processes. Rapidly applied stress treatments, intended to discover the first responding processes, have further exaggerated the independence of these two major components of photosynthetic response. When *Eucalyptus* seedlings were deprived of water and exposed to water stress over a period of several days, the decline in photosynthesis was attributable to a similar decline in both stomatal and nonstomatal conductance (Collatz et al., 1976). Bunce (1977) compared 12 woody species from habitats ranging from desert to streamside. When these plants were exposed to water stress the stomatal and nonstomatal conductance to CO_2 uptake began to decrease at the same water potential. After several decades of attempts to partition the response of whole leaf photosynthesis during stress it now seems likely that stomatal responses are controlled by changes in photosynthetic metabolism.

Many studies have now shown a straignt line relationship between assimilation (A) and stomatal conductance (C_s) over a wide range of conditions in which A has been manipulated by stage of development, light intensity or nutrition. Because the intercellular CO_2 concentration (p_i) is related to ambient CO_2 concentration (p_a) by the equation

$$p_i = p_a - A/C_s$$

a constant ratio A/C_s implies that intercellular CO_2 concentration remains more or less constant at a particular ambient external CO_2 concentration, temperature and humidity deficit. Wong et al. (1979) point out that a constant ratio A/C_s does not mean that stomata control assimilation but rather that stomata respond to the photosynthetic capacity of the underlying mesophyll cells. They suggest that should stomata be responsive to the photosynthetic capacity of the underlying mesophyll and maintain constant p_i, they could function so as to optimize the rate of assimilation in terms of the rate of water loss. The mechanisms responsible for the proposed linkage between photosynthetic capacity and stomatal conductance are obscure. Presumably these involve a coordination of biochemical and biophysical processes at least as complex as those underlying the cooperative metabolism in mesophyll and bundle sheath cells during C_4 photosynthesis.

Studies of the response of C_3 plants show that a linear relationship between A and C_s is also maintained in response to water stress. In Figure 2a I have plotted data from experiments with *Eucalyptus* (Collatz et al., 1976) and *Glycine* (Turner et al., 1978) which show that the ratio of A/C_s is more or less constant over a wide range of assimilation rates. In Figure 2b I have plotted data from experiments in which the responses of assimilation to salinity was examined in a number of C_3 species (Downton, 1977; De Jong, 1978;

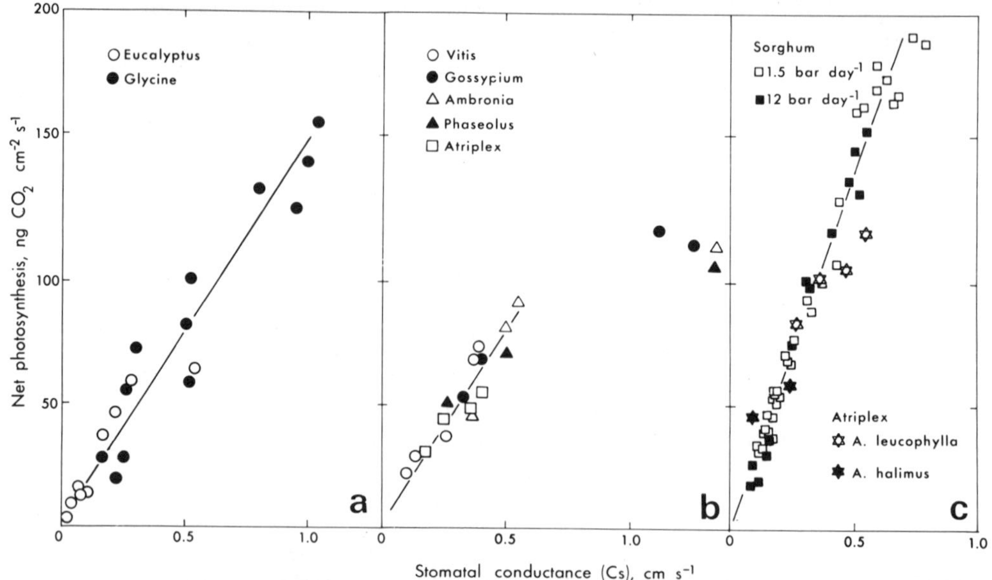

Fig. 2. Relationship between net photosynthesis in air (A) and stomatal conductance (C_s) in stressed plants. (a) Water stressed C_3 plants (Eucalyptus data calculated from Collatz et al., 1976; Glycine data calculated from Turner et al., 1978). (b) Salt stressed C_3 plants (Vitis data calculated from Downton, 1977; Gossypium and Phaseolus data calculated from Longstreth and Nobel, 1979; Ambronia and Atriplex data calculated from De Jong, 1978). (c) Water and salinity stressed C_4 plants (Sorghum data from Jones, 1979; Atriplex data from Kaplan and Gale, 1972; De Jong, 1978).

Longstreth and Nobel, 1979). Irrespective of whether the species studied were halophytes or salt sensitive crop plants, the relationship between A and C_s is more or less linear and has a slope comparable with that in Figure 2b. Some data points from control treatments with high conductance fall to the right of the line, indicating that stomata may partly close at low levels of salinity with little effect on assimilation. Similar direct responses of stomata, without effect on assimilation, have been observed if leaves are treated with ABA (Wong et al., 1979), or if leaves are exposed to sudden changes in leaf-air vapor pressure difference (Farquhar et al., 1980a).

Jones (1979) found that the relationship between A and C_s was the same in Sorghum (C_4) leaves which underwent osmoregulation when slowly stressed (1.5 bar day^{-1}) as in leaves which were rapidly stressed (12 bar day^{-1}) and did not osmoregulate (Figure 2c). As was previously noted (Osmond et al., 1980), when leaves of Zea (C_4) are subject to even more rapid stress (Lawlor and Fock, 1978), a curvilinear relationship between A and C_s is obtained. Few data are available for the responses of whole leaf photosynthesis in C_4 plants to salinity. However, Kaplan and Gale (1972) and De Jong (1978) measured the effects of salinity on gas exchange in the arid halophyte Atriplex halimus and the coastal halophyte Atriplex leucophylla and their data fit quite closely to Jones' curve for Sorghum (Figure 2c).

The causes of reduced photosynthetic capacity in response to stress cannot be considered in detail here. It should be noted, however, that the time of peak water stress commonly coincides with the peak of illumination. If water stress interferes with the normal transduction of photochemical energy, or its orderly dissipation in carbon metabolism, we anticipate that photoinhibition as a consequence of illumination may be responsible for the reduction in quantum yield and other component processes (Osmond et al., 1980). Furthermore unless leaf cells are able to accumulate the incoming salt into the vacuole and to make compatible solutes for the cytoplasm, salinity will also lead to a disruption of normal cell metabolism. In salt sensitive species such as Vitis, salinity substantially decreases nonstomatal components of leaf conductance (Downton, 1977), presumably as a result of metabolic interference. In leaves of halophytes, such as Atriplex, salinity has little effect on nonstomatal conductance (Gale and Poljakoff-Mayber, 1970).

Reconsideration of data in the above terms suggests that the coupling between stomatal conductance and assimilation capacity observed by Wong et al. (1979) persists in a wide range of species exposed to water stress. Essentially the same coupling is observed in response to salinity, and it is possible that salinity responses are largely due to effects on plant water relations. Osmoregulation does not seem to matter much in these relationships (Figure 2c). The constancy of A/C_s means that whole leaf CO_2 exchange proceeds at more or less constant intercellular CO_2 concentration; approximately 200 μl l^{-1} in C_3 plants and 100 μl l^{-1} in C_4 plants (Wong et al., 1979). Repeated exposure of Gossypium and Vigna plants to drying cycles appears to lower the steady state intercellular CO_2 concentration in these C_3 plants to between 90-150 μl l^{-1} (K. Winter, unpublished). The "homeostasis" of intercellular CO_2 which is a product of integrated stomatal physiology and mesophyll metabolism, presumably ensures that the balance of photosynthetic CO_2 metabolism is stabilized, within limits, during water and salinity stress.

THE DAILY COURSE OF PHOTOSYNTHESIS

If stomatal response is integrated with respect to photosynthetic

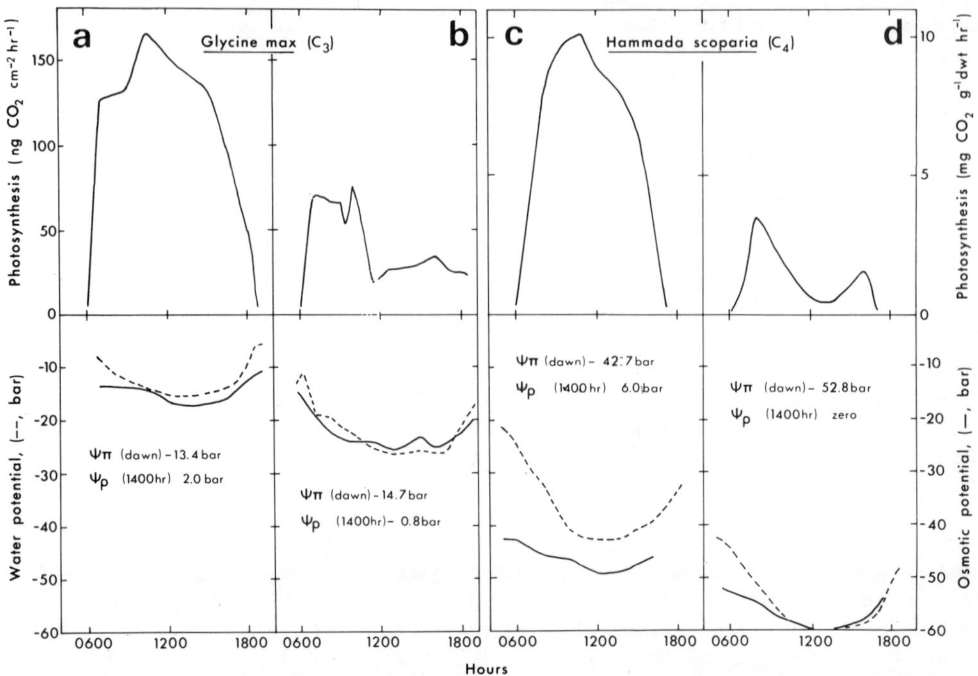

Fig. 3. Daily pattern of net photosynthesis and water relations parameters in Glycine (C_3) a mesic species which does not osmoregulate and in Hammada (C_4) and arid species which osmoregulates (Redrawn from Turner et al., 1978; Kappen et al., 1976).

capacity as outlined above, it follows that stomata may function so as to minimize the amount of water lost during the day for the amount of carbon absorbed. Predictions of optimal stomatal behavior (Cowan and Farquhar, 1977) are realized in the daily CO_2 exchange patterns of many species under stress conditions in which CO_2 exchange shows pronounced midday depression due to stomatal closure.

Figure 3a shows the daily course of photosynthesis in Glycine max (C_3) growing in the field, 6 days after irrigation ceased (Turner et al., 1978). These plants maintain positive turgor throughout the day and show a midday peak of CO_2 fixation. Figure 3b shows the daily course of CO_2 fixation and leaf water relations in the same crop 33 days after irrigation ceased. Glycine has a very small capacity for osmoregulation and consequently leaf turgor falls to zero for several hours at midday under these stress conditions. Associated with this, there is a pronounced midday depression of photosynthesis associated with stomatal closure. In line with the data discussed above, if photosynthetic capacity is reduced at zero turgor, stomata close in response to this. Stomata may also respond directly to reduced bulk leaf turgor and to the increasing water vapor difference at midday.

The daily courses of photosynthesis and water relations in leaves of Hammada scoparia (C_4) growing in the Negev desert under irrigated (Figure 3c) and control (Figure 3d) conditions is like that of Glycine (Kappen et al., 1976). This halophyte absorbs substantial amounts of salt and under irrigated conditions has a much lower osmotic potential than irrigated Glycine. Under irrigated conditions turgor can be maintained throughout the day and a normal midday peak of CO_2 fixation is observed (Figure 3c). As a consequence of salt absorption this halophyte can presumably maintain this performance in soils which dry to -30 bar or more. Hammada scoparia has the capacity for further osmoregulation in response to stress, as shown by dawn values of osmotic potential which are 10 bar more negative in the stressed, non irrigated plants (Figure 3d). In spite of the additional capacity for osmoregulation, this halophyte ultimately experiences zero turgor and a midday depression of CO_2 uptake which is qualitatively similar to that in Glycine.

Figure 3 appears to show that the presence or absence of the capacity for osmoregulation matters little to the daily course of CO_2 fixation. However we must bear in mind that the environmental conditions under which the experiments were done seem to have been vastly different. The Negev experiments were done under hotter conditions at lower relative humidity and to truly assess the significance of osmoregulation with respect to integrated photosynthesis in the same or different genotypes, comparisons must be made under the same conditions. In such a comparison, the greater capacity for solute uptake or synthesis and osmoregulation in Hammada could enable this species to remain active long after Glycine ceased to function, but such comparisons have yet to be made.

ASSIMILATE ALLOCATION AND GROWTH

All plant growth ultimately depends on the carbon assimilated in photosynthesis. Desirable genotypes in terms of biomass production tend to be those in which the carbon fixation and allocation systems have been optimally integrated for maximum growth rate in nutrient rich, warm and wet tropical habitats. In these the bulk of assimilated carbon can be devoted to synthesis of additional photosynthetic surface and reporduction can be assured by a small investment in organs of vegetative production. These plants are likely to be most useful in terms of biomass for fuel or as a source of chemicals. Desirable genotypes in terms of food production tend to be those in which carbon fixation and allocation systems are integrated for maximum reproductive effort. In these, selection for grain yield under cultivation has been at the expense of photosynthesis rate on a leaf area basis (Dunstone et al., 1973), presumably as a consequence of selection for the optimal carbon allocation system. Obviously the integration of photosynthetic and growth processes in different genotypes is likely to determine their responses to the stresses of different habitats, and may constrain our capacity to exploit different species for their biomass production or reproductive output.

Control of the allocation of assimilates between new photosynthetic, transpiring tissues and nonphotosynthetic water absorbing tissues is an important but poorly understood component of plant responses to stress. Observations suggest that annuals and perennials from mesic and arid habitats show a consistent increase in the root/shoot ratio in response to decreasing soil water potential (Fischer and Turner, 1978). Such a response seems "purposeful" in our integrated context because it effectively reduces the transpiring surface and increases the water absorbing surface. In salt sensitive species such as cotton, salinity has a similar effect on root/shoot ratio (Hoffman et al., 1971). More complex questions with respect to the allocation of assimilates remain to be explored. For example, does capacity to assimilate more carbon for a given amount of water permit C_4 plants to divert a greater proportion of assimilates to roots for water acquisition in response to stress? However, in C_3 and C_4 halophytes the stimulation of growth by low levels of salinity is largely achieved by a decrease in the root/shoot ratio.

The stimulation of growth of the halophyte *Atriplex halimus* in a dry atmosphere in response to salt provides a good example of the significance of osmoregulation for integrated photosynthetic metabolism (Gale and Poljakoff-Mayber, 1970). The presence of 10 bar NaCl in culture solutions enabled these plants to maintain turgor throughout the day when exposed to a dry atmosphere (Kaplan and Gale, 1972). In the absence of NaCl plants experienced zero turgor for several hours. Because the salt was effectively compartmented in leaf cells, photosynthetic capacity on a leaf area basis was only slightly reduced by salt treatment. Maintenance of turgor may have allowed these plants to avoid the midday depression of CO_2 uptake, although the daily course of CO_2 fixation was not measured. Turgor maintenance in the salt-treated plants presumably provided the driving force for leaf expansion and, with the additional carbon assimilated, led to an increased leaf area in salt treated plants (Gale and Poljakoff-Mayber, 1970). Greater leaf area was then available for further photosynthesis, leading ultimately to increased growth.

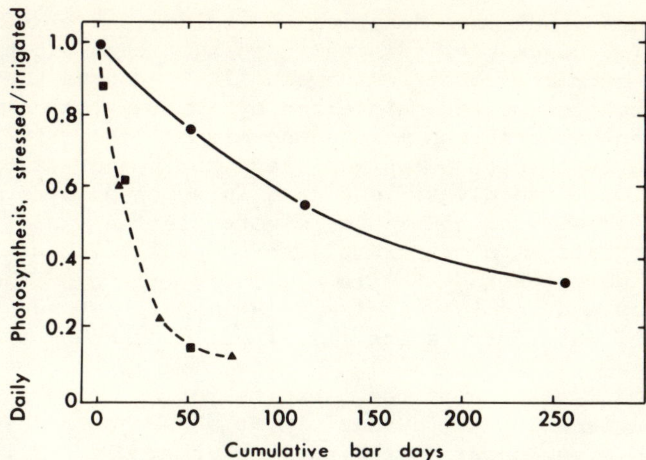

Fig. 4. Daily net photosynthetic CO_2 fixation, relative to irrigated controls in water stressed Sorghum at different rates of stress. Plants stressed at 1.5 bar day^{-1} (o) and 7 bar day^{-1} (Δ) showed osmoregulation whereas those stressed at 12 bar day^{-1} (■) did not (redrawn from Jones and Rawson, 1979).

CONCLUSIONS

Most of the experiments discussed here do not make a case for a significant role of osmoregulation in integrated photosynthetic metabolism. It seems unlikely that the diversion of photosynthetic products for osmoregulation is a very significant part of carbon metabolism. Moreover, it seems that the presence or absence of osmoregulation does not effect the capacity of leaves to integrate stomatal and nonstomatal components of CO_2 exchange during stress. Whether a plant is or is not capable of osmoregulation in response to stress ultimately makes little difference to the diurnal pattern of CO_2 exchange. However, these few observations are insufficient to evaluate the role of osmoregulation and suggest perhaps that other comparisons may be more meaningful.

Turgor maintenance through osmoregulation has the potential for maintaining photosynthesis and growth in the face of increasing soil water deficits. This potential is presumably realized by extending the duration of normal leaf turgor under stress conditions and hence extending the period of carbon acquisition. Fereres et al. (1980) showed, for example, that osmotic adjustment in non-irrigated Sorghum enabled these plants to maintain leaf turgor at the same levels as irrigated controls, although plant water potentials were 5 bar more negative in the dryland plants. Stomatal conductance (~0.30 cm s^{-1}) remained unchanged throughout the growing season. Furthermore, Jones and Rawson (1979) showed that *Sorghum* plants in which turgor was maintained by osmoregulation during slowly applied water stress maintained higher rates of photosynthesis at low leaf water potentials, and for longer periods, than plants which were rapidly stressed and in which no osmoregulation was observed (Figure 4). However, osmoregulation was also observed at an intermediate rate of stress (7 bar day^{-1}) without any impact on declining daily photosynthesis. Clearly, factors other than osmoregulation are involved in the relationship between rate of application of water stress and its effect on photosynthesis.

Should manipulation of the capacity for osmoregulation by means of genetic engineering prove feasible, the new genotype may be able to photosynthesize longer during periods of slowly applied stress. Hopefully, the improvement in photosynthesis may be translated into improved biomass production or harvestable yield. That osmoregulation in response to salinity actually stimulates the growth of *A. halimus* in dry atmospheres should stimulate us to contrive further experiments.

ACKNOWLEDGMENTS

I am grateful to Detlef Schulze for Figure 1 and for suggesting the format used in Figure 4. Klaus Winter, Madelaine Jones, Neil Turner and Graham Farquhar have contributed data and stimulating discussion during preparation of the manuscript.

REFERENCES

Berry, J. and Farquhar, G. D., 1978, The CO_2 concentrating function of C_4 photosynthesis, a biochemical model, *in*: "Photosynthesis 77," D. O. Hall, J. Coombs, and T. W. Goodwin, eds., The Bio-Chemical Society, London.

Björkman, O., 1971, Comparative photosynthetic gas exchange in higher plants, *in*: "Photosynthesis and Photorespiration," M. D. Hatch, C. B. Osmond, and R. O. Slatyer, eds., Wiley Interscience, New York.

Björkman, O., 1975, Environmental and biological control of photosynthesis: inaugural address, *in*: "Environmental and Biological Control of Photosynthesis," R. Marcelle, ed., Junk, The Hague.

Bunce, J. A., 1977, Nonstomatal inhibition of photosynthesis at low water potentials of intact leaves from a variety of habitats, *Plant Physiol*., 59:348.

Collatz, J. Ferrar, P. J., and Slatyer, R. O., 1976, Effects of water stress and differential hardening treatments on photosynthetic characteristics of a xeromorphic shrub *Eucalyptus socialis* F. Muell, Oecologia, 23:95.

Cornic, G., 1978, La photorespiration se déroulant dans un air sans CO_2 a-t-elle une fonction?, Can. J. Bot., 56:2128.
Cowan, J. R. and Farquhar, G. D., 1977, Stomatal function in relation to leaf metabolism and environment, in: "Integration of Activity in the Higher Plant," SEB Symposium 31, D. H. Jennings, ed., Cambridge University Press, Cambridge.
De Jong, T. M., 1978, Comparative gas exchange and growth responses of C_3 and C_4 beach species grown at different salinities, Oecologia, 36:59.
Downton, W. J. S., 1977, Photosynthesis in salt stressed grapevines, Aust. J. Plant Physiol., 4:183.
Dunstone, R. L., Gifford, R. M., and Evans, L. T., 1973, Photosynthetic characteristics of modern and primitive wheat species in relation to ontogeny and adaptation to light, Aust. J. Biol. Sci., 26:295.
Farquhar, G. D., Schulze, E.-D., and Küppers, U., 1980a, Responses to humidity by stomata of Nicotiana glauca L. and Corylus avellana L. are consistent with the optimization of CO_2 uptake with respect to water loss, (submitted for publication).
Farquhar, G. D., von Caemmerer, S., and Berry, J. A., 1980b, A biochemical model of photosynthetic CO_2 assimilation in leaves of C_3 species, Planta, (in press).
Fereres, E., Acevedo, E., Henderson, D. W., and Hsiao, T. C., 1978, Seasonal changes in water potential and turgor maintenance in sorghum and maize under water stress, Physiol. Plant., 44:261.
Fischer, R. A. and Turner, N. C., 1978, Plant productivity in the arid and semiarid zones, Ann. Rev. Plant Physiol., 29:277.
Gale, J. and Poljakoff-Mayber, A., 1970, Interrelations between growth and photosynthesis of saltbush (Atriplex halimus L.) grown in saline media, Aust. J. Biol. Sci., 23:937.
Hall, A. E. and Bjorkman, O., 1975. Model of leaf photosynthesis and respiration, in: "Perspectives in Biophysical Ecology," D. M. Gates, R. B. Schmerl, eds., Springer Verlag, Berlin.
Hatch, M. D. and Osmond, C. B., 1976, Compartmentation and transport in C_4 photosynthesis, in: "Transport in Plants III, Encyclopedia of Plant Physiology," (New Series) Vol. 3, U. Heber and C. R. Stocking, eds., Springer Verlag, Berlin.
Hoffman, G. J., Rawlins, S. T., Garber, M. J. and Cullen, E. M., 1971, Water relations and growth of cotton as influenced by salinity and relative humidity, Agron. J., 63:822.
Jones, M. M., 1979, "Physiological Responses of Sorghum and Sunflower to Leaf Water Deficity," Ph.D. Thesis, Australian National University, Canberra.
Jones, M. M. and Rawson, H. W., 1979, Influence of rate of development of leaf water deficits upon photosynthesis, leaf conductance, water use efficiency, and osmotic potential in sorghum, Physiol. Plant., 45:103.
Jones, M. M., Osmond, C. B., and Turner, N. C., 1980, Accumulation of solutes in leaves of sorghum and sunflower in response to water deficits, Aust. J. Plant Physiol., (submitted for publication).

Kaplan, A. M. and Gale, J., 1972, Effect of sodium chloride salinity on the water balance of _Atriplex halimus_, _Aust. J. Biol. Sci._, 25:895.

Kappen, L., Lange, O. L., Schulze, E.-D., Evenari, M., and Buschbom, U., 1976, Distributional pattern of water relations and net photosynthesis of _Hammada scoparia_ (Pomel) Illjin in a desert environment, Oecologia, 23:323.

Kluge, M. and Ting, I. P., 1978, "Crassulacean Acid Metabolism: Analysis of an Ecological Adaptation," Springer Verlag, Berlin.

Laing, W. A., Ogren, W. L., and Hageman, R. H., 1974, Regulation of soybean net photosynthetic CO_2 fixation by the interaction of CO_2, O_2 and ribulose-1,5-diphosphate carboxylase, _Plant Physiol._, 54:678.

Lawlor, D. W. and Fock, H., 1978, Photosynthesis, respiration and carbon assimilation in water-stressed maize at two oxygen concentrations, _J. Exp. Bot._, 29:579.

Longstreth, D. J. and Nobel, P. S., 1979, Salinity effects on leaf anatomy: consequences for photosynthesis, _Plant Physiol._, 63:700.

Lorimer, G. H., Woo, K. C., Berry, J. A., and Osmond, C. B., 1978, The C_2 photorespiratory carbon oxidation cycle in leaves of higher plants: pathway and consequences, _in_: "Photosynthesis 77," D. O. Hall, J. Coombs, T. W. Goodwin, eds., The Biochemical Society, London.

Lüttge, U. and Ball, E., 1977, Water relations parameters of the CAM plant _Kalanchoe daigremontiana_ in relation to diurnal malate oscillations, Oecologia, 31:85.

Nobel, P. S., 1977, Water relations of barrel cactus _Ferocactus acanthodes_ in the Colorado desert, Oecologia, 27:117.

Osmond, C. B., 1976, CO_2 assimilation and dissimilation in the light and dark in CAM plants, _in_: "CO_2 Metabolism and Plant Productivity," R. H. Burris and C. C. Black, eds., University Park Press, Baltimore.

Osmond, C. B., 1978, Crassulacean acid metabolism: a curiosity in context, _Ann. Rev. Plant Physiol._, 29:379.

Osmond, C. B. and Bjorkman, O., 1972, Simulataneous measurements of oxygen effects on net photosynthesis and glycolate metabolism in C_3 and C_4 species of _Atriplex_, _Carnegie Inst. Wash. Yearbook_, 71:141.

Osmond, C. B., Winter, K., and Powles, S. B., 1980, Adaptive significance of CO_2 cycling during photosynthesis in water stressed C_3, C_4 and CAM plants, _in_: "Adaptation of Plants to Water and High Temperature Stress," N. C. Turner and P. J. Kramer, eds., Wiley Interscience, New York.

Peisker, M., 1974, A model describing the influence of oxygen on photosynthetic carboxylation, _Photosynthetica_, 8:47.

Powles, S. B. and Osmond, C. B., 1978, Inhibition of the capacity and efficiency of photosynthesis in bean leaflets illuminated in a CO_2-free atmosphere at low oxygen: a possible role for photorespiration, _Aust. J. Plant Physiol._, 5:619.

Schulze, E.-D., Hall, A. E., Lange, O. L., Evenari, M., Kappen, L., and Buschbom, U., 1980, Long term effects of drought on wild and cultivated plants in the Negev Desert I. Maximal rates of net photosynthesis, Oecologia, (in press).

Turner, N. C., Begg, J. E., Rawson, H. M., English, S. D., and Hearn, A. B., 1978, Agronomic and physiological responses of soybean and sorghum crops to water deficit. III. Components of leaf water potential, leaf conductance, $^{14}CO_2$ photosynthesis and adaptation to water deficity, Aust. J. Plant Physiol., 5:179.

Waddington, C. H., 1972, The theory of evolution today, in: "Beyond Reductionism," A. Koestler and J. R. Smythies, eds., Hutchinson, London.

Winter, K. and Lüttge, U., 1976, Balance between C_3 and CAM pathway of photosynthesis, in: "Water and Plant Life," O. L. Lange, L. Kappen, and E.-D. Schulze, eds., Springer-Verlag, Berlin.

Winter, K., Lüttge, U., Winter, E., and Troughton, J. H., 1978, Seasonal shift from C_3 photosynthesis to Crassulacean acid metabolism in Mesembryanthemum crystallinum in its native environment, Oecologia, 34:225.

Wong, S. C., Cowan, I. R., and Farquhar, G., 1979, Stomatal conductance correlated with photosynthesis capacity, Nature, 282:424.

ENERGY COST OF ION TRANSPORT

B. W. Veen

Centre for Agrobiological Research
Wageningen
The Netherlands

INTRODUCTION

For a plant to survive in a saline environment it is necessary to have specific control of internal ion concentrations.

The influence of high salt concentrations in the root environment on plant growth is two-fold:

Because of the high salt concentrations in the environment, osmotic adjustment of the plant tissue is necessary so that the water potential does not result in outflow of water and dehydration of cells. Bernstein (1963, 1963) and Lagerwerf and Eagle (1963) demonstrated osmotic adjustment of bean plants, grown under saline conditions. When salt concentrations in the root environment were increased the plant took up ions in excess of the amount required for growth. However these ions were accumulated at levels optimal for adjustment of water potential. Under saline conditions nutrients required for growth (e.g., K^+) are present in small amounts relative to ions not required for physiological reactions (e.g., Na^+), and, dependent on the plant species, these excess ions can have a detrimental effect on the metabolism of the plant. Adaptation to saline conditions involve uptake mechanisms with high selectivity for K^+ relative to Na^+ (Rains, 1972). Pitman and Sadler (1967) suggest the existence of ion pumps at the plasmalemma transporting K^+ and Cl^- inward and Na^+ outward. Kylin and Quatrano (1975) conclude that active secretion into the vacuole protects the cytoplasm against too high a Na^+ concentration, and increases cell turgor. The adaptation of plants to saline conditions mentioned above involve transport of ions across cell membranes.

One of the responses of plants to saline conditions is an increase in respiration (Rains, 1972). In comparison with the respiration of the whole plant the respiration of the root system is relatively high. From data of Andre et al. (1978) it can be calculated that in the vegetative stage the root respiration of a maize plant amounts to 40% of the total plant respiration. Hansen and Jensen (1977) calculated that the relative energy consumption of the roots of Lolium multiflorum was about 40%. Because of the relative importance of root respiration in the consumption of energy by the plant, the relation between respiration and ion uptake, and the energy cost of uptake in relation to the energy cost of other root activities will be considered here.

MATERIAL AND METHODS

Maize plants var. "Avanti" were grown on a Hoagland nutrient solution in a controlled climate room. The day temperature was 22 C, the night temperature 20 C and the relative humidity about 70%. Four Philips HPI 375 W lamps provided a light intensity of about 70 W m^{-2}, with a daily interval of 16 h light and 8 h dark.

The system was used for the simultaneous measurement of nitrate and potassium uptake, water uptake, root growth and root respiration (Veen, 1977; Figure 1). A maize root was situated in the system through which an aerated nutrient solution with a temperature of 23 C was circulated. Concentrations of nitrate and potassium were measured with specific ion electrodes and recorded, together with the volume of the nutrient solution at short intervals on a data logger. From these data rates of nitrate and potassium uptake were calculated.

Every 4 hours aeration was stopped for one hour. From changes in oxygen concentration in the nutrient solution during that period root respiration was calculated.

The daily growth in volume of the root system was calculated as the difference of the volume of solution that the root vessel contained measured in one day time-intervals. The volume of the root system at any time during the experiments was obtained by subtracting the daily growth of the roots from the root volume measured at harvest of the plant.

To study the mutual relation between root respiration and ion uptake three types of experiments were performed: (a) short-time experiments where the rate of ion uptake was varied by changing the ion supply and measuring differences in ion uptake and root respiration, (b) long-time experiments where the total daily root respiration was measured during a number of consecutive days. Root respiration was separated in three components by multiple regression analyses, using root growth, root volume and ion uptake as the

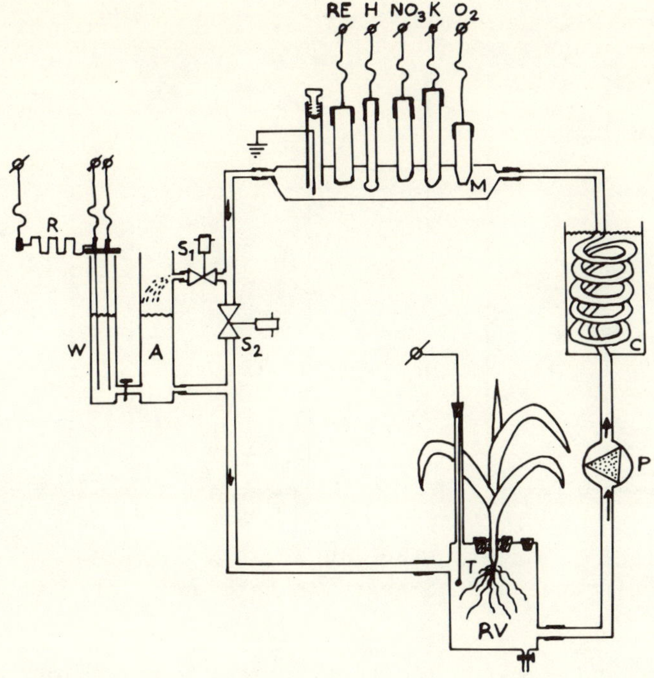

Fig. 1. A diagram of the system for the measurement of uptake of water, ions, and oxygen by plant roots from a circulated nutrient solution. RV = root vessel, P = pump, C = cooling unit, M = measuring chamber, O_2 = oxygen electrode, K = potassium electrode, NO_3 = nitrate electrode, H = pH electrode, RE = reference electrode, W = water measuring vessel, A = aeration vessel, S = solenoid valve, R = resistance, T = semiconductor for temperature measurement.

independent variables, and (c) long-term experiments where the energy supply of the roots is varied, the relative effects of energy deficiency on growth, maintenance and ion uptake were studied.

In all the experiments the NO_3^--uptake is used as a measure of the total ion uptake.

RESULTS

The influence of supply of a diluted Hoagland nutrient solution to maize roots, after a short period of depletion is shown in Figure 2. Increasing the ion supply results in greater ion uptake and a concomitant rise in respiration. After a few hours the ion concentration decreases, resulting in a decreased ion uptake and respiration. When change in respiration is attributed completely to

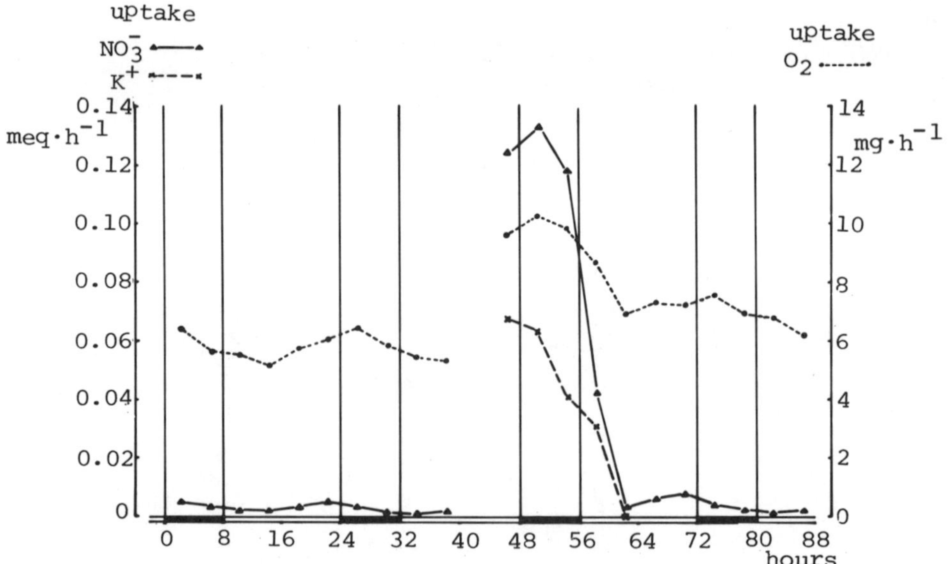

Fig. 2. Effect of nutrient supply after a short depletion period on nitrate- and potassium uptake and root respiration.

change in ion uptake, the relation between respiration and ion uptake can be calculated and amounts to 30.2 mg O_2 per meq NO_3^-, with a standard deviation of 4.8. Uncertainty about the effect of nutrient supply on the respiration associated with root growth decreases the significance of the calculations.

In a second series of experiments, root growth, the average root volume, the ion uptake and the daily root respiration have been measured for consecutive days.

Table I summarizes the root activities for 7 days. To increase the relative differences between growth respiration, uptake respiration and maintenance respiration, part of the root system was removed one day before the start of the experiments. For the same reason the light intensities were changed as indicated in Table I.

The results of multiple regression analysis are shown in Table II.

The uptake respiration amounts to 36.8 mg O_2 per meq NO_3^- (S_d 2.1). The reliability of these results is greater than the results of the short-time experiments, because here root growth and maintenance respiration have been taken into account. The NO_3^- uptake covers about 90% of the anion uptake of maize roots so that the uptake respiration amounts to 33 mg O_2 per meq anions.

Table I. Root weight and root activity on seven successive days with different light intensities.

period nr.	light int. $W.m^{-2}$	NO_3 uptake $meq.h^{-1}$	root growth $g.h^{-1}$	root weight g	O_2 uptake $mg.h^{-1}$
1	35	0.0065	0.059	4.8	1.957
2	35	0.0150	0.104	9.8	3.358
3	70	0.0590	0.119	16.3	5.761
4	70	0.1150	0.148	22.7	8.498
5	35	0.0850	0.102	29.2	6.402
6	70	0.1110	0.065	34.2	6.900
7	0	0.0400	0.037	37.1	3.550

Table II. Oxygen consumption of a maize root in relation to ion uptake, growth and root size.

energy cost of :		S_d
ion uptake $mg\ O_2.\ meq\ NO_3^{-1}$	36.8	2.1
root growth $mg\ O_2.\ g^{-1}$	24.5	1.2
maintenance resp. $mg\ O_2.\ g^{-1}.h^{-1}$	0.032	0.005

Table III. Relative energy consumption of uptake, growth and maintenance processes.

period nr.	light int. $W.m^{-2}$	% of total respiration used for		
		ion uptake	growth	maintenance
1	35	13	78	8
2	35	16	75	9
3	70	39	52	9
4	70	49	42	9
5	35	48	38	14
6	70	60	24	16
7	0	41	25	33

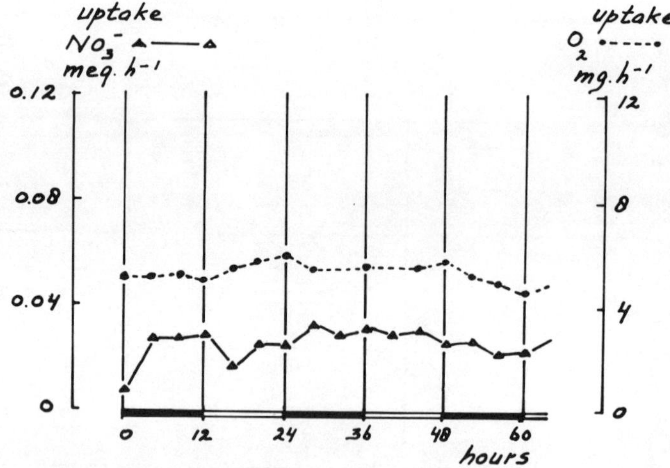

Fig. 3. NO_3^- uptake and root respiration in a 12 h light period.

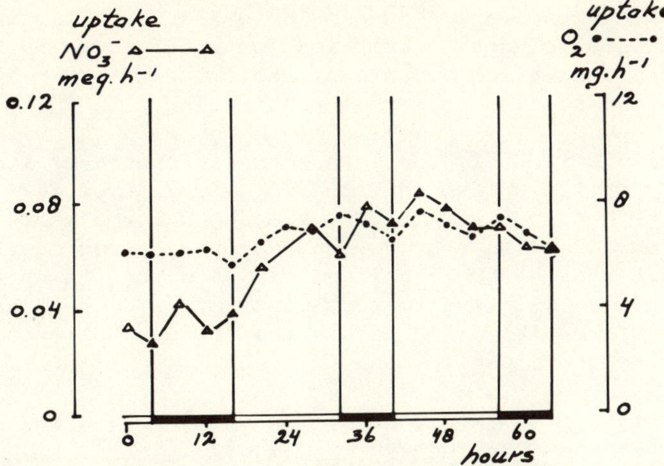

Fig. 4. Effect of changing the light period from 12 h to 16 h on NO_3^- uptake and root respiration.

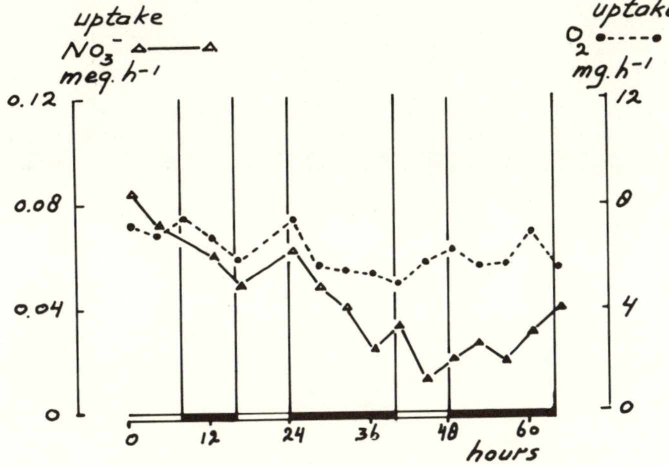

Fig. 5. Effect of changing the light period from 16 h to 8 h on NO_3^- uptake and root respiration.

The percentages of the total respiration used for ion uptake, growth and maintenance are shown in Table III. Shortly after the roots are excised, oxygen consumption for root growth is relatively high, but with increasing amounts of roots the uptake respiration increases to 60% of the total root respiration.

In a third series of experiments the influence of limiting energy supply on ion uptake was studied. In Figures 3, 4 and 5, nitrate uptake and root respiration are shown for a maize plant grown for a number of days at various day length regimes. After a change from 12 h light to 16 h light per 24 h the ion uptake is strongly increased relative to root respiration. When the light period is decreased from 12 h to 8 h the ion uptake is decreased relative to the root respiration.

A possible explanation is that ion uptake has a low priority when the root system has energy deficiencies.

DISCUSSION

Root respiration accounts for an appreciable amount of the total plant respiration. From the total root respiration of a maize plant, growing under normal conditions about 50% is due to ion uptake, and roughly 20% of the total plant respiration is used for uptake and transport of ions.

Salinity increases inward and outward transport of ions across cell membranes, possibly resulting in an increase of root respiration. In situations where the root system or part of it becomes energy deficient, the internal ionic environment cannot be controlled, resulting in death of the plant. An indication of the negative effect of salinity on the energy balance of a plant has been given by Rush and Epstein (1976) who observed a steep decrease in sugar level in roots of tomato plants under salt stress when compared with unsalinized controls. Livne and Levin (1967) and Hason-Porath and Poljakoff-Mayber (1970) found a decrease in ATP/ADP ratio when pea plants were salinized with NaCl.

In our experiments the oxygen consumption has been used as a measure of the energy availability for ion uptake. Experiments of Lüttge et al. (1971) indicate that salt transport is directly linked to ATP generated by respiration. The efficiency of respiration in producing ATP is however not always the same, but depends on the existence of an alternative non-phosphorylative electron transport chain, which is highly dependent on the plant species, variety or developmental stage (Lambers, 1979).

Selection for varieties generating ATP with a high efficiency may increase the possibility of increasing the tolerance of plants to salt stress.

REFERENCES

Andre, M., Massimino, D., and Daguenet, A., 1978, Daily patterns under the life cycle of a maize crop. II. Mineral nutrition, root respiration and root excretion, Physiol. Plant., 44:197.

Bernstein, L., 1961, Osmotic adjustment of plants to saline media. I. Steady state, Amer. J. Bot., 48:909.

Bernstein, L., 1963, Osmotic adjustment of plants to saline media. II. Dynamic phase, Amer. J. Bot., 50:360.

Hansen, G. K. and Jensen, C. R., 1977, Growth and maintenance respiration in whole plants, tops and roots of Lolium multiflorum, Physiol. Plant., 39:155.

Hason-Porath, E. and Poljakoff-Mayber, A., 1970, Effect of chloride and sulfate types of salinity on the nicotinamide-adenine-dinucleotide in pea root tips, J. Exptl. Bot., 21:300.

Kylin, A. and Quatrano, R. S., 1975, Metabolic and biochemical aspects of salt tolerance, in: "Plants in Saline Environments," A. Poljakoff-Mayber and J. Gale, eds., Springer-Verlag, New York.

Lagerwerf, J. V. and Eagle, H. E., 1961, Osmotic and specific effects of excess salts on beans, Plant Physiol., 36:472.

Lambers, H., 1979, Efficiency of root respiration in relation to growth rate, morphology and soil composition, Physiol. Plant., 46:194.

Livne, A. and Levin, N., 1967, Tissue respiration and mitochondrial oxidative phosphorylation of NaCl-treated pea seedlings, Plant Physiol., 42:407.

Lüttge, U., Cram, W. J., and Laties, G. G., 1971, The relationship of salt stimulated respiration to localized ion transport in carrot tissue, Z. Pflanzenphysiol., 64:418.

Pitman, M. G. and Saddler, H. D. W., 1967, Active sodium and potassium transport in cells of barley roots, Proc. Natl. Acad. Sci., 57:44.

Rains, D. W., 1972, Salt transport in relation to salinity, Ann. Rev. Plant Physiol., 23:367.

Rush, D. W. and Epstein, E., 1976, Genotypic responses to salinity. Differences between salt-sensitive and salt-tolerant genotypes of tomato, Plant Physiol., 57:162.

Veen, B. W., 1977, The uptake of potassium, nitrate, water and oxygen by a maize root system in relation to its size, J. Expt. Bot., 28:1389.

SECTION IV

MECHANISMS OF DROUGHT AND COLD TOLERANCE IN PLANTS

OSMOREGULATION IN PLANTS DURING DROUGHT

J. S. Boyer and R. F. Meyer

USDA/SEA, Department of Botany, and
Department of Agronomy
University of Illinois
Urbana, IL 61801

INTRODUCTION

For higher plants, water deficiency is more commonly caused by a lack of sufficient water than by exposure to high concentrations of solutes. Crop production is frequently limited by this factor and certain stages of development are more sensitive than others. A particularly sensitive time is during germination. The plant has little root and is located in a layer of soil that quickly dehydrates. Growth must continue, however, since the plant must become autotrophic. The possibility that osmoregulation may protect against the inhibitory effects of dry soil has not been explored during germination (Meyer and Boyer, 1972). In the present paper, we will describe a highly developed system of osmoregulation that permits seedling growth under dry soil conditions that would otherwise be lethal for the plant.

MATERIALS AND METHODS

We germinated soybean (*Glycine max* L. Merr.) seeds in a dark, humid environment in vermiculite containing adequate water and a few calcium and chloride ions (10^{-4} M $CaCl_2$). After 48 hours, the seedlings were transplanted to vermiculite containing various amounts of moisture. Growth and osmoregulatory behavior were observed after transplanting. The dark humid conditions simulated those occurring during early germination in soil. The transplanting to low moisture conditions simulated the dehydration that typically takes place in soil a short time after a rain.

RESULTS

Soybean seedlings had a well developed hypocotyl with an apical meristem, an elongation region of about 3 cm, and a mature basal region. The meristematic cells elongated rapidly in the zone of elongation and became approximately 13 times longer than their meristematic counterparts. Growth was rapid, and the cells made the transition from the meristematic region to the mature region in as little as 15 hours. Consequently, the cells doubled in length every 4 hours. Because of the large size of the elongation region, it was possible to measure the water potential, osmotic potential, and turgor (the difference between water potential and osmotic potential) in the region undergoing growth. Osmoregulation was considered to occur when the osmotic potential compensated for changes in the water potential so that turgor remained constant.

Table I shows that the growth rate of the hypocotyls was inhibited by the moisture deficits in the vermiculite. Rates of growth were constant for at least 72 hours after transplanting. Studies of longitudinal sections of the enlarging tissue showed that the profile of cell lengths was unaltered by the moisture treatments. Therefore, from the anatomical standpoint, the seedlings were identical and differed only in rates of growth.

The water potential of the growing region of the hypocotyls decreased from -2.3 bars in vermiculite having adequate water to -8.5 bars in vermiculite having 7% of the moisture contained in the control (Table I). The osmotic potential decreased from -7.8 bars to -13 bars under the same conditions (Table I). Therefore, turgor was 5.5 bars and 4.5 bars in the enlarging region of the control seedlings and low moisture seedlings, respectively, which indicates that the solute changes had compensated to a large extent for the decreases in tissue water potential (Table I). In these conditions, the seedlings appeared turgid and showed no sign of wilting.

If the cotyledons were removed from the seedlings, growth ceased at water potentials of -6 bars in the enlarging region and the plants appeared wilted. The osmotic potential of the enlarging region remained constant regardless of the moisture treatment (-7.6 bars). Therefore, the seedlings did not osmoregulate and considerable turgor had been lost. Evidently, in the absence of significant salts in the external medium, the soybean seedling was capable of compensating for changes in the external water supply using internally derived solutes, but only if the cotyledons were present.

When the seedlings were transplanted to water deficient medium, the osmotic potential changed to a new steady level within 20 hours. This was prevented by cool temperatures (4°). Measurements of soluble protein, free amino acids, glucose and fructose, and

Table I. Growth and Osmoregulation in Hypocotyls of Intact Soybean Seedlings

Moisture Levels[a]	Hypocotyl Growth Rates[b]	Elongation Zone[b]		
		Water Potential	Osmotic Potential	Turgor
1.00X	1.48 mm hr^{-1}	-2.3 bars	-7.8 bars	+5.5 bars
0.50X	0.90	-3.1	-8.4	+5.3
0.25X	0.59	-7.0	-11.3	+4.3
0.13X	0.22	-8.3	-12.4	+4.1
0.07X	0.14	-8.5	-13.0	+4.5

[a]Moisture levels were varied by supplying different amounts of water containing 10^{-4} M $CaCl_2$ to the vermiculite around the roots.

[b]Growth rates were measured in the hypocotyl and the mean water potential and osmotic potential were determined for the growing region of the hypocotyl. Turgor was calculated as the difference between the water potential and the osmotic potential. Matric effects were small and were ignored. Growth rates, water potentials, and osmotic potentials became constant about 15 hr after the treatments were imposed and remained constant for at least 72 hr.

sucrose contents in the enlarging tissue indicated that 70% of the change in solute content could be attributed to glucose and fructose plus free amino acids. Since amino acids and sucrose are likely to be the major translocation forms from the cotyledons, it appears that osmoregulation in the growing region consists primarily of those molecules or simple derivatives of them.

CONCLUSIONS

We conclude that soybean seedlings are capable of internal osmotic compensation for changes in water potential caused by the external water supply. This compensation occurs in the absence of any significant salts in the external medium and could happen frequently under natural conditions. The compensation is sufficiently large to result in a 70% increase in solute content of the enlarging region within 20 hours, which permits slow growth to continue under conditions where wilting, cessation of growth, and eventual death would otherwise take place. For soybean, the solutes involved are mainly glucose, fructose, and amino acids.

REFERENCE

Meyer, R. F. and Boyer, J. S., 1972, Sensitivity of cell division and cell elongation to low water potentials in soybean hypocotyls, *Planta*, 108:77.

MEMBRANE DYNAMICS: EFFECTS OF ENVIRONMENTAL STRESS

M. J. Saxton, R. W. Breidenbach, and J. M. Lyons*

Plant Growth Laboratory/Agronomy & Range Science Dept.
University of California, Davis, CA 95616

*Department of Vegetable Crops
University of California, Davis, CA 95616

INTRODUCTION

Many workers have studied the effects of temperature on the molecular ordering of lipids in cell membranes, the relation of this ordering to the activity of membrane enzymes, and the physiological consequences of this relation. The effects of low temperature on plant membranes were reviewed recently by Lyons, Raison and Steponkus (1980).

The molecular ordering of a membrane is a function of a number of variables, each of which interacts with the temperature response. These include the concentrations of ions in the aqueous phase, mechanical stress -- both pressure and tangential stress -- and various external forces. External forces arise from the electrical potential across the membrane, external electrical fields, the interaction of the cell wall with the membrane, and the interaction of cytoskeletal elements with components of the membrane.

Environmental stresses such as chilling, salinity and drought alter these variables and thus evoke the plant's response. We present here an outline of a unified view of environmental stress in terms of its effects on membrane dynamics. First, we discuss the effects of ion concentrations and mechanical stress on molecular ordering. We next consider the effects of the cell wall and the cytoskeletal elements on the membrane. Then, we consider the effects of environmental stress on the membrane. Finally, we summarize possible physiological mechanisms in the membrane, and their sensitivity to the molecular ordering of the membrane. A summary of the view then allows us to discuss some important topics for further research.

MOLECULAR ORDERING IN MEMBRANES

Phase transitions in lipid bilayers can be caused by changes in temperature, but they can also be brought about isothermally by changes in pH, or in the concentration of univalent or polyvalent cations. The viscosity of a given phase will also be a function of these variables.

In pure lipids, a phase transition involves change in the rotational freedom of the hydrocarbon chains, in the freedom of motion of the head groups, and in the average tilt of the head groups. Associated with these changes are alterations in the packing density and degree of hydration. In lipid mixtures there is the additional phenomenon of lateral phase separation which usually involves formation of "fluid-like" and "solid-like" domains. Wu and McConnell (1975) have suggested, however, that under some conditions immiscible liquid-like domains also may be formed.

A related phenomenon is the formation of small transient quasi-crystalline clusters of lipids "floating" in the liquid phase. The presence of such transient clusters would explain the abrupt changes in activation energies observed in certain systems well above the temperature of the gel-to-fluid transition (Lee et al., 1974; Wunderlich et al., 1975; Ting and Solomon, 1975; Bashford et al., 1976; but see Davis et al., 1976). For recent reviews of the general concepts of membrane structure, see Lee (1975, 1977a,b) and Melchior and Steim (1976).

The molecular ordering of a lipid bilayer depends on the Gibbs free energy of the various phases. The Gibbs free energy is determined by the interaction energy of the lipids, the entropy of the phase, and the pressure-volume relation for the lipids. For the purpose of this review it is sufficient to discuss the interaction energy.

One of the most significant contributions to the total energy is the interaction energy of the headgroups, involving hydrogen bonding and Coulomb forces. And, in turn, hydrogen bonding is sensitive to the pH of the aqueous phase (MacDonald et al., 1976), while the electrostatic interactions depend upon both the pH and the ionic composition of the aqueous phase. Thus, the effect of pH on the temperature of the phase transition can be large. The transition temperature of fully ionized DPPG is 17 C higher than that of neutral DPPG (Watts et al., 1978; see also Galla and Sackmann, 1975; Jacobson and Papahadjopoulos, 1975; and MacDonald et al., 1976). In DMPA, a phase transition can be induced isothermally by a pH change of one unit (Träuble and Eibl, 1974). The viscosity of a membrane also depends on pH; increased ionization leads to expansion of the membrane, lowering the viscosity (Watts et al., 1978).

The main effect of monovalent ions is through the ionic strength of the medium, which affects both the degree of ionization of the headgroups and the range of their electrostatic interaction. One of the larger shifts observed was in DPPS at pH 6; addition of 0.1 M NaCl or KCl shifted the transition temperature by 8 C (MacDonald et al., 1976; also see Träuble and Eibl, 1974; Jacobson and Papahadjopoulos, 1975; and Weller and Haug, 1977).

Specific interactions may be important when polyvalent ions are present. Calcium ions can shift the transition temperature to a much higher value (Jacobson and Papahadjopoulos, 1975). In mixed vesicles of DPPA and DPPC, Ca^{++} binds to DPPA, forming rigid clusters of lipid. Lateral phase separations can thus be induced at constant temperature (Ohnishi and Ito, 1973, 1974; Galla and Sackmann, 1975). Polylysine can have a similar effect (Galla and Sackmann, 1975; Träuble and Eibl, 1974) and the use of polylysine to bind cells to substrates (Mazia et al., 1975; McKeehan and Ham, 1976) may therefore alter the state of the membrane.

Another important variable affecting the molecular ordering of membranes is mechanical stress. The gel-to-fluid transition leads to decreased thickness and to lateral expansion. The transition is therefore sensitive to both the normal and tangential components of mechanical stress.

The influence of membrane curvature on molecular ordering has been demonstrated by comparison of phase transitions in large multilamellar liposomes with those in small unilamellar vesicles of radius 10-15 nm (see for example, Lentz et al., 1976a; Suurkuusk et al., 1976; Marsh et al., 1977). The transition temperature in the smaller vesicles is lower by 2-5 K, and is much broader. This broadening implies a decrease in the cooperativity of the transition. A further indication of the influence of membrane curvature can be found in the studies of Lentz et al. (1976b), who showed that different species of phosphatidylcholine mix less ideally in smaller vesicles than in liposomes. Theoretical treatments indicate that a change in the composition of the aqueous phase can induce a change in the curvature of a membrane; such changes are well known in erythrocytes (Evans, 1974; Gruler, 1975; Gebhardt et al., 1977; Evans and Hochmuth, 1978).

The coupling of mechanical variables, chemical variables, and molecular ordering by the membrane is essential to an understanding of turgor pressure sensing, as will be discussed in detail in the following section.

In this section, we have discussed membrane dynamics for

artificial lipid bilayers with one or two components. In these membranes, phase changes and lateral phase separations are well defined and take place over a relatively narrow temperature range. In complex membranes, such as those of higher plants, many types of lipids are present, sterols are present, and membrane proteins affect the conformation of the lipids. These factors broaden the temperature range of changes in molecular ordering (Melchior and Steim, 1976). For a review of the effects of sterols see Demel and De Kruyff (1977); for a theoretical discussion of protein effects on phase transitions, see Owicki et al. (1978) and Owicki and McConnell (1979).

INTERACTIONS WITH THE CELL WALL

The cell wall is not an inert container, but part of the physiological machinery of the cell. One would predict a strong interaction between the plasma membrane and the cell wall; but, because of experimental difficulties, this interaction has been little investigated (see Burgess, 1978, for a brief review of the physiological differences between protoplasts and intact cells.)

One indication of this interaction can be found in data showing that the cell wall appears to be involved in control of cell division (Lloyd, 1978). Protoplasts prepared from parenchyma cells of tobacco divide, even though the original cells with an intact cell wall do not (Nagata and Takebe, 1970). Apparently the process of regeneration of the cell wall is necessary for division; furthermore, addition of cellulase to the protoplast culture medium inhibits cell division (Schilde-Rentschler, 1977).

Another example of the interaction is that the electrical potential of a cell is affected by the presence of the cell wall. Protoplasts are electrically positive with respect to the medium, even though the cells from which they are derived are electrically negative (Heller et al., 1974; Racusen et al., 1977; Rona et al., 1977). This difference in potential is attributed to the diffusion potential resulting from fixed charges in the cell wall (Rona et al., 1977). The potential of a cell with respect to the medium changes during plasmolysis or deplasmolysis. In tobacco protoplasts, the sign and magnitude of this change depend upon the age of the protoplasts, and therefore may depend upon the extent of regeneration of the cell wall (Racusen et al., 1977).

The interaction of the cell wall with the plasma membrane may also be involved in osmoregulation. In the marine alga _Valonia_, K^+ transport is stimulated when the turgor pressure is lowered (Gutknecht, 1968). This stimulation is due to the turgor pressure, not the absolute pressure (Hastings and Gutknecht, 1974) and several mechanisms have been suggested to explain this observation (Hastings and Gutknecht, 1974; Gutknecht et al., 1978; Zimmermann et al., 1976; Zimmermann, 1978; see also the discussion in Zimmermann and Dainty, 1974. First, transport may be affected by compression of the membrane. Second, the interactions of negative charges in the cell wall with charged membrane components may control K^+ transport. Third, transport may be controlled by the interaction of the plasma membrane with cellulose microfibrils. At normal turgor pressure, the membrane is pressed around the microfibrils, which are approximately 30 nm in diameter. This imposed curvature may lead to lateral phase separation in the lipids. When turgor pressure drops, the membrane relaxes, and the lipids may redistribute themselves.

Zimmermann and co-workers proposed thinning of a membrane under pressure as the mechanism of sensing turgor pressure (Coster et al.; 1977; Zimmermann, 1978). In *Valonia*, a change in turgor pressure of 1 bar is estimated to change the thickness of the membrane by 2% (Zimmermann et al., 1977). They present a model of electrical breakdown of membranes, requiring as the condition of mechanical equilibrium that the sum of the forces from turgor pressure and from the electrical potential across the membrane be equal to the restoring force due to the elasticity of the cell wall (Coster et al., 1977). Their model of the force due to the electrical potential is oversimplified, since it assumes that the membrane components are electrically neutral, but the basic idea is correct: the turgor pressure and the electrical potential are interconnected. If the electrical potential affects the thickness, then the thickness is sensitive to the ionic composition of the medium (Zimmermann, 1977).

Similarly it was shown that the electrical resistance of the membrane has a maximum as a function of turgor pressure, and the position of the maximum depends on the volume of the cell (Zimmermann et al., 1976). Addition of IAA increased the maximum resistance and shifted the maximum to higher pressure. Since the elastic modulus of the cell wall is unchanged, the shift was not due to cell wall loosening by IAA (Zimmermann et al., 1976; Zimmermann and Steudle, 1977).

The treatment of Zimmermann and co-workers is in terms of macroscopic variables, and little connection has been made to molecular properties of the membrane. As we will discuss later, one of the characteristics of the lipid phase transition is a high compressibility, so that near the transition the permeability will be especially sensitive to pressure changes.

There is not yet enough evidence to identify the mechanism of turgor pressure effects on transport. Lateral stretching of the membrane, thinning of the membrane under pressure, and interaction of the membrane with the cell wall are all possible, as is volume sensing by cytoskeletal elements. In protoplasts it is possible to isolate the effect of stretching on the membrane, though the magnitude of the stretching is quite small. Fluorescence depolarization measurements on rose petal protoplasts showed that the viscosity of the membrane decreased on swelling (Borochov and Borochov, 1979).

As indicated at the outset, while the existence of an interaction between the cell wall and the plasma membrane is probable, investigating such an interaction is extremely difficult. Ultrastructural studies are hampered by the incompatibility of the techniques used for preparing cell wall specimens with those used for preparing specimens of plasma membranes (Roland, 1973). In addition, fixatives and cryoprotectants have been shown to destroy the interface between the cell wall and the plasma membrane (Willison, 1976; Brown and Montezinos, 1976; Mueller et al., 1976). We are planning experiments to test the effect of the cell wall on membrane dynamics. Fluorescence photobleaching recovery will be used to measure the lateral diffusion constants of membrane proteins in a small region of the cell, enabling us to compare the lateral diffusion rates in protoplasts with those in intact cells at various turgor pressures. In fluorescence photobleaching recovery, membrane proteins are labeled with a fluorescent substance. A small area of the cell surface, about 1 μm in diameter, is irreversibly photobleached with a pulse of laser light. The fluorescence of that area is then monitored to measure the diffusion rate of unbleached protein back into the bleached area (Poo and Cone, 1974; Peters et al., 1974; Axelrod et al., 1976; Koppel et al., 1976; Edidin et al., 1976; Jacobson et al., 1976a,b; Zagyansky and Edidin, 1976). These experiments should provide some estimate of the influence of the cell wall on the plasma membrane.

INTERACTION WITH CYTOSKELETAL ELEMENTS

A key feature of eukaryotic cells, distinguishing them from prokaryotes, is the cytoskeleton. Porter et al. (1979) have categorized three filament subclasses that can be visualized as part of this system: 24 nm microtubules, containing tubulin; 10 nm intermediate filaments; and 5 to 7 nm microfilaments, containing actin and associated myosin.

In animals it is well established that these elements are involved in maintaining cell shape, in cell motility, in cell surface topography, in transport and secretion processes, in organelle stratification, and in cytokinesis and cyclosis. In plants, firm evidence exists only for their role in the latter three processes, and most of this evidence has been obtained from algae, especially the giant charean algae such as *Nitella* and *Chara* (for review, see Hepler and Palevitz, 1974; and Palevitz, 1976). There is evidence for at least one important and unique role for these cytoskeletal elements in plants: that of organizing the distribution and orientation of cellulose fibrils in the secondary cell wall. Brown and Willison (1977) have shown that the orientation of these fibrils coincides precisely with that of intracellular (cortical) microtubules.

Microtubules have no contractile properties but are thought to provide a structural framework for the contractile action of other elements. They are often observed in close association with the cell surface in both plants and animals; in animals they are implicated in all surface activities such as patching and capping of mobile receptors. Intermediate filaments are also thought to serve as structural elements (Porter et al., 1979) but their composition and function are not well known.

Microfibrils arranged in cables or bundles are clearly involved in mechanochemical work in both plants and animals. In animal cells these actin filaments are associated with myosin and high-molecular-weight actin-binding proteins which can crosslink actin and cause it to gel. Such actin-binding proteins have been isolated from sea urchin eggs (Kane, 1975), _Acanthamoeba_ (Pollard and Korn, 1973) and pulmonary macrophages (Hartwig and Stossel, 1975; Stossel and Hartwig, 1976).

Evidence for myosin or actin-binding proteins is lacking for plants, but heavy meromysin decoration has been used to establish the actin-like nature of microfilaments in both algae (Palevitz et al., 1974; Kersey et al., 1976; Williamson, 1974) and higher plants (Forer and Jackson, 1975; Condeelis, 1974). Preparations with the biochemical and electrophoretic behavior of actin have been isolated from _Nitella_ by Palevitz (1976) and we have recently obtained from wheat germ a partially purified actin which was immunologically cross-reactive with antibodies against turkey gizzard actin, and contained proteins with electrophoretic mobilities comparable to rabbit and sea urchin actins (Ilker et al., 1979).

The final element in a complete description of the cytoplasmic morphology is the three-dimensional network of microtubeculae comprising the protein-rich gel-like phase of the ground cytoplasm surrounding the so-called cytoskeletal elements described above. Porter et al. (1979) suggest that the material of this lattice coats and interconnects all of the cell structures except the mitochondria and extends to the undersurface of the cell membrane.

Our interest in the cytoskeletal elements and the microtubecular system stems from their response to temperature, pressure and mechanical disturbance, their association with membrane and membrane-mediated responses, and therefore their potential involvement in stress responses or stress sensing in higher plants.

The gelation of actin, myosin, and actin-binding proteins is dependent on temperature and pressure, and the gel is destroyed by cold (e.g., low temperature above 0 C). In many kinds of cells, movement and mechanical work appear to depend on the presence of a gel-like cytoplasm, because low temperatures or extreme pressure destroy the gel and this destruction coincides with a cessation of cycloses and cytokinesis (Pollard, 1976). The effect of temperature on cycloses in plants has been investigated extensively (Kamiya, 1962; Lewis, 1956; Das et al., 1966; Patterson and Graham, 1977).

Allen (1976), in her study of the streaming of *Nitella* endoplasm, stated "the best way to observe, count, or photograph endoplasmic filaments is in cells that have been mechanically stimulated by tapping on the slide. Within a second after cessation of streaming it can be seen that the cytoplasm contains a dozen or more filaments at every distinguishable focal plane within the endoplasm. As streaming resumes, the particles associate with and saltate along endoplasmic filaments as well as along the subcortical fibrils."

This rapid and sensitive response to mechanical forces suggests that a possible role of these cytoskeletal elements in volume (turgor pressure) sensing should be considered.

We have begun a preliminary investigation of possible roles of the cytoskeletal system in the low temperature response. Figure 1 shows the results of an experiment (A. D. Keith, unpublished) in which the viscosity of the cytoplasm of cultured tomato cells was measured as a function of temperature using the water-soluble spin label Tempone. At temperatures below about 10 C the viscosity of the cytoplasm parallels the viscosity of water; but, as the temperature increases, the viscosity of the cytoplasm remains relatively constant whereas that of water continues a marked decrease. The difference curve for these viscosities exhibits a sharp change at about 10 C coinciding with the threshold temperature for chilling injury in these cells (Figure 1).

Finally, the interaction of the cytoskeletal system with the membrane and its influence upon membrane components is notable. In animals the involvement of microtubules and microfilaments in patching and capping of membrane surface receptors is clearly demonstrated by immunofluorescence studies (Ash, Louvard and Singer, 1977; Bourguignon and Singer, 1977; Gabbiani et al., 1977; and Toh and Hard, 1977).

EFFECTS OF ENVIRONMENTAL STRESS ON MEMBRANE DYNAMICS

Temperature

The viscosity of the membrane is an important variable to a poikilothermic organism, as is shown by the existence of mechanisms regulating viscosity. Modification of lipid composition to maintain a certain range of viscosity has been demonstrated in a variety of organisms since the work of Sinensky (1974) on *E. coli*.

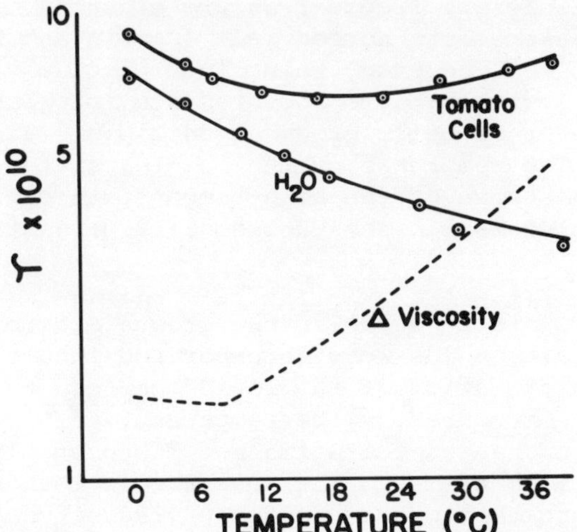

Fig. 1. The effect of temperature on the viscosity of the aqueous cytoplasm of tomato cells compared to water. Viscosity measured as rotational correlation time (τ) for the water soluble spin-label Tempone (2,2,6,6-tetramethylpiperidine-4-keto-N-oxyl).

Acholeplasma laidlawii can incorporate into its membrane lipids a variety of fatty acids supplied in the growth medium. The presence of exogenous fatty acids shifts the chain length of the fatty acids synthesized by the organism so as to maintain the viscosity within the required range for growth (Silvius et al., 1977). Hsung et al. (1974) varied the ratio of saturated to unsaturated fatty acyl groups in total membrane lipid between 0.5 and 12.5. Despite this variation in composition, the viscosity at the growth temperature of 37 C was essentially constant. Butler et al. (1978) obtained similar results. Possible regulatory mechanisms are discussed by Silvius et al. (1977) and by Melchior and Steim (1977).

Control of membrane viscosity in the protozoan Tetrahymena pyriformis has been studied in great detail (reviewed by Thompson and Nozawa, 1977). When cells grown at 39 C were chilled to 15 C, the fractions of myristic acid (14:0) and palmitic acid (16:0) in microsomal membranes decreased, and the fraction of linolenic acid (18:3) increased. When membrane properties were measured as a function of time after the temperature shift, it was found that the increase in the degree of unsaturation of the fatty acids paralleled the change in viscosity. Viscosity was measured both by fluorescence depolarization of DPH and by observation of the distribution of membrane particles by freeze-fracture electron microscopy. The two viscosity measurements agreed well (Martin and Thompson, 1978). Tetrahymena does not, however, maintain a precisely constant viscosity. Experiments with 5-nitroxide stearate spin label showed that the order parameter of cells grown at 34 C was 87% of that of cells grown at 15 C. The membranes of the low-temperature cells are thus more viscous than those of the high-temperature cells (Nozawa et al., 1974; see also Martin and Thompson, 1978; Martin and Foyt, 1978).

The effect of fatty acids in the growth medium was also studied. When exogenous fatty acids were incorporated into the phospholipids, the degree of unsaturation of the lipids changed rapidly, so that the viscosity (as measured by the temperature of incipient phase separation) was unchanged (Kitajima and Thompson, 1977). The response of the mechanisms to the anaesthetic methoxyflurane was similar. The anaesthetic decreased the viscosity of the membrane, and the fraction of unsaturated fatty acids decreased to compensate for this (Nandini-Kishore et al., 1977).

Hence, Tetrahymena responds to temperature changes, addition of fatty acids to the growth medium, and a fluidizing anaesthetic by altering the degree of unsaturation of the fatty acids to maintain constant viscosity. These results support the hypothesis that fatty acid desaturase is regulated by the viscosity of the membrane (Thompson and Nozawa, 1977; Kitajima and Thompson, 1977). If the enzyme simply responds to the viscosity of the membrane, then any variable affecting the viscosity will change the degree of unsaturation of fatty acids (Thompson, 1980). Such a shift in unsaturation was observed as a response to high salinity (Thompson, 1980; Thompson and Nozawa, 1977), though details of this work have not yet been published.

Raison and Berry (1978) examined the lipids of Nerium oleander chloroplasts and the thermophilic alga Synechococcus lividus. N. oleander grown at temperatures of 20/15 C (day/night) had a higher viscosity than those grown at 45/30 C, but there was partial compensation for the change of viscosity with temperature. Chlorophyll fluorescence increased sharply at 53 C for the high-temperature plants and at 43 C for the low-temperature plants. This increase in fluorescence was attributed to a disruption of energy transfer in the chloroplast, caused by thermal instability of the membrane. The viscosity of the lipids from the low-temperature plant at 43 C was equal to that of lipids from the high-temperature plant at 53 C, suggesting that there is a minimum viscosity required to maintain stable, functional membranes (Pike et al., 1980). This conclusion is supported by the observation that the viscosity of S. lividus lipids at its upper temperature limit of 55 C was equal to that of N. oleander lipids near its upper temperature limit (Raison and Berry, 1978).

Viscosity is one of the quantities affected by the molecular ordering of membranes. Another manifestation of molecular ordering is lateral phase separation.

Melchior and Steim (1976) concluded that lateral phase separations have no physiological role because the phase transition temperature is typically well below the growth temperature of the cell. The examples they cite, however, are from mammalian cells and prokaryotes; clearly this is insufficient evidence to conclude that lateral phase separation is unimportant in all organisms. In cell membranes, the temperature range of lateral phase separation is very large on account of the variety of lipids and integral proteins. Some poikilotherms are exposed to a range of environmental temperatures broad enough to encompass such a transition range, and these organisms are precisely the ones in which lateral phase separation could be physiologically significant.

The existence of abrupt changes in plots of rate functions of enzymes, membranes, cells, and plants due to changes in temperature is controversial (Bagnall and Wolfe, 1978; Silvius et al., 1978; Raison et al., 1980; Willcox, 1980; Wolfe, 1980; Potter, 1980). Clearly some Arrhenius plots are better fit by curves than by sets of straight lines (Bagnall and Wolfe, 1978; Wolfe and Bagnall, 1980). As pointed out by Raison et al. (1980), the temperature response of secondary events might be better represented by a curve than by a straight line, even though the primary event shows a discontinuity at a critical temperature. It must be emphasized, however, that there _are_ abrupt changes in the slope in Arrhenius plots of enzyme activities, and there _are_ abrupt changes in plant response to temperature. For example, in the experiments of Bagnall and Wolfe (1978; see also Raison et al., 1980), morning glory (_Pharbitis nil_) was grown at 28 C and then transferred to a lower temperature. Those held at 10.8 C gained in dry weight; those held at 9.8 C died. Raison et al. (1980) found that an Arrhenius plot of succinate oxidase activity of maize root mitochondria shows vertical displacements as well as changes in slope at 12 and 27 C; motion of spin label in the membranes showed changes in slope at the same temperatures.

Wunderlich et al. (1975) studied temperature effects on endoplasmic reticulum membranes in _Tetrahymena_. Freeze-etch electron micrographs showed that particle-free zones began to appear between 15 and 20 C, indicating the onset of lateral phase separation. Activity of glucose-6-phosphatase showed a slight discontinuity at 15 C. Binding of 8-anilino-1-napthalenesulfonate by the membrane showed a distinct change in slope between 16 and 20 C. Mobility of a spin-labeled fatty acid changed slope at 18 C. In NMR measurements, the height of methyl and methylene signals changed abruptly at 17 C. The agreement of all of those independent measurements is strong evidence for an abrupt change in molecular ordering around 17 C.

Salinity

Salinity is expected to influence the membrane in two ways. First, a change in the ionic strength of the aqueous phase could affect the ionization of the phospholipid headgroups and the range of their electrostatic interaction. Second, loss of turgor pressure changes the mechanical stress on the membrane and the interactions of the membrane with the cell wall. Salinity may thus have an important effect on the molecular ordering of the membrane.

Qualitative correlations are known between lipid composition and salt resistance. For example, as barley adapts to high salinity, the proportion of galactolipids in the chloroplasts decreases (Müller and Santarius, 1978). This result is consistent with the results of Kuiper (1968) on grape root stocks, showing that the fraction of monogalactose diglycerides decreased as the salt resistance of the root stock increased.

These changes in the lipid headgroups may be a means of maintaining the viscosity within an optimum range despite variations in the ionic environment. This hypothesis has not yet been tested in higher plants, but experiments with *Dunaliella* support the hypothesis. Spin label measurements of cells grown in various salt concentrations between 1.7 mM and 5.0 M showed that the viscosity of the membrane is constant (Fontana and Haug, 1979).

We plan to examine the effect of salinity on the plasma membrane by the use of fluorescence photobleaching techniques described in a previous section. Lateral diffusion rates of membrane proteins will be compared for salt-tolerant and salt-resistant strains of alfalfa (Croughan et al., 1978) and barley (Epstein and Norlyn, 1977). To distinguish between changes in the lipid composition and changes in the membrane proteins, we will measure the diffusion rates of a fluorescent lipid analog, and compare the proteins by gel electrophoresis.

Drought

The primary effect of drought on the membrane is a loss of turgor pressure, changing the mechanical stress on the membrane and its interaction with the cell wall. The loss of turgor pressure may also be sensed by cytoskeletal elements (Cram, 1976). Unless ionic effects are shown to be significant, drought and salinity will have similar immediate effects on the membrane.

In his discussion of mechanisms of response to water stress, Hsiao (1973) suggests that "changes in pressure differential across the wall may also affect spatial relations of enzyme reaction systems located at the plasmalemma." We can now make this suggestion more specific: a pressure-sensitive lateral phase separation affects either the diffusion and interaction of mobile membrane proteins, or the conformation of a membrane protein, or both.

The accumulation of certain organic compounds (for example, proline, betaine, and polyhydric alcohols) is frequently associated with stress: salinity, drought, heat stress, chilling stress (see, for example, Flowers et al., 1977; Hellebust, 1976; Singh et al., 1973; Chu et al., 1978; Yelenosky, 1979; Tal et al., 1979). These compounds enable the cell to increase the osmotic pressure in the cytoplasm to respond to the stress without disrupting the structure of salt-sensitive proteins (Stewart and Lee, 1974; Wyn Jones et al., 1977). It would be interesting to test the effect of these compounds on membrane viscosity. Perhaps these compounds, or others found in plants subject to salinity fluctuations, increase the range of salinity over which a membrane maintains constant viscosities, just as sterols broaden the temperature range of constant viscosity (Demel and De Druyff, 1976).

Freezing

We will not discuss freezing in this review, but we point out that the molecular ordering of the membrane is important, since some aspects of resistance to freezing injury depend in part on the permeability of the plasma membrane to water.

Air Pollutants

Air pollution is an environmental stress and the plant's response to it has characteristics in common with the response to other stresses. Furthermore, the interaction of air pollution with other environmental stresses is of significant concern. This interaction can be manifest through modification by environmental stress of the response of plants to air pollutants and, conversely, modification by air pollutants of the response of the plant to temperature, salt, or drought stress (Weinstein and McCune, 1979).

Studies with the oxidant air pollutants such as ozone and peroxyacetyl nitrate (PAN) have shown that the membrane permeability changes upon exposure of the membrane to these phytotoxicants. Permeability of whole cells, tissues and isolated organelles can be altered by exposure to these oxidants (Ting et al., 1974). Alterations in membrane properties could be measured in plant tissues in which visible symptoms had not yet appeared. It was concluded that cellular membranes are the initial target for oxidant injury, and that subsequent symptoms, such as altered respiration, photosynthesis, and metabolism, are secondary effects resulting from early changes in membrane properties.

While the specific mechanism of membrane alteration by these air pollutants is not clear, it is undoubtedly related to the oxidizing properties of these compounds. For example, ozone is known to attack double bonds of lipids and specific amino acids of proteins (Goldstin and Balchum, 1967). Studies in our laboratory with thin films at an air/water interface (unpublished results) have shown that ozone altered the film primarily by oxidizing the lipid double bonds and it had a lesser effect on lipoprotein complexes. PAN, on the other hand, was most effective against protein films, which can be explained by the reactivity of PAN with sulfhydryl groups and possible oxidation to the disulfide (Mudd, 1966). Hence, it would appear that ozone causes its stress on plant tissues by oxidizing unsaturated bonds of lipids and aromatic amino acid side chains of protein, thereby disrupting the molecular ordering of the membrane, while PAN perturbs -SH groups of specific enzymatic proteins and cofactors. As discussed under drought, it would be interesting to test the ability of those organic compounds associated with other stresses to maintain membrane viscosity in the presence of air pollutants.

PROCESSES IN THE PLANE OF THE MEMBRANE

We have seen various ways in which environmental stresses can affect the molecular ordering of the membrane. Next, we consider the effects of the molecular ordering on physiological processes in the membrane. In this section, we summarize those processes directly dependent on motion of the proteins in the plane of the membrane. In the following section, we discuss transport mechanisms.

Patching and Capping

When a multivalent ligand -- a lectin or an antibody -- binds to mobile receptors on a cell surface, the receptors collect into patches. This clustering does not require metabolic energy. Later, the patches collect into a cap, in a process requiring metabolic energy (see, for example, DePetris and Raff, 1973; Bourguinon and Singer, 1977). Both patching and capping have been observed in various mammalian cells; only patching has been observed in plant cells (Williamson et al., 1976; Burgess and Linstead, 1977; Williamson, 1979).

Observations of patching and capping are significant in two respects. First, patching demonstrates the mobility of membrane proteins in the plane of the membrane. Second, capping provides evidence that cytoskeletal elements are involved in membrane dynamics, at least in mammalian cells. Double fluorescence labeling experiments showed that actin and myosin are concentrated immediately under the caps (Toh and Hard; 1977; Gabbiani et al., 1977; Bourguignon and Singer, 1977; Ash et al., 1977).

A patching mechanism has been shown to be involved in the action of insulin on rat adipocytes (Kahn et al., 1978). An antibody to the insulin receptor was isolated from a patient with insulin-resistant diabetes, and this antibody had insulin-like effects. A monovalent fragment of the antibody was a competitive antagonist to insulin, but had no biological effect. When an antibody to the monovalent fragment was added, the insulin-like activity was restored, showing that crosslinking of receptors is involved in the action of insulin.

Mobile Receptor Hypothesis

In some mammalian cells, adenylate cyclase activity is regulated by several hormones. The hormones do not compete for receptors, and their effects are not additive. In fat cells, at least eight different receptors are coupled to adenylate cyclase, and it is difficult to understand how so many receptors could be arranged so that they would be simultaneously exposed at the outer surface of the cell and in contact with the adenylate cyclase.

According to the mobile receptor hypothesis (Cuatrecasas, 1974; Cuatrecasas and Hollenberg, 1976; De Haën, 1976), the hormone recepors are not attached to the adenylate cyclase molecule, but diffuse freely in the membrane. When a hormone binds to its receptor, the receptor is activated. When this activated receptor encounters an adenylate cyclase molecule, the receptor activates the adenylate cyclase. Several independent receptors can thus activate the same enzyme.

Evidence for this model was provided by the experiments of Rimon et al. (1978). The viscosity of turkey erythrocyte membranes was modified by addition of cis-11-octadecenoic acid. Activation of adenylate cyclase by epinephrine was a linear function of reciprocal viscosity, implying diffusional control. Activation by adenosine was independent of viscosity, indicating that the adenosine receptor is coupled to adenylate cyclase.

Lateral Electrophoresis

If the membrane contains charged, mobile components, then an external electric field will produce lateral motion of these components. Eventually, a steady state will be reached between diffusion and electrophoresis, and membrane components will be distributed asymmetrically. Lateral electrophoresis was proposed by Jaffe (1977) and observed by Poo and co-workers (Poo and Robinson, 1977; Poo et al., 1978, 1979; Orida and Poo, 1978). When *Xenopus* larval muscle cells were exposed to an electric field and then labeled with a fluorescent lectin, an asymmetric distribution of fluorescence was observed. The development of asymmetry was not affected by metabolic inhibitors or microtubule inhibitors. Typically, an electric field of 10 V/cm was used, corresponding to a potential difference across a cell of approximately 30 mV. The threshold field was approximately 1.5 V/cm, giving a voltage difference of 2-3 mV across a cell. Exposure times ranged between 10 minutes and 3 hours.

Lateral electrophoresis may be involved in the control of plant development (see Jaffe and Nuccitelli, 1977). Raven (1979) proposed lateral electrophoresis as a mechanism to produce a greater concentration of IAA^- carriers at the basal end of a cell than at the apical end, as required by the chemiosmotic hypothesis of polar transport.

Cell Wall Formation

Heath (1974) proposed that the cellulose synthetase complexes required for cell wall formation are located in the plasma membrane and move freely in the plane of the membrane. Cytoskeletal elements move the enzyme complex through the membrane, so that the cellulose fibrils are oriented. Brown and co-workers have obtained electron micrographs consistent with this hypothesis in the unicellular green alga *Oocystis* (Brown and Montezinos, 1976) and in maize (Mueller et al., 1976). For further discussion of the role of the plasma membrane and cytoskeletal elements in cellulose synthesis, see, for example, Brown and Willison (1977), Palevitz and Hepler (1976), and Hepler and Palevitz (1974).

TRANSPORT PROCESSES

Permeability

Below the temperature of the gel-to-liquid transition, the permeability of a lipid bilayer increases with temperature. The permeability reaches a maximum at the transition temperature, and decreases at higher temperatures. This peak in permeability was observed, for example, in the efflux of Na^+ from DPPG vesicles (Papahadjopoulos et al., 1973), in valinomycin-mediated K^+ conductivity in DPPC membranes (Wu and McConnell, 1973), in the efflux of K^+ from DMPC liposomes (Blok et al., 1975), and in penetration of DMPC vesicles by TEMPO-choline (Marsh et al., 1976).

One-component lipid bilayers are likely to be near their critical point at ordinary temperatures (Hui et al., 1975). Near a critical point, density fluctuations increase, and the lateral compressibility becomes very large. Similar behavior is expected in the temperature range of lateral phase separation of a lipid mixture. The increase in permeability is attributed to the increase in density fluctuations, and the observed decrease at higher temperatures is attributed to the decrease in density fluctuations as the temperature moves away from the critical temperature (a theoretical treatment was presented by Marčelja and Wolfe, 1979).

The increase in lateral compressibility of the membranes facilitates penetration of the membrane, and may affect transport, insertion of proteins into the membrane, conformational changes in proteins, and the action of enzymes on membrane components (see Shimshick and McConnell, 1973; Linden et al., 1973; Linden and Fox, 1975; Phillips et al., 1975; Nagle and Scott, 1978). Op den Kamp et al. (1974, 1975) showed that pancreatic phospholipase A_2 does not hydrolyze lecithin liposomes above or below their transition temperature, but at the transition temperature, hydrolysis occurs.

The permeability is thus highly sensitive to the temperature range of the phase transition, and is expected to be equally sensitive to the range of other variables which may induce a phase transition.

Gibberellic acid lowers the temperature of the gel-to-liquid transition in various lipid bilayers (Wood and Paleg, 1974), thus increasing the permeability of the bilayers to glucose, sucrose, and chromate (Wood and Paleg, 1972). The specificity of this effect is limited; the biologically-inactive compound GA_8 produced as strong a response as GA_3, but the inactive methyl ester of GA_3 had no effect on the transition temperature. NMR measurements showed that GA_3 forms a complex with lecithin in a deuterochloroform solution (Wood et al., 1974).

Active Transport

The activity of Na^+, K^+-ATPase has been shown to depend upon the identity and the molecular ordering of the membrane lipids (see, for example, Grisham and Barnett, 1973; Hsung et al., 1974; Kimelberg and Papahadjopoulos, 1974; Palatini et al., 1977). Likewise, the activity of Ca^{2+}-ATPase from sarcoplasmic reticulum depends upon the presence and identity of membrane lipids. The presence of approximately 30 phospholipid molecules is required for full enzyme activity; reconstitution experiments show that the activity of the enzyme depends strongly on the headgroup of the phospholipids added (Hesketh et al., 1976; Metcalfe et al., 1976; Dean and Tanford, 1977; Warren and Metcalfe, 1977). Regulation of membrane enzymes by lipids was reviewed recently by Sandermann (1978); interactions of proteins with lipids were reviewed by Gennis and Jonas (1977).

Ionophores

Ions may also be transported by carrier ionophores and channel ionophores, though little is known yet about the importance of these mechanisms in plants. Carrier ionophores bind to an ion at the membrane surface, diffuse across the membrane, and release the ion at the other surface. The ionophores are cyclic compounds with a hydrophobic exterior and a hydrophilic interior, able to replace the solvation shell of the ion. The compounds are therefore highly selective. For example, the conductivity of a DOPC membrane containing valinomycin is 10^4 times higher for K^+ than for Na^+. The conductivity of the bilayer is proportional to the concentration of valinomycin over a range of at least three orders of magnitude (Läuger, 1972). Since the carriers must diffuse across the membrane, the conductivity is sensitive to the viscosity of the membrane.

Channel ionophores, such as gramicidin A and alamethicin, polymerize to form transient channels across the membrane. In the case of gramicidin A, a linear hydrophobic peptide made up of fifteen amino acids, the channel is a dimer with a mean lifetime of approximately one second, and a pore diameter of 40 nm (Bamberg and Läuger, 1974). The dimerization constant depends upon the lipid composition of the membrane (Veatch et al., 1975), but once the channel is formed, ion transport is insensitive to the viscosity of the membrane. The conductance of the membrane is unchanged by the gel-to-liquid transition, but the activation energy for ion transport is changed (Krasne et al., 1974).

Lea and Collins (1979) showed that ABA forms conducting channels in lecithin-chloresterol membranes in 1 M KCl. The conductance per channel is comparable to that of gramicidin A, but the mean lifetime is much shorter, 0.2 ms.

SUGGESTED AREAS OF INVESTIGATION

We have reviewed a number of aspects of membrane dyanmics and outlined the variables likely to affect the molecular ordering of a membrane and the changes in those variables produced by environmental stresses. We have summarized possible transport mechanisms and possible physiological mechanisms based on lateral motion of membrane proteins, and the sensitivity of these mechanisms to the molecular ordering of the membrane. In conclusion, we suggest some important topics for further investigation:

1) Tests of the general hypothesis of correlation of physiological functions with constant membrane viscosity. Experiments in which both salinity and temperature are varied would be informative.

2) Measurements of lateral diffusion of membrane proteins in plants. Labeled proteins could be used as physiologically unobtrusive probes of membrane viscosity. These investigations would provide the groundwork for experiments to see if the mobile receptor mechanism is used in plants.

3) Theoretical and experimental study of pressure effects on the molecular ordering of a membrane constrained by a cell wall. Even if the cell wall were structureless and inert, it would be likely to affect the molecular ordering of the membrane. A free membrane may alter its curvature to minimize the free energy, but a membrane in a plant cell at full turgor is constrained to a certain shape. The effects of this constraint on the structure and

function of the membrane is of fundamental interest since the presence of a cell is one of the essential differences between plant and animal cells.

4) Specific interactions of the plasma membrane with the cell wall. Like the previous problem, this is an unexplored area of fundamental importance. Plant cells are not just animal cells that happen to live in an apartment building of cellulose.

5) Investigations of cytoskeletal elements in plants.

6) Behavior of proteins in lateral phase separation. Various fluorescent probes behave differently in a lateral phase separation: pyrenedecanoic acid partitions into the fluid phase (Galla and Sackmann, 1975) while trans-parinaric acid partitions into the solid phase (Sklar et al., 1979). Do different membrane proteins prefer different lipid phases?

7) Plant hormone effects on membranes. Several suggestions have already been mentioned: the decrease in viscosity caused by GA_3 (Wood and Paleg, 1972), ABA as a channel ionophore (Lea and Collins, 1979), lateral electrophoresis of IAA^- receptors (Raven, 1979). Several natural and artificial growth regulators, however, were shown to have no effect on membrane viscosity (Parups and Miller, 1978; Leshem and Inbar, 1978; but see Helgerson et al., 1976 and Robenek, 1979). A major problem with membrane-based mechanisms for hormone action is specificity. It would be useful to test natural lipid mixtures as well as synthetic lipid bilayers in experiments on hormone interactions with membranes, to see if there is specific binding to a minor lipid. Much more work is needed on hormone-membrane interactions to test the ideas of Kende and Gardner (1976) and of Trewavas (1976, 1979) that the primary response to hormones may be in the membranes. Ultimately this could lead to a unified view of growth regulation and environmental stress.

The investigation of these areas of membrane dynamics in plants is important. The questions deal with fundamental problems of biophysics and biochemistry, and are directly related to some of the major constraints on world food production. Answering these questions is important now; in the event of changes in global patterns of temperature and rainfall, answering them may become imperative.

<u>Abbreviations</u>

ABA	abscisic acid
DPH	1,6-diphenylhexatriene
DMPA	dimyristoylphosphatidic acid
DMPC	dimyristoylphosphatidylcholine
DOPC	dioleylphosphatidylcholine
DPPA	dipalmitoylphosphatidic acid

DPPC	dipalmitoylphosphatidylcholine
DPPG	dipalmitoylphosphatidylglycerol
DPPS	dipalmitoylphosphatidylserine
IAA	indoleacetic acid
PAN	peroxyacetyl nitrate
TEMPO-choline	2,2,6,6-tetramethylpiperidinyl-1-oxycholine

For fatty acids, the length of the carbon chain and the number of double bonds is indicated:

linolenic acid	18:3
myristic acid	14:0
oleic acid	18:1
palmitic acid	16:0

REFERENCES

Allen, N. S., 1976, Undulating filaments in *Nitella* endoplasm and motive force generation, *in*: "Cell Motility B," R. Goldman, T. Pollard and J. Rosenbaum, eds., Cold Springs Harbor Laboratory.

Ash, J. F., Louvard, D., and Singer, S. J., 1977, Antibody-induced linkages of plasma membrane proteins to intracellular actomyosin-containing filaments in cultured fibroblasts, *Proc. Natl. Acad. Sci. USA*, 74:5584.

Axelrod, D., Koppel, D. E., Schlessinger, J., Elson, E., and Webb, W. W., 1976, Mobility measurement by analysis of fluorescence photobleaching recovery kinetics, *Biophys. J.*, 16:1055.

Bagnall, D. J. and Wolfe, J. A., 1978, Chilling sensitivity in plants: do the activation energies of growth processes show an abrupt change at a critical temperature?, *J. Expt. Bot.*, 29:1231.

Bamberg, E. and Läuger, P., 1974, Temperature-dependent properties of gramicidin A channels, *Biochim. Biophys. Acta*, 367:127.

Bashford, C. L., Morgan, C. G., and Radda, G. K., 1976, Measurement and interpretation of fluorescence polarizations in phospholipid dispersions, *Biochim. Biophys. Acta*, 426:157.

Blok, M. C., Van Der Neut-Kok, E. C. M., Van Deenen, L. L., and De Gier, J., 1975, The effect of chain length and lipid phase transitions on the selective permeability properties of liposomes, *Biochim. Biophys. Acta*, 406:187.

Borochov, A. and Borochov, H., 1979, Increase in membrane fluidity in liposomes and plant protoplasts upon osmotic swelling, *Biochim. Biophys. Acta*, 550:546.

Bourguignon, L. Y. W. and Singer, S. J., 1977, Transmembrane interactions and the mechanism of capping of surface receptors by their specific ligands, *Proc. Natl. Acad. Sci. USA*, 74:5031.

Brown, Jr., R. M. and Montezinos, D., 1976, Cellulose microfibrils: visualization of biosynthetic and orienting complexes in association with the plasma membrane, Proc. Natl. Acad. Sci. USA, 73:143.

Brown, Jr., R. M. and Willison, J. H. M., 1977, Golgi apparatus and plasma membrane involvement in secretion and cell surface deposition, with special emphasis on cellulose biogenesis, in: "International Cell Biology 1976-1977," B. R. Brinkley and K. R. Porter, eds., Rockefeller Univ. Press, New York.

Burgess, J., 1978, Plant cells without walls?, Nature, 275:588.

Burgess, J. and Linstead, P. J., 1977, Membrane mobility and the concanavalin A binding system of the plasmalemma of higher plant protoplasts, Planta, 136:253.

Butler, K. W., Johnson, K. G., and Smith, I. C. P., 1978, Acholeplasma laidlawii membranes: an electron spin resonance study of the influence on molecular order of fatty acid composition and cholesterol, Arch. Biochem. Biophys., 191:289.

Condeelis, J. S., 1974, The identification of F actin in the pollen tube of Amaryllis belladonna, Exptl. Cell Res., 88:435.

Chu, T. M., Jusaitis, M., Aspinall, D., and Paleg, I. G., 1978, Accumulation of free proline at low temperatures, Physiol. Plant., 43:254.

Coster, H. G. L., Steudle, E., and Zimmermann, U., 1977, Turgor pressure sensing in plant cell membranes, Plant Physiol., 58: 636.

Cram, W. J., 1976, The maintenance of turgor, volume and nutrient supply, in: "Transport in Plants II, Part A, Cells," U. Lüttge and M. G. Pitman, eds., Springer-Verlag, New York.

Croughan, T. P., Stavarek, S. J., and Rains, D. W., 1978, Selection of a NaCl tolerant line of cultured alfalfa cells, Crop Sci., 18:959.

Cuatrecasas, P., 1974, Membrane receptors, Ann. Rev. Biochem., 43: 169.

Cuatrecasas, P. and Hollenberg, M. D., 1976, Membrane receptors and hormone action, Adv. Protein Chem., 30:251.

Das, T. M., Hildebrandt, A. C., and Riker, A. J., 1966, Cinephotomicrography of low temperature effects on cytoplasmic streaming, nucleolar activity and mitosis in single tobacco cells in microculture, Amer. J. Bot., 53:253.

Davis, D. G., Inesi, G., and Gulig-Krzywicki, T. G., 1976, Lipid molecular motion and enzyme activity in sarcoplasmic reticulum membrane, Biochim., 15:1271.

Dean, W. L. and Tanford, C., 1977, Reactivation of lipid-depleted Ca^{2+}-ATPase by a nonionic detergent, J. Biol. Chem., 252:3551.

de Haën, C., 1976, The non-stoichiometric floating receptor model for hormone sensitive adenylyl cyclase, J. Theor. Biol., 58: 383.

de Petris, S. and Raff, M. C., 1973, Normal distribution, patching and capping of lymphocyte surface immunoglobulin studied by electron microscopy, Nature New Biol., 241:257.

Demel, R. A. and De Kruyff, B., 1976, The function of sterols in membranes, Biochim. Biophys. Acta, 457:109.
Edidin, M., Zagyansky, Y., and Lardner, T. J., 1976, Measurement of membrane protein lateral diffusion in single cells, Science, 191:466.
Epstein, E. and Norlyn, J. D., 1977, Seawater-based crop production: a feasibility study, Science, 197:249.
Evans, E. A., 1974, Bending resistance and chemically induced moments in membrane bilayers, Biophys. J., 14:923.
Evans, E. A. and Hochmuth, R. M., 1978, Mechanochemical properties of membranes, in: "Current Topics in Membrane Transport," Vol. 10, F. Bronner and A. Kleinzeller, eds., Academic Press, New York.
Flowers, T. J., Troke, P. F., and Yeo, A. R., 1977, The mechanism of salt tolerance in halophytes, Ann. Rev. Plant Physiol., 28:89.
Fontana, D. B. and Haug, A., 1978, Salt adaptation and the membrane of Dunaliella primolecta, Plant Physiol., 61:93 (Suppl.).
Forer, A. and Jackson, W. T., 1975, Distribution of actin in the spindle of a higher plant Haemanthus katherinae, J. Cell. Biol., 67:117a.
Gabbiani, G., Chaponnier, C., Zumbe, A., and Vassalli, P., 1977, Actin and tubulin co-cap with surface immunoglobulins in mouse B lymphocytes, Nature, 269:697.
Galla, H.-J. and Sackmann, E., 1975, Chemically induced phase separation in mixed vesicles containing phosphatidic acid. An optical study, J. Am. Chem. Soc., 97:4114.
Gebhardt, C., Gruler, H., and Sackmann, E., 1977, On domain structure and local curvature in lipid bilayers and biological membranes, Z. Naturforsch., 32c:581.
Gennis, R. B. and Jonas, A., 1977, Protein-lipid interactions, Ann. Rev. Biophys. Bioeng., 6:195.
Goldstein, B. D. and Balchum, O. J., 1967, Effect of ozone on lipid peroxidation in the red blood cell, Proc. Soc. for Exptl. Biol. and Med., 126:356.
Grisham, C. M. and Barnett, R. E., 1973, The role of lipid-phase transitions in the regulation of the (sodium + potassium) adenosine triphosphatase, Biochem., 12:2635.
Gruler, H., 1975, Chemoelastic effect of membranes, Z. Naturforsch., 30c:608.
Gutknecht, J., 1968, Salt transport in Valonia: inhibition of potassium uptake by small hydrostatic pressures, Science, 160:68.
Gutknecht, J., Hastings, D. F., and Bisson, M. A., 1978, Ion transport and turgor pressure regulation in giant algal cells, in: "Transport Across Multi-Membrane Systems," G. Giebisch, ed., Springer-Verlag, Berlin.
Hartwig, J. H. and Stossel, T. P., 1975, Isolation and properties of actin, myosin and a new actin binding protein in rabbit alveolar macrophages, J. Biol. Chem., 250:5696.

Hastings, D. F. and Gutknecht, J., 1974, Turgor pressure regulation: modulation of active potassium transport by hydrostatic pressure gradients, in: "Membrane Transport in Plants," U. Zimmermann and J. Dainty, eds., Springer-Verlag, New York.

Heath, I. B., 1974, A unified hypothesis for the role of membrane bound enzyme complexes and microtubules in plant cell wall synthesis, J. Theor. Biol., 48:445.

Helgerson, S. L., Cramer, W. A., and Morré, D. J., 1976, Evidence for an increase in microviscosity of plasma membranes from soybean hypocotyls induced by the plant hormone, indole-3-acetic acid, Plant Physiol., 58:548.

Hellebust, J. A., 1976, Osmoregulation, Ann. Rev. Plant Physiol., 27:485.

Heller, R., Grignon, C., and Rona, J.-P., 1974, Importance of the cell wall in the thermodynamic equilibrium of ions in free cells of Acer pseudoplatanus L., in: "Membrane Transport in Plants," U. Zimmermann and J. Dainty, eds., Springer-Verlag, New York.

Hepler, P. K. and Palevitz, B. A., 1974, Microtubules and microfilaments, Ann. Rev. Plant Physiol., 25:309.

Hesketh, T. R., Smith, G. A., Houslay, M. D., McGill, K. A., Birdsall, N. J. M., Metcalfe, J. C., and Warren, G. B., 1976, Annular lipids determine the ATPase activity of a calcium transport protein complexed with dipalmitoyllecithin, Biochem., 15:4145,

Hsiao, T. C., 1973, Plant responses to water stress, Ann. Rev. Plant Physiol., 24:519.

Hsung, J.-C., Huang, L., Hoy, D. J., and Haug, A., 1974, Lipid and temperature dependence of membrane-bound ATPase activity of Acholeplasma laidlawii, Can. J. Biochem., 52:974.

Hui, S. W., Cowden, M., Papahadjopoulos, D., and Parsons, D. F., 1975, Electron diffraction study of hydrated phospholipid single bilayers: effects of temperature, hydration and surface pressure of the "precursor" monolayer, Biochim. Biophys. Acta, 382:265.

Ilker, R., Breidenbach, R. W., and Murphy, T. M., 1979, Partial purification of actin from wheat germ, Phytochem., 18:1781.

Jacobson, K. and Papahadjopoulos, D., 1975, Phase transitions and phase separations in phospholipid membranes induced by changes in temperature, pH, and concentration of bivalent cations, Biochem., 14:152.

Jacobson, K., Wu, E., and Poste, G., 1976a, Measurement of the translational mobility of concanavalin A in glycerol-saline solutions and on the cell surface by fluorescence recovery after photobleaching, Biochim. Biophys. Acta, 433:215.

Jacobson, K., Derzko, Z., Wu, E.-S., Hou, Y., and Poste, G., 1976b, Measurement of the lateral mobility of cell surface components in single, living cells by fluorescence recovery after photobleaching, J. Supramol. Struc., 5:565.

Jaffe, L. F., 1977, Electrophoresis along cell membranes, Nature, 265:600.

Jaffe, L. F. and Nuccitelli, R., 1977, Electrical controls of development, Ann. Rev. Biophys. Bioeng., 6:445.

Kahn, C. R., Baird, K. L., Jarrett, D. B., and Flier, J. S., 1978, Direct demonstration that receptor crosslinking or aggregation is important in insulin action, Proc. Natl. Acad. Sci. USA, 75:4209.

Kamiya, N., 1962, Protoplasmic streaming, in: "Encyclopedia of Plant Physiology," XVII, 2, W. Rhuland, ed., Springer-Verlag, Berlin.

Kane, R. E., 1975, Preparation and purification of polymerized actin from sea urchin egg extracts, J. Cell Biol., 66:305.

Kende, H. and Gardner, G., 1976, Hormone binding in plants, Ann. Rev. Plant Physiol., 27:267.

Kersey, Y. M., Hepler, P. K., Palevitz, B. A., and Wessells, N., 1976, Polarity of actin filaments in characean algae, Proc. Natl. Acad. Sci. USA, 73:165.

Kimelberg, H. K. and Papahadjopoulos, D., 1974, Effects of phospholipid acyl chain fluidity, phase transitions, and cholesterol on ($Na^+ + K^+$)-stimulated adenosine triphosphatase, J. Biol. Chem., 249:1071.

Kitajima, Y. and Thompson, Jr., G. A., 1977, Self-regulation of membrane fluidity: the effect of saturated normal and methoxy fatty acid supplementation on Tetrahymena membrane physical properties and lipid composition, Biochim. Biophys. Acta, 468:73.

Koppel, D. E., Axelrod, D., Schlessinger, J., Elson, E. L., and Webb, W. W., 1976, Dynamics of fluorescence marker concentration as a probe of mobility, Biophys. J., 16:1315.

Krasne, S., Eisenman, G., and Szabo, G., 1971, Freezing and melting of lipid bilayers and the mode of action of nonactin, valinomycin, and gramicidin, Science, 174:412.

Kuiper, P. J. C., 1968, Lipids in grape roots in relation to chloride transport, Plant Physiol., 13:1367.

Läuger, P., 1972, Carrier-mediated ion transport, Science, 178:24.

Lea, E. J. A. and Collins, J. C., 1979, The effect of the plant hormone abscisic acid on lipid bilayer membranes, New Phytol., 82:11.

Lee, A. G., 1975, Functional properties of biological membranes: a physical-chemical approach, Prog. Biophys. Molec. Biol., 29:3.

Lee, A. G., 1977a, Lipid phase transitions and phase diagrams. I. Lipid phase transitions, Biochim. Biophys. Acta, 472:237.

Lee, A. G., 1977b, Lipid phase transitions and phase diagrams. II. Mixtures involving lipids, Biochim. Biophys. Acta, 472:285.

Lee, A. G., Birdsall, N. J. M., Metcalfe, J. C., Toon, P. A., and Warren, G. B., 1974, Clusters in lipid bilayers and the interpretation of thermal effects in biological membranes, Biochem., 13:3699.

Lentz, B. R., Barenholz, Y., and Thompson, T. E., 1976a, Fluorescence depolarization studies of phase transitions and fluidity in phospholipid bilayers. I. Single component phosphatidylcholine liposomes, Biochem., 15:4521.

Lentz, B. R., Barenholz, Y., and Thompson, T. E., 1976b, Fluorescence depolarization studies of phase transitions and fluidity in phospholipid bilayers. II. Two-component phosphatidylcholine liposomes, Biochem., 15:4529.

Leshem, Y. Y. and Inbar, M., 1978, Resistance to gibberellin-induced changes of lipid fluidity in wheat embryo mitochondrial membranes as assessed by the fluorescent probe 1,6-diphenyl-1,3,5-hexatriene, J. Exptl. Bot., 29:671.

Lewis, D. A., 1956, Protoplasmic streaming in plants sensitive and insensitive to chilling temperatures, Science, 124:75.

Linden, C. D., Wright, K. L., McConnell, H. M., and Fox, C. F., 1973, Lateral phase separations in membrane lipids and the mechanism of sugar transport in Escherichia coli, Proc. Natl. Acad. Sci. USA, 70:2271.

Linden, C. D. and Fox, C. F., 1975, Membrane physical state and function, Accounts of Chem. Res., 8:321.

Lloyd, C., 1978, The first division, Nature, 276:562.

Lyons, J. M., Raison, J. K., and Steponkus, P. L., 1980, The plant membrane in response to low temperature: an overview, in: "Low Temperature Stress in Crop Plants: The Role of the Membrane," J. M. Lyons, D. Graham, and J. K. Raison, eds., Academic Press, New York.

MacDonald, R. C., Simon, S. A., and Baer, E., 1976, Ionic influences on the phase transition of dipalmitoylphosphatidylserine, Biochem., 15:885.

Marčelja, S. and Wolfe, J., 1979, Properties of bilayer membranes in the phase transition or phase separation region, Biochim. Biophys. Acta, 557:24.

Marsh, D., Watts, A., and Knowles, P. F., 1976, Evidence for phase boundary lipid. Permeability of Tempo-choline into dimyristoylphosphatidylcholine vesicles at the phase transition, Biochem., 15:3570.

Marsh, D., Watts, A., and Knowles, P. F., 1977, Cooperativity of the phase transition in single- and multibilayer lipid vesicles, Biochim. Biophys. Acta, 465:500.

Martin, C. E. and Foyt, D. C., 1978, Rotational relaxation of 1,6-diphenylhexatriene in membrane lipids of cells acclimated to high and low growth temperatures, Biochem., 17:3587.

Martin, C. E. and Thompson, Jr., G. A., 1978, Use of fluorescence polarization to monitor intracellular membrane changes during temperature acclimation. Correlation with lipid compositional and ultrastructural changes, Biochem., 17:3581.

Mazia, D., Schatten, G., and Sale, W., 1975, Adhesion of cells to surfaces coated with polylysine, J. Cell Biol., 66:198.

McKeehan, W. L. and Ham, R. G., 1976, Stimulation of clonal growth of normal fibroblasts with substrata coated with basic polymers, J. Cell. Biol., 71:727.

Melchior, D. L. and Steim, J. M., 1976, Thermotropic transitions in biomembranes, Ann. Rev. Biophys. Bioeng., 5:204.

Melchior, D. L. and Steim, J. M., 1977, Control of fatty acid composition of *Acholeplasma laidlawii* membranes, Biochim. Biophys. Acta, 466:148.

Metcalfe, J. C., Bennett, J. P., Hesketh, T. R., Houslay, M. D., Smith, G. A., and Warren, G. B., 1976, The lateral organization of lipids around a calcium transport protein: evidence for a phospholipid annulus that modulates function, in: "The Structural Basis of Membrane Function," Y. Hatefi and L. Djavadi-Ohaniance, eds., Academic Press, New York.

Mudd, J. B., 1966, Reaction of peroxyacetyl nitrate with glutathione, J. Biol. Chem., 241:4077.

Müller, M. and Santarius, K. A., 1978, Changes in chloroplast membrane lipids during adaptation of barley to extreme salinity, Plant Physiol., 62:326.

Mueller, S. C., Brown, R. M., and Scott, T. K., 1976, Cellulosic microfibrils: nascent stages of synthesis in a higher plant cell, Science, 194:949.

Nagata, T. and Takebe, I., 1970, Cell wall regeneration and cell division in isolated tobacco mesophyll protoplasts, Planta, 92:301.

Nagle, J. F. and Scott, Jr., H. L., 1978, Lateral compressibility of lipid mono- and bilayers: theory of membrane permeability, Biochim. Biophys. Acta, 513:236.

Nandini-Kishore, S. G., Kitajima, Y., and Thompson, Jr., G. A., 1977, Membrane fluidizing effects of the general anesthetic methoxyflurane elicit an acclimation response in *Tetrahymena*, Biochem. Biophys. Acta, 471:157.

Nozawa, Y., Iida, H., Fukushima, H., Ohki, K., and Ohnishi, S., 1974, Studies on *Tetrahymena* membranes: temperature-induced alterations in fatty acid composition of various membrane fractions in *Tetrahymena pyriformis* and its effect on membrane fluidity as inferred by spin-label study, Biochim. Biophys. Acta, 367:134.

Ohnishi, S. and Ito, T., 1973, Clustering of lecithin molecules in phosphatidylserine membranes induced by calcium ion binding to phosphatidylserine, Biochem. Biophys. Res. Commun., 51:132.

Ohnishi, S. and Ito, T., 1974, Calcium-induced phase separations in phosphatidylserine-phosphatidylcholine membranes, Biochem., 13:881.

Op Den Kamp, J. A. F., De Gier, J., and Van Deenen, L. L. M., 1974, Hydrolysis of phosphatidylcholine liposomes by pancreatic phospholipase A_2 at the transition temperature, Biochim. Biophys. Acta, 345:253.

Op Den Kamp, J. A. F., Kauerz, M. Th., and Van Deenen, L. L. M., 1975, Action of pancreatic phospholipase A_2 on phosphatidylcholine bilayers in different physical states, Biochem. Biophys. Acta, 406:169.

Orida, N. and Poo, M., 1978, Electrophoretic movement and localization of acetylcholine receptors in the embryonic muscle cell membrane, Nature, 275:31.

Owicki, J. C., Springgate, M. W., and McConnell, H. M., 1978, Theoretical study of protein-lipid interactions in bilayer membranes, Proc. Natl. Acad. Sci. USA, 75:1616.

Owicki, J. C. and McConnell, H. M., 1979, Theory of protein-lipid and protein-protein interactions in bilayer membranes, Proc. Natl. Acad. Sci. USA, 76:4750.

Palatini, P., Dabbeni-Sala, F., Pitotti, A., Bruni, A., and Mandersloot, J. C., 1977, Activation of ($Na^+ + K^+$)-dependent ATPase by lipid vesicles of negative phospholipids, Biochim. Biophys. Acta, 466:1.

Palevitz, B. A., 1976, Actin cables and cytoplasmic streaming in green plants, in: "Cell Motility B," R. Goldman, T. Pollard, and J. Rosenbaum, eds., Cold Springs Harbor Laboratory.

Palevitz, B. A., Ash, J. F., and Hepler, P. K., 1974, Actin in the green algae Nitella, Proc. Natl. Acad. Sci. USA, 71:363.

Palevitz, B. A. and Hepler, P. K., 1976, Cellulose microfibril orientation and cell shaping in developing guard cells of Allium: the role of microtubules and ion accumulation, Planta, 132:71.

Papahadjopoulos, D., Jacobson, K., Nir, S., and Isac, T., 1973, Phase transitions in phospholipid vesicles: fluorescence polarization and permeability measurements concerning the effect of temperature and cholesterol, Biochim. Biophys. Acta, 311:330.

Parups, E. V. and Miller, R. W., 1978, Investigation of effects of plant growth regulators on liposome fluidity and permeability, Physiol. Plant., 42:415.

Patterson, B. C. and Graham, D., 1977, Effect of chilling temperature on the protoplasmic streaming of plants from different climates, J. Exptl. Bot., 28:736.

Peters, R., Peters, J., Tews, K. H., and Bahr, W., 1974, A microfluorimetric study of translational diffusion in erythrocyte membranes, Biochim. Biophys. Acta, 367:282.

Phillips, M. C., Graham, D. E., and Hauser, H., 1975, Lateral compressibility and penetration into phospholipid monolayers and bilayer membranes, Nature, 254:155.

Pike, C. S., Berry, J. A., and Raison, J. K., 1980, Fluorescence polarization studies of membrane phospholipid phase separations in warm and cool climate plants, in: "Low Temperature Stress in Crop Plants: The Role of the Membrane," J. M. Lyons, D. Graham, and J. K. Raison, eds., Academic Press, New York.

Pollard, T. P., 1976, The role of actin in the temperature-dependent gelation and contraction of extracts of Acanthamoeba, J. Cell Biol., 68:579-601.

Pollard, T. D. and Korn, E. D., 1973, The "contractile" proteins of Acanthamoeba castellanii, Cold Springs Harbor Symposium Quant. Biol., 37:573.

Poo, M. and Cone, R. A., 1974, Lateral diffusion of rhodospin in the photoreceptor membrane, Nature, 247:438.

Poo, M. and Robinson, K. R., 1977, Electrophoresis of concanavalin A receptors along embryonic muscle cell membrane, Nature, 265:602.

Poo, M., Poo, W.-J. H., and Lam, J. W., 1978, Lateral electrophoresis and diffusion of concanavalin A receptors in the membrane of embryonic muscle cell, J. Cell Biol., 76:483.

Poo, M., Lam, J. W., Orida, N., and Chao, A. W., 1979, Electrophoresis and diffusion in the plane of the cell membrane, Biophys. J., 26:1.

Porter, K. R., Byers, H. R., and Ellisman, M. H., 1979, The cytoskeleton, in: "The Neurosciences: Fourth Study Program," F. O. Schmitt and F. G. Worden, eds.

Potter, F. and Ross, G. J. S., 1980, Maximum likelihood estimation of breakpoints and the comparison of the goodness of fit with that of conventional curves, in: "Low Temperature Stress in Crop Plants: The Role of the Membrane," J. M. Lyons, D. Graham, and J. K. Raison, eds., Academic Press, New York.

Racusen, R. H., Kinnersley, A. M., and Galston, A. W., 1977, Osmotically induced changes in electrical properties of plant protoplast membranes, Science, 198:405.

Raison, J. K. and Berry, J. A., 1978, The physical properties of membrane lipids in relation to the adaptation of higher plants and algae to contrasting thermal regimes, Carnegie Institution Yearbook, 77:276.

Raison, J. K., Chapman, E. A., Jacobs, S. W. L., and Wright, L. C., 1980, Membrane lipid transitions: their correlation with the climatic distribution of plants, in: "Low Temperature Stress in Crop Plants: The Role of the Membrane," J. M. Lyons, D. Graham, and J. K. Raison, eds., Academic Press, New York.

Raven, J. A., 1979, The possible role of membrane electrophoresis in the polar transport of IAA and other solutes in plant tissues, New Phytol., 82:285.

Rimon, G., Hanski, E., Braun, S., and Levitzki, A., 1978, Mode of coupling between hormone receptors and adenylate cyclase elucidated by modulation of membrane fluidity, Nature, 276:394.

Robenek, H., 1979, Der Einfluß von Indonyl(3)essigsäure (IES) auf die Verteilung der intramembranösen Partikel des Plasmalemmas isolierter Sproßkallusprotoplasten von Skimmia japonica Thunb., Z. Pflanzenphysiol., 93:317.

Roland, J.-C., 1973, The relationship between the plasmalemma and plant cell wall, Intl. Rev. Cytol., 36:45.

Rona, J. P., Cornel, L. D., and Heller, R., 1977, Determination and interpretation of the electrical profile of free cells of Acer pseudoplatanus, in: "Echanges Ioniques Transmembranaires Chez Les Vegetaux," M. Thellier, A. Monnier, M. Demarty, and J. Dainty, eds., C.N.R.S., Paris.

Sandermann, Jr., H., 1978, Regulation of membrane enzymes by lipids, Biochim. Biophys. Acta, 515:209.

Schilde-Rentschler, L., 1977, Role of the cell wall in the ability of tobacco protoplasts to form callus, Planta, 135:177.

Shimshick, E. J. and McConnell, H. M., 1973, Lateral phase separation in phospholipid membranes, Biochem., 12:2351.

Silvius, J. R., Saito, Y., and McElhaney, R. N., 1977, Membrane lipid biosynthesis in *Acholeplasma laidlawii* B., Arch. Biochem. Biophys., 182:455.

Silvius, J. R., Read, B. D., and McElhaney, R. N., 1978, Membrane enzymes: artifacts in Arrhenius plots due to temperature dependence of substrate-binding affinity, Science, 199:902.

Sinensky, M., 1974, Homeoviscous adaptation -- a homeostatic process that regulates the viscosity of membrane lipids in *Escherichia coli*, Proc. Natl. Acad. Sci. USA, 71:522.

Singh, T. N., Paleg, L. G., and Aspinall, D., 1973, Stress metabolism. III. Variations in response to water deficit in the barley plant, Aust. J. Biol. Sci., 26:65.

Sklar, L. A., Miljanich, G. P., and Dratz, E. A., 1979, Phospholipid lateral phase separation and the partition of *cis*-parinaric acid and *trans*-parinaric acid among aqueous, solid lipid, and fludi lipid phases, Biochem., 18:1707.

Stewart, G. R. and Lee, J. A., 1974, The role of proline accumulation in halophytes, Planta, 120:279.

Stossel, T. P. and Hartwig, J. H., 1976, Interaction of actin, myosin and a new actin-binding protein in rabbit pulmonary macrophages. II. Role of cytoplasmic movement and phagocytosis, J. Cell Biol., 68:602.

Suurkuusk, J., Lentz, B. R., Barenholz, Y., Biltonen, R. L., and Thompson, T. E., 1976, A calorimetric and fluorescent probe study of the gel-liquid crystalline phase transition in small single-lamellar dipalmitoylphosphatidylcholine vesicles, Biochem., 15:1391.

Tal, M., Rosental, I., Abramovitz, R., and Forti, M., 1979, Salt tolerance in *Simmondsia chinensis*: water balance and accumulation of chloride, sodium and proline under low and high salinity, Ann. Bot., 43:701.

Thompson, G., 1980, Molecular control of membrane fluidity, *in*: "Low Temperature Stress in Crop Plants: The Role of the Membrane," J. M. Lyons, D. Graham, and J. K. Raison, eds., Academic Press, New York.

Thompson, Jr., G. A. and Nozawa, Y., 1977, *Tetrahymena*: a system for studying dynamic membrane alterations within the eukaryotic cell, Biochim. Biophys. Acta, 472:55.

Ting, I. P., Perchorowicz, J., and Evans, L., 1974, Effect of ozone on plant cell membrane permeability, *in*: "Air Pollution Effects on Plant Growth," H. Dugger, ed., ACS Symposium Series 3.

Ting, P. and Solomon, A. K., 1975, Temperature dependence of N-phenyl-1-naphthylamine binding in egg lecithin vesicles, Biochim. Biophys. Acta, 406:447.

Toh, B. H. and Hard, G. C., 1977, Actin co-caps with concanavalin A receptors, Nature, 269:695.

Träuble, H. and Eibl, H., 1974, Electrostatic effects on lipid phase transitions: membrane structure and ionic environment, Proc. Natl. Acad. Sci. USA, 71:214.

Trewavas, A. J., 1976, Plant growth substances, in: "Molecular Aspects of Gene Expression in Plants," J. A. Bryant, ed., Academic Press, New York.

Trewavas, A., 1979, What is the molecular basis of plant hormone action?, Trends in Biochim. Sci., 4:N199.

Veatch, W. R., Mathies, R., Eisenberg, M., and Stryer, L., 1975, Simultaneous fluorescence and conductance studies of planar bilayer membranes containing a highly active and fluorescent analog of gramicidin A, J. Mol. Biol., 99:75.

Warren, G. B. and Metcalfe, J. C., 1977, What is the phospholipid specificity of a reconstituted calcium pump?, Biochem. Soc. Trans., 5:517.

Watts, A., Harlos, K., Maschke, W., and Marsh, D., 1978, Control of the structure and fluidity of phosphatidylglycerol bilayers by pH titration, Biochim. Biophys. Acta, 510:63.

Weinstein, L. H. and McCune, D. C., 1979, Air pollution stress, in: "Stress Physiology in Crop Plants," H. Mussell and R. C. Staples, eds., John Wiley & Sons, New York.

Weller, H. and Haug, A., 1977, Effects of Ca^{2+} and K^{+} on the physical state of the membrane lipids in Thermoplasma acidophila, J. Gen. Micro., 99:379.

Willcox, M. E. and Patterson, B. D., 1980, Breaks or curves?, A visual aid to the interpretation of data, in: "Low Temperature Stress in Crop Plants: The Role of the Membrane," J. M. Lyons, D. Graham, and J. K. Raison, eds., Academic Press, New York.

Williamson, F. A., Fowke, L. C., Constabel, F. C. and Gamborg, O. L., 1976, Labelling of concanavalin A sites on the plasma membrane of soybean protoplasts, Protoplasma, 89:305.

Williamson, F. A., 1979, Concanavalin A binding sites on the plasma membrane of leek stem protoplasts, Planta, 144:209.

Williamson, R. E., 1974, Actin in the alga Chara corallinia, Nature, 248:801.

Willison, J. H. M., 1976, An examination of the relationship between freeze-fractured plasmalemma and cell-wall microfibrils, Protoplasma, 88:187.

Wolfe, J., 1980, Some physical properties of membranes in the phase separation region and their relation to chilling damage in plants, in: "Low Temperature Stress in Crop Plants: The Role of the Membrane," J. M. Lyons, D. Graham, and J. K. Raison, eds., Academic Press, New York.

Wolfe, J. and Bagnall, D., 1980, Statistical tests to decide between straight line segments and curves as suitable fits to Arrhenius plots or other data, in: "Low Temperature Stress in Crop Plants: The Role of the Membrane," J. M. Lyons, D. Graham, and J. K. Raison, eds., Academic Press, New York.

Wood, A. and Paleg. L. G., 1972, The influence of gibberellic acid on the permeability of model membrane systems, Plant Physiol., 50:103.

Wood, A. and Paleg, L. G., 1974, Alteration of liposomal membrane fluidity by gibberellic acid, Aust. J. Plant Physiol., 1:31.

Wood, A., Paleg, L. G., and Spotswood, T. M., 1974, Hormone-phospholipid interaction: a possible hormonal mechanism of action in the control of membrane permeability, Aust. J. Plant Physiol., 1:167.

Wu, S. H. and McConnell, H. M., 1973, Lateral phase separations and perpendicular transport in membranes, Biochem. Biophys. Res. Commun., 55:484.

Wu, S. H. and McConnell, H. M., 1975, Phase separations in phospholipid membranes, Biochem., 14:847.

Wunderlich, F., Ronai, A., Speth, V., Seelig, J., and Blume, A., 1975, Thermotropic lipid clustering in Tetrahymena membranes, Biochem., 14:3730.

Wyn Jones, R. G., Storey, R., Leigh, R. A., Ahmad, N., and Pollard, A., 1977, A hypothesis on cytoplasmic osmoregulation, in: "Regulation of Cell Membrane Activities in Plants," E. Marré and O. Ciferri, eds., Elsevier North Holland, Amsterdam.

Yelenosky, G., 1979, Accumulation of free proline in citrus leaves during cold hardening of young trees in controlled temperature regimes, Plant Physiol., 64:425.

Zagyansky, Y. and Edidin, M., 1976, Lateral diffusion of concanavalin A receptors in the plasma membrane of mouse fibroblasts, Biochim Biophys. Acta, 433:209.

Zimmermann, U., 1977, Cell turgor pressure regulation and turgor pressure-mediated transport processes, in: "Integration of Activity in the Higher Plant," D. H. Jennings, ed., Cambridge University Press, Cambridge.

Zimmermann, U., 1978, Physics of turgor and osmoregulation, Ann. Rev. Plant Physiol., 29:121.

Zimmermann, U. and Dainty, J., eds., 1974, Membrane Transport in Plants, Springer-Verlag, New York, pp. 102-103.

Zimmermann, U. and Steudle, E., 1977, Action of indoleacetic acid on membrane structure and transport, in: "Regulation of Cell Membrane Activities in Plants," E. Marré and O. Ciferri, eds., Elsevier North Holland, Amsterdam.

Zimmermann, U., Steudle, E., and Lelkes, P. I., 1976, Turgor pressure regulation in Valonia utricularis, Plant Physiol., 58:608.

Zimmermann, U., Beckers, F., and Steudle, E., 1977, Turgor sensing in plant cells by the electro-mechanical properties of the membrane, in: "Echanges Ioniques Transmembranaires Chez Les Vegetaux," M. Thellier, A. Monnier, M. Demarty, and J. Dainty, eds., C.N.R.S., Paris.

A UNIFIED CONCEPT OF STRESS IN PLANTS?

Peter L. Steponkus

Department of Agronomy
Cornell University
Ithaca, New York 14853

INTRODUCTION

Consideration of a unified concept of stress resistance in plants is often prompted by the numerous examples where seasonal variation in resistance to a particular environmental stress is paralleled by increased resistance to other stresses. As early as 1929, Maximov discussed Walter's (1925) suggestion that there is a close analogy between freezing and drought resistance. Scarth (1941), drawing upon data of Levitt and Siminovitch (1940), reported that the relative resistance to desiccation, freezing and plasmolysis was equal in such diverse tissues as cabbage petioles and cortex tissue of *Catalpa* and *Cornus* twigs. Pisek and Larcher (1954) have documented that several conifers and broad-leaved evergreens exhibit parallel trends in drought and freezing resistance, while Parker (1972) indicates that seasonal trends in heat resistance of various species of conifers and broad-leaved evergreens resemble those for freezing resistance. Coffman (1957) demonstrated a parallelism between freezing resistance and heat resistance of twelve varieties of two *Avena* species, while Sullivan and Ross (1979) reported positive correlations between levels of cellular desiccation tolerance and heat tolerance.

Further support for a unified concept of stress resistance is obtained from observations that "hardening" (acclimating or conditioning) plants to a particular environmental stress results in increased resistance to other stresses or the corollary that resistance to a particular stress may be elicited by any one of several hardening stimuli. The most comprehensive analysis of this was presented by Chandler (1913), in which the influence of cold acclimation, withholding water, and irrigating with various salt and sugar solutions on low temperature survival was considered in an extensive array of horticultural plants. Similarly, Rosa (1921) observed that the increase in cold hardiness of cabbage elicited by low temperatures could also be obtained by insufficient watering or introduction of salt solutions to the soil, and Whiteside (1941) increased the desiccation tolerance of several wheat cultivars by either drought or cold hardening. Such observations have led to the consideration that cold acclimation results in a parallel increase in tolerance to freezing, drought, and osmotic stress (Siminovitch and Briggs, 1953). For example, cold acclimation of barley plants increases their drought resistance (Levitt, Sullivan and Krull, 1960) while water stress increases the cold hardiness of redosier dogwood (Chen et al., 1975, 1977; Chen and Li, 1977; Chen and Li, 1978).

Finally, the fact that physiological changes which occur during acclimation to one environmental stress can be elicited by acclimation to another stress, further warrants the consideration of a unified concept of stress resistance. The many changes and the subsequent inferences regarding their significance have been extensively discussed (Chandler, 1913; Maximov, 1929; Scarth, 1941; Levitt, 1951, 1958, 1972).

Cumulatively, such observations have led to speculation that plants may possess a general resistance to several or even all environmental stresses (Levitt, 1972). While Maximov (1929) considered this as a "physiological resistance", Biebl (1952) preferred the phrase "ecological resistance", and Parker (1972) invoked the term "protoplasmic resistance". Levitt (1951) originally contended that "... frost, drought (i.e., desiccation), and heat resistance are all basically similar, and that resistance to one of these factors carried with it a resistance to the others", but subsequently (Levitt, 1958) sought refuge in semantics and referred to the general phenomenon as "environment resistance" and most recently (Levitt, 1972) addressed the question more specifically but with more carefully defined considerations. Consideration of a general resistance to environmental stresses has influenced some individuals to advocate and pursue selection for resistance to one stress as a criterion for selecting for resistance to other stresses. For instance, the ability to withstand plasmolysis and subsequent deplasmolysis has been used to select for cold hardiness (Scarth and Levitt, 1937; Siminovitch and Levitt, 1941; Siminovitch and Briggs, 1953; Levitt, 1972). The ability to germinate in the presence of high concentrations of osmoticum has been used for inferring relative drought resistance (Williams et al., 1967), and a measure of high temperature tolerance has been used to infer relative drought resistance of cereals (Sullivan and Ross, 1979).

While such approaches may be justified and valid in some instances, the attitude that there is a general resistance to all or even several environmental stresses may, however, be misleading or erroneous. In fact, Lange (1967) does not subscribe to Levitt's proposal (1958) that there is a "clear cut parallel between frost, drought, and heat tolerance" and that the general increase in resistance is caused by a single protoplasmic factor -- most notably the sulfhydryl-disulfide hypothesis (Levitt, 1962), which has been extensively invoked in all forms of stress resistance (Levitt, 1972). Lange (1967) cites examples where the relation between heat- and frost-resistance occurs only periodically in most investigated species and is even totally absent in others. Similarly, while Santarius (1973) considered certain similarities between frost, desiccation, and heat resistance, he acknowledged that they do not always change in parallel. Specifically, heat resistance reaches a second maximum in summer when cold hardiness is at a minimum. Secondly, although Scarth (1941) contends that drought-hardening and cold-hardening result in the same degree of desiccation resistance; Levitt (1972) reports that the increase in cold hardiness resulting from drought hardening is usually much less than that obtained by cold acclimation. Thus, the drought-induced increase in cold hardiness of redosier dogwood only increased cold hardiness to -11 C while cold acclimation results in survival to -196 C (Chen et al., 1975, 1977). In contrast, drought hardening of cabbage resulted in the same degree of cold hardiness achieved by cold acclimation (Cox and Levitt, 1976), however, the maximum extent of hardiness in this species was only -13 C.

Emerging is the dilemma that although it is possible to garner evidence suggestive of a common basis for resistance to many environmental stresses -- inconsistencies emerge which diminish the validity of any such generalizations. It should be emphasized that each environmental stress results in a myriad of cellular stresses each of which may result in specific lesions either structural or metabolic. Some of the cellular stresses may be unique to a particular environmental stress, while some may be common to several environmental stresses. Any such commonalities may be considered as the basis for a certain degree of similarity in resistance to various environmental stresses. Such a consideration is only appropriate, however, if an additional proviso is stipulated -- that the particular common cellular stress arises in a similar priority sequence with respect to injury. In other words, a cellular lesion resulting from a common cellular stress which is elicited by any one of several different environmental stresses could not be prempted by a lesion resulting from a cellular stress unique to any particular environmental stress. A further elaboration would allow that the common cellular stress is primary and can result in an array of secondary cellular stresses which are a function of the intensity of the primary cellular stress. Furthermore, cellular stresses unique to a particular environmental stress may interact with either the common primary or secondary cellular stresses to cause injury. If such a common cellular stress can be demonstrated to be compatible with the above provisos, then observations of similarities in resistance to various environmental stresses may be explained without invoking a universal resistance to environmental stresses. The following discussion will attempt to document and demonstrate that:

1) cellular dehydration is a primary cellular stress which is common to drought, salt, and freezing stress;

2) various secondary stresses occur as a result of cellular dehydration, but as different functions;

3) osmotic adjustment can effectively ameliorate the extent of cellular dehydration and may have additional consequences in relation to specific secondary stresses;

4) osmotic adjustment may be elicited by exposure to various environmental stresses.

CELLULAR DEHYDRATION

Drought, salt, and freezing stresses all result in cellular dehydration. While cellular dehydration resulting from drought or salt stress is readily envisioned, that resulting from freezing may require further elaboration and has been discussed by Mazur (1969, 1970) and Steponkus and Wiest (1980). As early as 1880 and 1886, Muller-Thurgau proposed that cellular dehydration is the cause of death of cells subjected to frost (Maximov, 1929). Each of these environmental stresses results in a lower chemical potential of water external to the cell, and due to the semi-permeable nature of the plasma membrane, the cells behave as osmometers. Boyle-van't Hoff behavior predicts that cell volume will vary linearly with osmolality^{-1} as shown in Figure 1. The application of the Boyle-van't Hoff relation for describing osmometric behavior of plant cells and organelles and its limitations have been discussed by Nobel (1970). While plant cells exhibit such behavior over a certain range, their upper limit is constrained by the cell wall. Volumetric responses of individual cells are often characterized by plasmometric techniques (e.g., Stadelmann, 1966), while those of excised tissues may be inferred from pressure-volume methodology (Scholander et al., 1964; Tyree and Hammel, 1972; e.g., Cutler et al., 1979). A reasonable approximation is presented in Figure 1, where the cell volume asymptotically approaches a maximal value and is influenced by the elasticity of the cell wall (Dainty, 1972, 1976).

Starting at maximal hydration, decreases in the chemical potential of the external solution will result in disequilibrium between the chemical potential of the intra- and extracellular

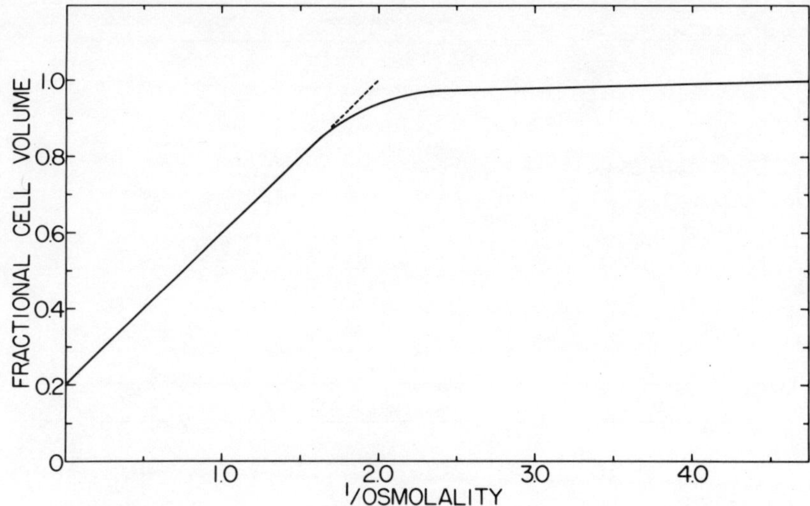

Fig. 1. Volumetric response of a cell with an internal solute concentration of 500 mosmolal at maximal hydration and an osmotically inactive fractional volume of 0.2, when subjected to increasing osmotic concentrations.

water. Efflux of water from the cell results in a decrease in both the turgor and solute potential of the cell. Since small volumetric decreases will result in large decreases in turgor potential, initial decreases in the chemical potential of water external to the cell will be principally offset by decreases in turgor potential and relatively small decreases in solute potential. With the dimunition of turgor potential, progressively greater decreases in solute potential are required to achieve thermodynamic equilibrium and hence a greater extent of cellular dehydration will result. At the point of incipient plasmolysis, when turgor potential is zero, further decreases in external water potential are entirely offset by decreases in solute potential. When turgor potential reaches zero, the change in cell volume per change in external water potential is maximal and subsequently declines. At extreme intensities of dehydration, a third component of water potential, matric potential may become significant (Hsiao, 1973).

Although drought, salt, and freezing all result in cellular dehydration, injury is manifested at different intensities (Figure 2). For instance, drought stress in many mesophytic crop species is usually manifested in the range of -1 to -20 bars, which corresponds to a loss of 0 to 15 percent of tissue water (Hsiao, 1973). In contrast, a mild-freezing stress of -5 C results in extensive cell dehydration (Steponkus and Wiest, 1980). At -20 C which corresponds to a water potential of nearly -250 bars, the majority

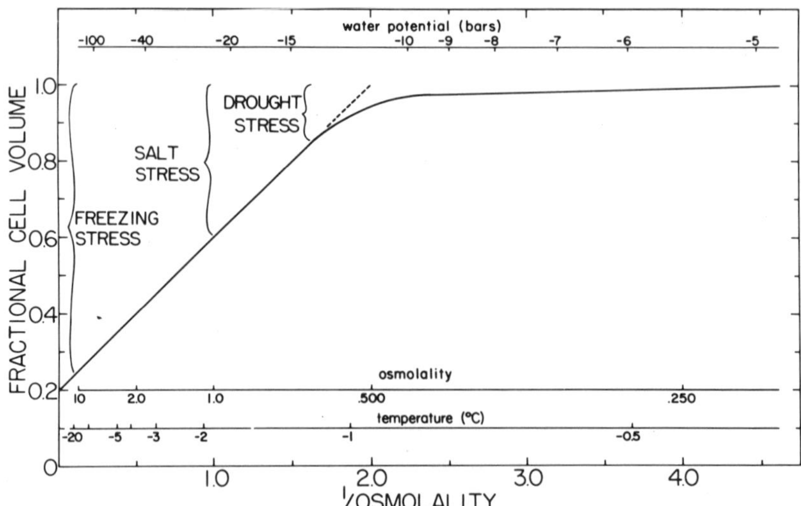

Fig. 2. Extent of cellular dehydration, expressed as a fractional cell volume, commonly associated with injury due to drought, salt, or freezing stress, assuming an internal solute concentration of 500 mosmolal at maximum hydration.

of the osmotically-active water is removed, assuming an initial, internal solute concentration of 500 mosmolal. Salt stress resulting from exposure to salt solutions of 0.5 to 1.5 osmolal will result in intermediate degrees of dehydration.

There are several physico-chemical events which result from cellular dehydration and may be considered as secondary stresses. In the case of relatively mild cellular dehydration resulting from drought stress, Hsiao (Hsiao, 1973; Hsiao et al., 1976a) has identified five subsequent effects: 1) reduced chemical potential or activity of cellular water, 2) decreases in turgor potential, 3) concentration of solutes and macromolecules, 4) altered spatial relations in cellular membranes, and 5) removal of water of hydration of macromolecules. In the case of extreme dehydration resulting from freezing stress, Mazur (1969) has identified six subsequent effects: 1) concentration of solutes, 2) precipitation of solutes, 3) reduction in cell water content, 4) cell shrinkage, 5) changes in pH, and 6) reduction in spatial separation of macromolecules. A similar analysis has been presented by Heber and Santarius (1973). Many investigators tend to consider these effects as singular events and elect one as the basis for a hypothesis on the mechanism of damage. Cellular dehydration, however, results in a concentration of events (Mazur, 1977). Therefore, it would be more appropriate to view the situation as a sequence of events (Steponkus, 1978) and to order the sequence as a function of intensity or incidence (Figure 3). A reasonable sequence would be:

1. reduction of turgor potential
2. reduction in cell volume
3. reduction in cell surface area
4. altered spatial relations in cellular membranes
5. concentration of solutes
6. precipitation of solutes
7. changes in pH
8. reduced chemical activity of water
9. removal of water of hydration of macromolecules.

Not included are the various stresses which are unique to the particular environmental stress. These will not be elaborated on since a demonstration of commonality based on cellular dehydration requires that they do not take precedence over the common stress in causing injury. For instance, in crop plants drought resistance is judged in terms of plant performance rather than survival.

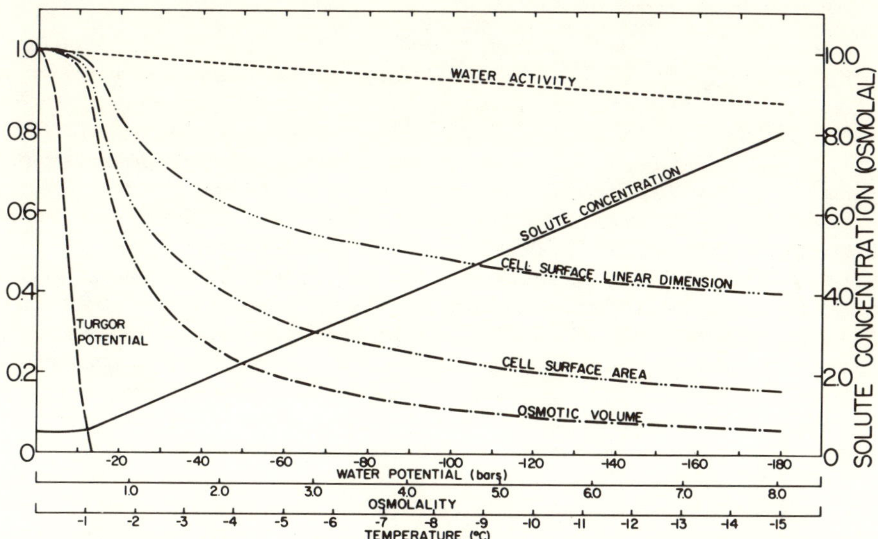

Fig. 3. Secondary cellular stresses resulting from cellular dehydration. All are presented as a fraction of their value at maximum hydration except solute concentration, which is presented in absolute units assuming a concentration of 500 mosmolal at maximum hydration.

Reduction of Turgor Potential

Relatively mild cellular dehydration (0 to 15 percent water loss) results in maximal losses of turgor potential before any appreciable change in the intensity of the other stresses occurs (Fig. 3). Hsiao (Hsiao, 1973; Hsiao et al., 1976a) therefore reasons that a reduction in turgor potential is the primary mechanism for the transduction of the effects of drought stress on mesophytic crop productivity. Furthermore, in a very comprehensive analysis, Hsiao argues that because of the strong dependency of cell growth on turgor potential, reductions in turgor potential can have far reaching repercussions on crop plant performance and productivity (Hsiao and Acevedo, 1974; Hsiao et al., 1976b). While his synthesis is reasonable and most appropriate when considering drought stress, it obtains from the premise that turgor pressure is "... the physical force needed to sustain enlargement" (Hsiao et al., 1976a). Cram (1974) suggests that while turgor pressure is the driving force for expansion, it does not necessarily follow that turgor regulates the rate of expansion. Since drought stress most frequently affects the rate of enlargement, the question of whether turgor potential per se bears a unique relationship to the rate of cell enlargement should be considered.

Reduction in Cell Volume

A reduction in cell volume inextricably results from cellular dehydration, albeit at very modest extents when the cell is in a turgid state. Directly associated with cellular volumetric decreases are decreases in the cell surface area and altered spatial relations in cellular membranes, most notably the plasma membrane (Figure 3). The impact of such changes is minimal in the case of drought stress, but may be of significance in salt stress and, most certainly, in the case of freezing stress. In fact, Meryman (1968, 1971) has proposed the "minimum critical volume hypothesis" to account for freeze-induced injury in a wide spectrum of biological cells. Wiest and Steponkus (1979a) have questioned the basic premise for this hypothesis, i.e., anomalous osmometric behavior and have discussed the application of this hypothesis to plant cells (Steponkus and Wiest, 1980). While a reduction in cell volume is the appropriate cell parameter when considering cellular stresses associated with freezing stress, it is more appropriate to consider reductions in cell surface area, or even more appropriately the plasma membrane, when considering cellular lesions. Wiest and Steponkus (1978a) have shown that a reduction in the plasma membrane surface area resulted in an alteration of the resilience of the plasma membrane. A physical deletion of plasma membrane components occurs upon contraction (Wiest and Steponkus, 1978b) and subsequently decreases the expansion potential during thawing. Such an alteration is probably a direct result of altered spatial relations in the plasma membrane which result in an increase in the global free energy of the plasma membrane (see Steponkus and Wiest, 1980, for a discussion of theoretical considerations).

Concentration of Solutes

The concentration of cellular solutes is frequently invoked in consideration of the mechanism of injury resulting from salt and freezing stresses. Following the reduction in turgor potential, solute concentration will increase linearly with further decreases in external water potential and proportionally smaller decreases in cell water will result in equal increases in solute concentration (Figure 3). As solute concentration increases, the solubility limits of some species may be exceeded and they will precipitate from the solution. If the buffering capacity of the cell solutes is dependent upon salts of differing solubilities, differential precipitation of buffering salts will result in pH changes. As early as the beginning of this century, Gorke (1906), Lidforss (1907), Harvey (1918) and Newton (1924) proposed that freezing injury was primarily attributable to the increased concentration of cellular solutes, especially salts. Lovelock (1953a,b, 1957) is often cited for his definitive work in this area and the proposals of Mazur (1977) rely very heavily on solute concentration as a major factor contributing to freezing injury.

Recently, the direct effects of salt concentration on the freeze-induced inactivation of light-induced proton uptake in chloroplast thylakoids have been considered (Heber and Santarius, 1973; Steponkus et al., 1977). Additionally, the high salt concentrations experienced during freezing have detrimental effects on the plasma membrane of isolated protoplasts (Wiest and Steponkus, 1979b). In the case of salt stress, the preferential uptake and concentration of specific salts is a primary cause of injury in salt-sensitive species (Levitt, 1972).

Removal of Water of Hydration

The removal of water of hydration in addition to altered spatial relations of macromolecules has been considered as a cause of freezing injury (Levitt, 1972) but is considered unlikely in the case of drought stress (Hsiao et al., 1976a). These events are, in part, invoked in the sulfhydryl-disulfide hypothesis of Levitt (1972) which, because of a paucity of unequivocal experimental support, has been often discounted (cf. Heber and Santarius, 1973; Mazur, 1969).

Thus, the primary cellular stress of dehydration is common to the environmental stresses of drought, salt, and freezing; and the various secondary stresses which result can be shown to result in injury due to different lesions. While all of the secondary stresses occur simultaneously, they vary with the degree of cellular dehydration so that their relative intensity varies greatly at any given extent of dehydration. Nonetheless, commonality in the primary factor responsible for injury can be demonstrated. Therefore, any factor which would ameliorate the extent of cellular dehydration should be mutually beneficial in conferring resistance to environmental stresses resulting in cellular dehydration.

OSMOTIC ADJUSTMENT IN RESPONSE TO ACCLIMATION

It is appropriate for this sympsoium that one cellular alteration which can ameliorate cellular dehydration is an increase in internal solute concentration referred to as osmoregulation or osmotic adjustment. The effect of increasing the internal solute concentration on cellular volumetric responses is shown in Figure 4. A doubling of the internal solute concentration decreases the extent of dehydration at any given water potential and thus correspondingly ameliorates the secondary stresses.

Although osmotic adjustment in response to drought conditioning enjoys recently acquired renown (Hsiao, 1973; Hsiao et al., 1976a; Boyer and McPherson, 1975; Begg and Turner, 1976; Turner, 1979), increases in cellular osmotic potential were recognized as early as 1921 by Rosa. He established that cold hardiness could be increased by either cold, drought, or salt conditioning and was associated with an increase in the sugar and amino acid contents. In 1929, Iljin concluded from studies of 122 species of plants, that plants growing in dry habitats contain more sugar than those

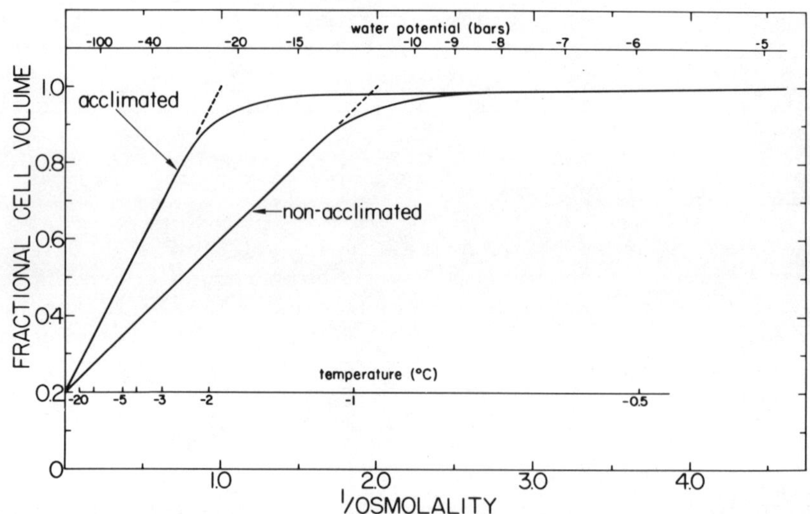

Fig. 4. Volumetric response of cells differing in their internal solute concentration at maximum hydration (500 vs. 100 mosmolal) and assuming zero turgor at the same fractional osmotic volume and equal fractions non-osmotic volumes.

in moist climates. In 1939, Schmidt demonstrated that osmotic pressure increased with drought. While Whiteside (1941) observed increased solute concentrations in two wheat varieties, Bartel (1947) demonstrated that drought decreased the osmotic potential in four wheat varieties, with the decrease in the order of the reputed order of drought resistance. Iljin (1957) discusses several other reports of alterations in osmotic potential in response to drought. Hsiao et al. (1976a) discount some of these early observations because they failed to demonstrate that the increase in solute concentration were not due solely to lower water contents. Recently, there has been widespread documentation of osmotic adjustment in response to drought in a wide range of crop species, e.g., Meyer and Boyer, 1972; Hsiao and Acevedo, 1974; Simmelsgaard, 1976; Morgan, 1977; Jones and Turner, 1978; Cutler and Rains, 1978; Cutler et al., 1980a,b; Acevedo et al., 1979. In retrospect, the early reports of increases in solute concentration were probably, in part, attributable to osmotic adjustment resulting from a net increase in solutes.

An increase in internal solute concentration in response to salt stress has been frequently observed and referred to as osmoregulation. Hellebust (1976) and Flowers et al. (1977) provide comprehensive reviews of osmoregulation in response to saline stress. In contrast to the response in drought hardening, where the increased solute concentration is generated from within the plant; osmoregulation in response to salt stress is frequently attributable to the uptake and accumulation of salts (Bernstein and Hayward, 1958). There are examples, however, where salt-induced increases in organic solutes occur (Greenway and Sims, 1974). A tentative generalization which may be offered is that osmoregulation in halophytes is a result of salt uptake while glycophytes may depend on synthesis of organic compounds for osmoregulation. In the latter case, increases in organic acids are frequently observed, e.g., Steiner and Eschrich, 1958; MacLennan et al., 1963; Jacoby and Laties, 1971.

The earliest observations of solute accumulation in response to environmental stresses can be found in the area of freezing resistance. In 1898, Lidforss documented the increase in soluble carbohydrates, and in 1907 proposed that the accumulation of sugars was related to cold acclimation. In the ensuing years, elaboration of the nature and quantification of the increased solutes resulting from cold acclimation has been an intensive area of activity of research in cold hardiness research. Countless examples can be found in the reviews of Maximov (1929), Levitt (1951, 1972), and Alden and Hermann (1971). The majority of these reports are subject to the reservations of Hsiao et al. (1976a). It should be emphasized that a doubling or tripling in the concentration of a particular solute cannot be automatically inferred as indicative of osmotic adjustment. If the solute in question is located solely in the cytoplasm of a cell whose osmotic volume is primarily (80 to 90%) in the vacuole, then such an increase will have a small impact on the osmotic properties of the cell. A rather unique, but fundamental, demonstration of osmotic adjustment in response to cold acclimation can be seen in Figure 5. Depicted is the osmometric behavior of protoplasts isolated from rye seedlings in various stages of acclimation.

CELLULAR CONSEQUENCES OF OSMOTIC ADJUSTMENT

The primary cellular consequence of osmotic adjustment in relation to cellular dehydration is that the extent of dehydration at any given water potential will be decreased. It follows that the intensity of the secondary stresses will also be diminished accordingly. Thus, the major consequence of osmotic adjustment with respect to drought stress is most often attributed to the capacity for maintenance of turgor at lower water potentials which may be the result of either diminishing soil water availability or transpirationally-induced diurnal fluctuations. Interestingly, the process of osmotic adjustment is responsive to both (Acevedo et al., 1979; Shahan et al., 1979), even though the former generally occurs rather slowly in comparison to rather rapid transpirationally-induced diurnal deficits. The major effect of osmotic adjustment is then to effectively increase the turgor potential or "turgor reservoir" of the cell. Additionally, the relationship between the extent of cell dehydration (cell volume) and wall elasticity will also be influenced by osmotic adjustment (Turner, 1979), since elasticity is turgor dependent (Steudle et al., 1977).

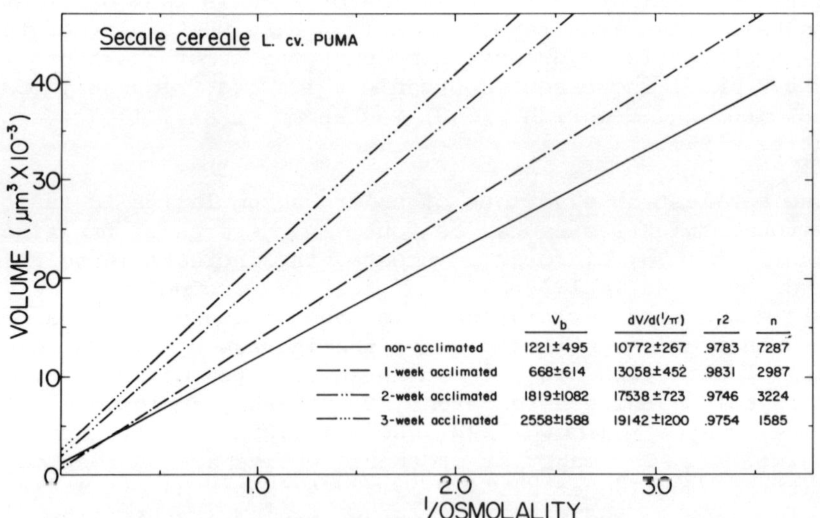

Fig. 5. Osmometric behavior of rye protoplasts isolated from tissue in different stages of acclimation.

In the area of salt stress, maintenance of turgor potential to lower water potentials is equally beneficial. Additionally, osmotic stresses imposed on the root systems result in decreased water potential gradients between the roots and aerial portions and reduce water flux. Under heavy transpirational demands, reduced water flux contributes to the incidence of dehydration in the aerial portions. Osmotic adjustment which allows for maintenance of turgor to lower water potentials would then allow for a greater reduction in water potential in the aerial portions which would favor increased water flux while turgor was maintained. In addition, salt stress may result in toxic internal solute concentrations due to salt uptake. Osmotic adjustment via organic solutes would minimize the uptake of inorganic solutes, by minimizing osmotic disequilibrium.

With respect to cold hardiness, early observations of increased solute concentrations were only viewed as being influential due to depression of the freezing point of the sap. Such considerations neglect the fact that under normal conditions ice formation only occurs extracellularly, and any deferral in the freezing point of the sap per se would not be of direct concern anyway. Maximov (1929), however, contended that the increased solute concentration influenced freezing resistance in some other manner. Chandler (1913), and in a very similar fashion, Akerman (1927) elaborated on this view and detailed the influence of solutes on reducing the extent of ice formed at any particular subzero temperature. At subzero temperatures, the water potential of the intracellular solution will equilibrate with that of the partially frozen external solution which will be determined by the subzero temperature. When the initial internal solute concentration is increased, less water will have to be removed from the cell to achieve thermodynamic equilibrium. Hence, the extent of cellular dehydration is reduced accordingly. Specifically, a doubling of the internal concentration will decrease the extent of cellular dehydration by 50 percent at any given subzero temperature (Figure 6). Thus, secondary stresses such as reductions in plasma membrane surface area which have been shown to result in freezing injury are effectively ameliorated at any particular temperature.

At extreme levels of cell dehydration where further reductions in temperature only result in minimal changes in cell volume, solute concentration continues to increase linearly. In such instances, osmotic adjustment has an additional consequence -- the concentration of any potentially toxic compounds that might arise during the general concentration of solutes is also ameliorated. In the absence of any turgor potential, thermodynamic equilibrium of the intra- and extracellular compartments will be achieved primarily by osmotic equilibration (the contribution of matric potential being poorly characterized). While the chemical potential of a partially frozen solution is determined by the subzero temperature and independent of the initial solute concentration, the initial solution

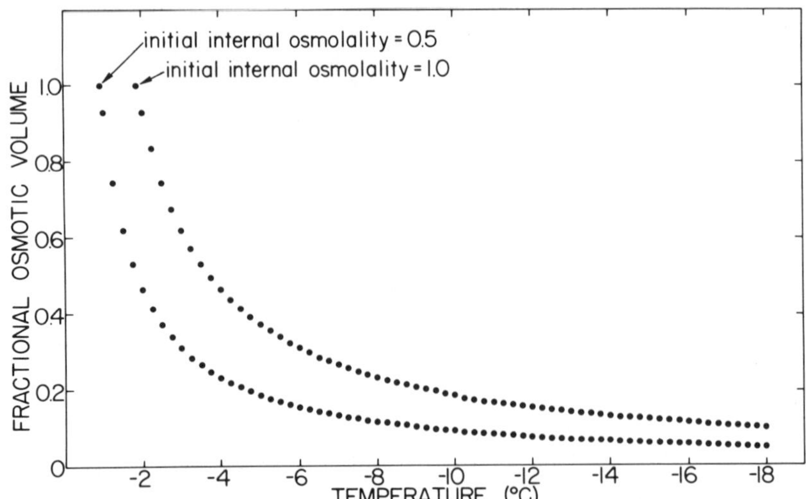

Fig. 6. Fractional osmotic volume at subzero temperatures in cells differing in their initial solute concentration (500 vs. 1000 mosmolal).

composition will influence the solution composition at the subzero temperature. Thus, if initially composed of low concentrations of both neutral and potentially toxic solutes, an increase in the neutral solutes prior to freezing will effectively diminish the concentration of toxic solutes attained during freezing. The requirement for the initial increase to be in neutral or protective solutes is not even necessary. That is, moderate increases in the concentration of several solutes, which individually may be toxic at the high concentrations incurred at subzero temperatures, may colligatively serve to preclude the attainment of a critical concentration of any one species (Heber and Santarius, 1973). Such an effect has been termed colligative protection (Lovelock, 1953). Heber and Santarius (1973) infer a large influence of colligative protection due to neutral solutes which accumulate during cold acclimation.

The mechanism by which solute accumulation contributes to membrane protection has been extended even further by Santarius (1973). In view of observations that certain sugars are more effective (trisaccharides>disaccharides>monosaccharides) in conferring protection against freezing damage to isolated chloroplast thylakoids (Heber and Santarius, 1973; Steponkus et al., 1977), Santarius (1973) considers that sugars contribute to increased protection by two mechanisms. One is via a colligative reduction of toxic solutes. However, since colligative properties are non-specific, differences elicited by equimolar concentrations of sugars suggests a more specific mechanism may also be involved in membrane stabilization, perhaps, as envisioned by Steponkus (1971). Recent work (Lineberger and Steponkus, 1980), however, indicates that the differential cryoprotection afforded to chloroplast thylakoids against freeze-induced uncoupling of cyclic photophosphorylation by equimolar concentrations of glucose, sucrose and raffinose appears to be due to non-ideal activity-concentration relations of sugars at the extremely high concentrations experienced during freezing. When cryoprotection is analyzed as a function of the mole fraction of NaCl to which the membranes are exposed during freezing, protection of cyclic photophosphorylation and its component reactions is not dependent upon the chemical identity of the protective solute. Such anomalous behavior appears to be a factor at extremely high solute concentrations as resulting from a freezing stress and is probably not significant at low to moderate solute concentrations resulting from drought stress.

Thus, it can be demonstrated that osmotic adjustment can serve directly as an effective moderator of cellular dehydration and hence ameliorate the ensuring secondary stresses associated with volume reduction and solute concentration. As such it functions primarily to diminish the intensity of the primary stress and hence may be inferred to increase the stress resistance. Osmotic adjustment is often considered as a "tolerance mechanism" in stress resistance terminology, especially with respect to drought resistance (Turner, 1979) in that it allows the plant to "tolerate" lower soil water potentials. It is equally appropriate to view it as an "avoidance" mechanism from the cellular perspective in that it enables the cell to avoid cellular dehydration (Steponkus et al., 1979). Levitt (1972) concluded that no one avoidance mechanism provides resistance against all kinds of water stress. In fact, he concluded that "there is no mechanism for avoiding osmotic stress, since this would involve impermeability of the cell to water." It would appear entirely appropriate to view osmotic adjustment as such a mechanism in that it specifically enables the cell to avoid an osmotic stress whether it be elicited by drought, salt, or freezing stress. In a final summation addressing the interrelationships between resistances to these stresses, Levitt (1972) invoked the sulfhydryl hypothesis (pp. 544-568) without any mention of the commonalities associated with cellular dehydration and osmotic adjustment even though the influence of solute concentration on dehydration was previously addressed (pp. 107-109).

CONCLUSIONS

Do the demonstrations that the three environmental stresses of drought, salt, and freezing all result in cellular dehydration; that osmotic adjustment can effectively ameliorate the extent of cellular dehydration; and that osmotic adjustment may be elicited by various hardening treatments justify the consideration of a unified concept of stress resistance? In my opinion, they do not. Cellular dehydration is but one of the multitude of stresses resulting from these environmental stresses. While it is common to all three, other cellular stresses unique to each environmental stress cannot be discounted. Cellular dehydration and the attendant secondary stresses must be viewed in relation to other stresses confronting the cell. Consider the diverse insults imposed on a plant during drought versus freezing.

It might be argued that drought, salt, and freezing stresses only result in greater intensities of cellular dehydration. While cold acclimation increases the drought resistance of certain species and drought acclimation increases the cold resistance of others, the cold hardiness accrued is, in most cases, only a fraction of the total potential which can be achieved by cold acclimation. The analsis might suggest that since drought only results in a mild extent of dehydration, whereas freezing results in extensive dehydration, perhaps cold acclimation results in a maximal elicitation of drought resistance while drought hardening only results in partial cold acclimation due to a difference in the extent of osmotic adjustment. Early reports, however, are not consistent with such a consideration (Scarth, 1941). Alternatively, just as the various environmental stresses result in a multitude of cellular stresses, the individual acclimation of hardening processes probably result in a multitude of cellular alterations. This suggests that just as cellular dehydration is a common consequence of several environmental stresses, osmotic adjustment may be a common consequence of acclimation to the environmental stresses.

There are reports which suggest that hardening also increases the tolerance capacity to the cellular stresses. In drought-hardened plants, for instance, there are reports of altered turgor-growth relations (Bunce, 1977; Steponkus et al., 1980). In cold acclimation, alterations in the tolerance of the plasma membrane to the osmotically induced stresses of contraction and expansion can be demonstrated (Steponkus et al., 1979). Whether these two responses are unique to the particular acclimation regimes requires elucidation. Once these and other consequences of the individual acclimation processes are fully elucidated, it is quite possible that unique aspects will be revealed. Of particular interest is whether drought-hardening and cold acclimation of a particular cultivar result not only in the same extent of osmotic adjustment, but also contribute to the acclimation of similar solutes and whether cellular tolerance to the various secondary stresses increases in parallel. Conversely, it might be considered more interesting if they were not similar!

REFERENCES

Acevedo, E., Fereres, E., Hsiao, T. C., and Henderson, D. W., 1979, Diurnal growth trends, water potential, and osmotic adjustment of maize and sorghum leaves in the field, Plant Physiol., 64:476.

Akerman, A., 1927, Studien über den Kältetod und die Kälteresistanz der Pflanzen nebst Untersuchungen über die Winterfestigkeit des Weizens, Lund. (Cited by Maximov, 1929; original not seen).

Alden, J. and Hermann, R. K., 1971, Aspects of the cold-hardiness mechanism in plants, Bot. Review, 37:37.

Bartel, A. T., 1947, Some physiological characteristics of four varieties of spring wheat presumably differing in drought resistance, J. Agr. Res., 74:97.

Begg, J. E. and Turner, N. C., 1976, Crop water deficits, Adv. Agron, 28:161.

Bernstein, L. and Hayward, H. E., 1958, Physiology of salt tolerance, Ann. Rev. Plant Physiol., 9:25.

Biebl, R., 1952, Ecological and nonenvironmental constitutional resistance of the protoplasm of marine algae, J. Marine Biol. Assoc. U.K., 31:307.

Boyer, J. S. and McPherson, H. G., 1975, Physiology of water deficits in cereal crops, Adv. Agron., 27:1.

Bunce, J. A., 1977, Leaf elongation in relation to leaf water potential in soybean, J. Expt. Bot., 28:156.

Chandler, W. H., 1913, The killing of plant tissue by low temperature, Mo. Agr. Expt. Sta. Res. Bull., 8:141.

Chen, H. H. and Li, P. H., 1978, Interactions of low temperature, water stress and short days in the induction of stem frost hardiness in red osier dogwood, Plant Physiol., 62:833.

Chen, P. M. and Li, P. H., 1977, Induction of frost hardiness in stem cortical tissues of Cornus stolonifera Michx. by water stress. 2. Biochemical changes, Plant Physiol., 59:240.

Chen, P., Li, P. H., and Weiser, C. H., 1975, Induction of frost hardiness in red osier dogwood stems by water stress, HortScience, 10:372.

Chen, P. M., Li, P. H., and Burke, M. J., 1977, Induction of frost hardiness in stem cortical tissues of Cornus stolonifera Michx. by water stress. 1. Unfrozen water in cortical tissues and water status in plants and soil, Plant Physiol., 59:236.

Coffman, F. A., 1957, Cold resistant oat varieties also resistant to heat, Science, 125:1298.

Cox, W. and Levitt, J., 1976, Interrelations between environmental factors and freezing resistance of cabbage leaves, Plant Physiol., 57:553.

Cram, W. J., 1974, The regulation of concentration and hydrostatic pressure in cells in relation to growth, in: "Mechanisms of Regulation of Plant Growth," R. L. Bieleski, A. R. Ferguson, and M. M. Cresswell, eds., Bull. Roy. Soc. New Zealand, 12:183.

Cutler, J. M. and Rains, D. W., 1978, Effects of water stress and hardening on the internal water relations and osmotic constituents of cotton leaves, Physiol. Plant., 42:261.

Cutler, J. M., Shahan, K. W., and Steponkus, P. L., 1979, Characterization of internal water relations of rice by a pressure-volume method, Crop. Sci., 19:681.

Cutler, J. M., Shahan, K. W., and Steponkus, P. L., 1980a, Alteration of the internal water relations of rice in response to drought hardening, Crop Sci., (in press).

Cutler, J. M., Shahan, K. W., and Steponkus, P. L., 1980b, Dynamics of osmotic adjustment in rice, Crop Sci., (in press).

Dainty, J., 1972, Plant cell water relations: the elasticity of the cell wall, Proc. Roy. Soc. Edin. (A), 70:89.

Dainty, J., 1976, Water relations of plant cells, in: "Encyclopedia of Plant Physiology," V. Luttge and M. G. Putman, eds., Springer-Verlag, New York.

Flowers, T. J., Troke, P. F., and Yeo, A. R., 1977, The mechanism of salt tolerance in halophytes, Ann. Rev. Plant Physiol., 28:89.

Gorke, H., 1906, Uber chemische Vorgänge beim Erfrieren der Pflanzen, Landw. Versuchs., 65:149.

Greenway, H. and Sims, A. P., 1974, Effects of high concentrations of KCl and NaCl on responses of malate dehydrogenase (decarboxylating) to malate and various inhibitors, Aust. J. Plant Physiol., 1:15.

Harvey, R. B., 1918, Hardening process in plants and developments from frost injury, J. Agr. Res., 15:83.

Heber, U. and Santarius, K. A., 1973, Cell death by cold and heat and resistance to extreme temperatures. Mechanisms of hardening and dehardening, in: "Temperature and Life," H. Precht, J. Christopherson, H. Hensel, and W. Larcher, eds., Springer-Verlag, New York.

Hellebust, J. A., 1976, Osmoregulation, Ann. Rev. Plant Physiol., 27:485.

Hsiao, T. C., 1973, Plant responses to water stress, Ann. Rev. Plant Physiol., 24:519.

Hsiao, T. C. and Acevedo, E., 1974, Plant responses to water deficits, water use efficiency, and drought resistance, Ag. Meteorol., 14:59.

Hsiao, T. C., Acevedo, E., Fereres, E., and Henderson, D. W., 1976a, Stress metabolism. Water stress, growth, and osmotic adjustment, Phil. Trans. R. Soc. Lond. B., 273:479.

Hsiao, T. C., Fereres, E., Acevedo, E., and Henderson, D. W., 1976b, Water stress and dynamics of growth and yield of crop plants, in: "Water and Plant Life," O. L. Lange, L. Kappen, and E. D. Schultze, eds., Springer-Verlag, New York.

Iljin, W. S., 1929, Der Einfluss der Standortsfeuchtigkeit auf den osmotischen Wert bei Pflanzen, Planta, 7:45.

Iljin, W. S., 1957, Drought resistance in plants and physiological processes, Ann. Rev. Plant Physiol., 8:257.

Jacoby, B. and Laties, G. G., 1971, Bicarbonate fixation and malate compartmentation in relation to salt-induced stoichiometric synthesis of organic acid, Plant Physiol., 47:525.

Jones, M. M. and Turner, N. C., 1978, Osmotic adjustment in leaves of sorghum in response to water deficits, Plant Physiol., 61:122.

Lange, O. L., 1967, Investigations on the variability of heat resistance in plants, in: "The Cell and Environmental Temperature," A. S. Troshin, ed., Pergamon Press, Oxford.

Levitt, J., 1951, Frost, drought and heat resistance, Ann. Rev. Plant Physiol., 2:245.

Levitt, J., 1958, Frost, drought and heat resistance, in: "Protoplasmatologia," L. V. Heilbrumn and F. Weber, eds., Springer, Vienna.

Levitt, J., 1962, A sulfhydryl-disulphide hypothesis of frost injury and resistance in plants, J. Theoret. Biol., 3:355.

Levitt, J., 1972, "Responses of Plants to Environmental Stresses," Academic Press, New York.

Levitt, J. and Siminovitch, D., 1940, The relation between frost resistance and the physical state of the protoplasm. I. The protoplasm as a whole, Canad. J. Res., 18:550.

Levitt, J., Sullivan, C. Y., and Krull, E., 1960, Some problems in drought resistance, Bull. Council Israel, 80:173.

Lidforss, B., 1907, Die wintergrune Flora. Lunds Universitats Arsskrift, N.F. 2, Afd. 2, No. 13, (Cited by Maximov, 1929; original not seen).

Lineberger, R. D. and Steponkus, P. L., 1980, Cryoprotection by glucose, sucrose and raffinose to chloroplast thylakoids. Plant Physiol., 65:(in press).

Lovelock, J. E., 1953a, The haemolysis of human red blood cells by freezing and thawing, Biochim. Biophys. Acta, 10:414.

Lovelock, J. E., 1953b, The mechanism of the protective action of glycerol against haemolysis by freezing and thawing, Biochim. Biophys. Acta, 11:28.

Lovelock, J. E., 1957, The denaturation of lipid-protein complexes as a cause of damage by freezing, Proc. Roy. Soc., 147:427.

MacLennan, D. H., Beevers, H., and Harley, J. L., 1963, 'Compartmentation' of acids in plant tissues, Biochem. J., 89:316.

Maximov, N. A., 1929, Internal factors of frost and drought resistance in plants, Protoplasma, 7:259.

Mazur, P., 1969, Freezing injury in plants, Ann. Rev. Plant Physiol., 20:419.

Mazur, P., 1970, Cryobiology: The freezing of biological systems, Science, 168:939.

Mazur, P., 1977b, Slow-freezing injury in mammalian cells, in: "The Freezing of Mammalian Embryos," Proc. Ciba Foundation Symposium, London.

Meryman, H. T., 1968, Modified model for the mechanism of freezing injury in erythrocytes, Nature, 218:333.

Meryman, H. T., 1971, Osmotic stress as a mechanism of freezing injury, Cryobiology, 8:488.

Meyer, R. F. and Boyer, J. S., 1972, Sensitivity of cell division and cell elongation to low water potentials in soybean hypocotyls, Planta, 108:77.

Morgan, J. M., 1977, Differences in osmoregulation between wheat genotypes, Nature, 270:234.

Newton, R., 1924, Colloidal properties of winter wheat plants in relation to frost resistance, J. Agr. Sci., 14:178.

Nobel, P. S., 1970, "Plant Cell Physiology. A Physicochemical Approach," W. H. Freeman Co., San Francisco.

Parker, J., 1972, Protoplasmic resistance to water deficits, in: "Water Deficits and Plant Growth," T. T. Kozlowski, ed., Academic Press, New York.

Pisek, A. and Larcher, W., 1954, Zusammenhang zwischen Austrocknungsresistenz und Frosthärte bei immergrünen Pflanzen, Protoplasma, 44:30. .

Rosa, J. T., 1921, Investigation of the hardening process in vegetable plants, Miss. Agr. Exp. Sta. Res. Bull., 48:1.

Santarius, K. A., 1973, The protective effect of sugars on chloroplast membranes during temperature and water stress and its relationship to frost, desiccation and heat resistance, Planta, 113:105.

Scarth, G. W., 1941, Dehydration injury and resistance, Plant Physiol., 16:171.

Scarth, G. W. and Levitt, J., 1937, The frost-hardening mechanism of plant cells, Plant Physiol., 12:51.

Schmidt, H., 1939, Plasmazustand und Wasserhaushalt bei Laminium maculatum, Protoplasma, 33:25.

Scholander, P. F., Hammel, H. T., Hemmingsen, E. A., and Bradstreet, E. D., 1964, Hydrostatic pressure and osmotic potential in leaves of mangroves and some other plants, Proc. Nat. Acad. Sci., 52:119.

Shahan, K. W., Cutler, J. M., and Steponkus, P. L., 1979, Influence of stress regime on osmotic adjustment in rice, Plant Physiol. 63:(abstract in press).

Siminovitch, D. and Levitt, J., 1941, The relationship between frost resistance and the physical state of protoplasm. II. The protoplasmic surface, Can. J. Res., 19:9.

Siminovitch, D. and Briggs, D. R., 1953a, Studies on the chemistry of the living bark of the black locust in relation to its frost hardiness. III. The validity of plasmolysis and dessication tests for determining the frost hardiness of tissues, Plant Physiol., 28:15.

Simmelsgaard, S. E., 1976, Adaptation to water stress in wheat, Physiol. Plant., 37:167.

Stadelmann, E. J., 1966, Evaluation of turgidity, plasmolysis, and deplasmolysis of plant cells, in: "Methods in Cell Physiology," D. M. Prescott, ed., Academic Press, New York.

Steiner, M. and Eschrich, W., 1958, Die osmotische Bedeutung der Mineralstoffe, in: "Encyclopedia of Plant Physiology," W. Ruhland, ed., Springer-Verlag, Berlin.

Steponkus, P. L., 1971, Cold acclimation of Hedera helix: Evidence for a two phase process, Plant Physiol., 47:175.

Steponkus, P. L., 1978, Cold hardiness and freezing injury of agronomic crops, Adv. Agron., 30:51.

Steponkus, P. L., 1979, Effects of freezing and cold acclimation

on membrane structure and function, in: "Stress Physiology of Crop Plants," H. Mussell and R. C. Stables, eds., Wiley-Interscience, New York.
Steponkus, P. L., Cutler, J. M., and O'Toole, J. C., 1980, Adaptation to water deficits in rice, in: "Adaptation of Plants to Water and High Temperature Stress," P. J. Kramer and N. C. Turner, eds., Wiley Interscience, New York, (in press).
Steponkus, P. L., Dowgert, M. F., and Roberts, S. R., 1979, Cryobiology of isolated plant protoplasts: VI. Influence of cold acclimation, Cryobiology, 16:(in press).
Steponkus, P. L., Garber, M. P., Myers, S. P., and Lineberger, R. D., 1977, Effects of cold acclimation and freezing on structure and function of chloroplast thylakoids, Cryobiology, 14:303.
Steponkus, P. L. and Wiest, S. C., 1978, Plasma membrane alterations following cold acclimation and freezing, in: "Plant Cold Hardiness and Freezing Stress -- Mechanisms and Crop Implications," P. H. Li and A. Sakai, eds., Acadamic Press, New York.
Steponkus, P. L. and Wiest, S. C., 1980, Freeze-thaw induced lesions in the plasma membrane, in: "Low Temperature Stress in Crop

PANEL DISCUSSION

OSMOREGULATION IN HIGHER PLANTS

Chair: R. H. Nieman

Participants: J. S. Boyer, R. W. Breidenbach, R. Jefferies, C. B. Osmond, P. Steponkus, and R. G. Wyn Jones

P. Steponkus: It may be appropriate to consider the transduction mechanism for osmotic adjustment in higher plants in response to various environmental stresses, but could be start with something at a more elementary level and ask if osmoregulation is some covert active process of higher plants or whether it is a passive impact that results from the incidence of other events? From yesterday's discussion I think that there is good evidence that is a very active covert process. I wonder though whether that can be extended to the incidence of osmotic adjustment in response to various environmental hardening, conditioning, and acclimating regimes? Is it something that is elicited in response to a specific cellular situation, or is it something that results as a result of other limitations? Is it just a by-product, down the line, or is it something that is very primary in a stress resistant strategy?

C. B. Osmond: I think that it would be simple to regard it as a response to loss of turgor. In single large vacuolated plant cells, loss of turgor means the loss of capacity for extension growth which results in a reduction of utilization of substrates that would otherwise be used for making raw materials. That's one extreme view and this would be a passive response.

A. D. Brown: I was going to start my paper yesterday by defining what I thought osmoregulation was, but I thought that it was totally unnecessary in this context. But since this has been initiated let me say what I would have said. All of this may be just words, and it would seem to me that at least from a cellular point of view that it is possibly more complicated for a higher plant than single cell organism. Surely, osmoregulation is no more or no less than maintaining approximately constant turgor, pressure and volume in the face of a changing extracellular water potential. Now is it any more than that? Any less than that?

C. B. Osmond: It's fair to say that, presumably, volume and pressure are equivalents in this sort of situation. It depends on the elasticity of the wall surrounding the organism.

R. W. Breidenbach: Is everyone willing to accept that there are no specific properties of those compounds which would stabilize any cellular component, albeit a simple macromolecule to a more complicated cellular structure?

P. Steponkus: What I was saying was that there can be many beneficial consequences of even a passive process of osmotic adjustment that are not necessarily related to bulk maintenance of turgor potentials by increased cellular concentrations. Let me put it in different terms. I think that John Boyer's system is probably one of the most elementary systems for integrating at a whole-plant level. This is a rather well defined system with certain constraints that are not as elaborate as an intact plant growing in the field and more organized than what one could approach with protoplasm. John, in your opinion, do you think that the increase in solute concentration observed is the result of some positive action, elicited by either a state of turgor or some other positive mechanisms for transducing a loss of water or a state of dehydration? Or could it result from solute accumulation and merely accrue because the cotyledons in their function of pumping out stored reserve materials, no longer have a sink in terms of a rapidly expanding meristematic region. These solutes, the products of the cotyledons, are merely accumulating in these regions and appear as osmotic adjustment.

J. Boyer: Until we can demonstrate experimentally I can only take a guess. My guess would be, at this stage, there is some sort of environmental signal, a positive environmental signal that brings about these events rather than a passive response. One of the reasons for this suggestion is that the turgor in these growing regions remains approximately as high in the osmoregulating system, the dry system and the system in dry media as in the controls. If one thinks of a turgor-mediated process such as cell enlargement then cell enlargement should be slowing down because of loss of turgor; of course, that's not happening in this system. In other words, turgor-driven growth should be occurring just as rapidly in the seedlings in dry medium as it is in the seedlings in moist medium. Furthermore, it's quite clear that the fundamental materials coming from the cotyledons are much more available in the cells, and there are more of them in the cells of the seedlings in the dry medium than in the control. So from the standpoint of turgor and cell extension, or from the standpoint of the availability of substrates for growth, both are sufficient to drive growth rapidly; but it's not occurring and not occurring nearly as rapidly as in the controls. I tend to think there has been some change in the control of growth resulting from this state. Now, during the transplant process the first two or three hours seedlings are impacted by this dry environment without the benefit of osmoregulatory changes. This is a complex dehydration that occurs in transient and the plant may receive signals generated at that time.

P. Steponkus: If one accepts the promiscuity of various environmental aclimating regimes for eliciting osmotic adjustment the acceptance of turgor potential as the transducer is weakened by the fact that when one co-aclimates plants, that is, grow them at +5 C under optimum water, nutrient and all other conditions, they will respond and will be drought hardy. They will be salt hardy, but they have not experienced any loss of turgor or any stress. What is common across all three of these hardening regimes is that there is a suppression of growth.

J. Boyer: That's certainly true in our case. In view of the fact that even though this initial transitory growth takes place without the benefit of osmoregulation, it is not certain that these plants have never experienced turgor change.

P. Steponkus: If one uses a winter cereal, such as winter wheat or winter rye, which can adapt to low temperatures, and can also be drought hardened; and if the maximal increase in drought resistance, as manifested by an increase in solute potential, is elicited by cold aclimation, it is not feasible to argue that turgor changes are the transducers for cell dehydration.

J. Boyer: There is one other issue which I think needs to be brought out at this time. The idea of turgor-transducing system for bringing about changes in osmoregulation is quite popular in the algal systems. It is visualized by the algal physiologist as perhaps arising as a membrane compression against the external pressure which is close to atmospheric. But in higher plants, a positive pressure exists on the inside acting toward the cell wall and a negative pressure on the outside which makes up the remainder of the increment that can be generated osmotically. The pressure difference across the membrane is, in fact, a positive pressure on the inside and a negative pressure on the outside or roughly equivalent to the osmotic potential of the cell contents. Now when the environment brings about a change in the pressure (the turgor) in these cells there's a compensating rise in the tension on the outside. Consequently the pressure difference across the membrane changes very little. It's difficult to see how the algal system, where the outside pressure on the system is always atmospheric, could be applied directly to the plant system. Biological organisms have marvelous ways of doing things which I have not been able to outguess before and so I'm certainly willing to be proven wrong on this guess, but because of this constancy of the pressure, rough constancy of the pressure difference across the membrane, I am reluctant to think that the turgor pressure is really the transducing mechanism myself. So, in that sense, I'm in agreement with you.

R. W. Breidenbach: You're saying that there is no compressive force across the membrane.

J. Boyer: Well, I would say that there is a compressive force, but I would think that it would be roughly the same regardless of the internal turgor. A change in the internal turgor should be compensated roughly by a change in the tension on the outside. I would have to say roughly because this tension, of course, is generated in the xylem, and when a change in the rate of water flow through the plant takes place there is a change in the osmotic content of the xylem outside. Presumably part of the reason is that the roots pump solutes into the xylem at a certain rate and the water flow through the system is changed so that the xylem delivers solution to cells in the shoot at a different concentration. In other words, as the water flow rate through the plant slows down the concentration of solutes in the xylem stream increases.

R. W. Breidenbach: That really takes into account one of the hypotheses for a sensing element. The other one is Zimmerman's hypothesis. This hypothesis suggests that there is not only a pressure but an electrical potential across the membrane which may be quite different. There is a compressive force if there are two capacitor plates and a dielectric in between with a charge on either side. This compressive force tries to pull the two plates together but is resisted by a restoring mechanical force and that compression is the thing that Zimmermann, of course, is looking at in terms of electrical breakdown -- the electrical breakdown hypothesis.

The other pressure is much more difficult to understand, at least in my view. The only other thing that, it seems to me that one can possibly consider is some kind of volume sensing, that doesn't have anything to do with pressure. I can be really wildly speculative and say that if there are some receptor sites on the surface of the membrane and some molecules in the cell wall which interact with those receptors on the membrane then you would have a distance sensing element, basically, because the turgor drops a little bit and the membrane relaxes away at a molecular dimension and reduces the concentration of interacting elements between the cell wall and the membrane. There's absolutely no evidence whatsoever for this but it's certainly something that can't be excluded until somebody looks at the surface of the membrane.

G. Wyn Jones: There is some evidence of turgor-regulated fluxes in higher plants. There is, I think, quite reasonable evidence in *Halophila* which is a marine angiosperm, that there are turgor-regulated fluxes in that system. In John Cram's review he, in fact, suggests that there may be evidence of it in beet discs as well. But I don't think that the evidence is generally strong; as far as I am aware there is no evidence of turgor-regulated fluxes in any Graminae. There is some evidence that in some higher plants there are turgor-regulated fluxes which are analogous to the Valonia type situation in marine algae.

R. C. Valentine: I would like to ask Dr. Wyn Jones a question regarding his experiments dealing with the Koch's postulate where he added back the compound that is osmoregulating. That concept of a compound that is isolated from the cell and then the pure compound added back to the system to observe its function is common in microbial systems. Do you think these kinds of experiments, in general, are very useful with plants? If you were going to do several of these kinds of experiments, what would you do?

G. Wyn Jones: We did this simple experiment to test whether there was any increase in tolerance. Clearly, in many ways a very unsatisfactory system, because the compound is supplied externally when, in fact, presumably, the synthesis is internal, whether it's due to a decreasing growth or whether it's a response to a specific signal. Our motives were to determine whether any compound applied externally could increase tolerance. Clearly if one could produce a growth habit of some sort which increased tolerance this would be quite valuable in a number of contexts. I don't think, however, that this has any great commercial benefits for a number of reasons. First of all, the quantities that you would have to apply are quite out of any commercial realm to increase tolerance on any scale. It's just an interesting observation which is important because it suggests that there is some interaction between the cytoplasmic solutes and ion fluxes and therefore, it implies that there may be some interaction between cytoplasmic events and vacuolar events.

R. Jefferies: When you applied the betaines to the leaves had you any knowledge that this actually got into the leaves?

G. Wyn Jones: Yes, a certain amount gets in, but again I think that the rate of entry is very low and we tried various surfactants to get more to enter. It was done as a one off shot to see if it worked and not for any great biochemical purpose.

R. Jefferies: I think another point is that the applications of these substances externally through multicellular systems such as whole plants where it has to transfer through the root and into the xylem system creates a range of problems that doesn't apply when you're dealing with intact cells.

R. C. Valentine: Without agronomic value or economic value, do you think that this approach is really a reasonable thing to do at this stage with whole plants, plant tissues, or anything?

G. Wyn Jones: I don't think there's an answer to your question *per se*. I would not like to give the impression that betaine is a magic compound. I think that in terms of the whole plant response to salt stress, the ion relations, the control of ion fluxes and the distribution of ions between various compartments are probably more important than the ability to generate a specific cytoplasmic solute. We don't have strong evidence and I think that the group in Michigan State concur with the view that betaine is factor, that glycinebetaine is a factor, but there isn't really strong evidence that this is the crucial factor in salt tolerance in barley.

R. W. Breidenbach: Is there a reason why a relatively small number of compounds display this characteristic? Is there any meaning to the fact that there is a relatively small number of compounds found in many different organisms and there seems to be some specificity as to which compound is used?

G. Wyn Jones: There appear to be a number of classes of compounds that occur in organisms from the marine invertebrates to the algae. These include glycerol, arabitol, mannitol, sorbitol, and other polyol compounds. The evidence is fairly strong that the reason they can be accumulated at high concentrations according to the Schobert model is that they fit into the structure of water in such a way as they cause minimum perturbation. Amino acids and their derivatives are also related to stress phenomenon. Proline is a specific example and glycinebetaine is a specific example. These compounds occur in marine invertebrates and insects, in algae and higher plants. Now, the physics is far less clearly defined. It is quite apparent from biochemical experiments that enzymes can withstand high concentrations of these compounds. What the physical basis is for this property is a moot question. If you look at the physics of glycinebetaine it would appear to be a molecule of minimum interaction with water. It is a Zwitterion in which the point charge of the nitrogen has a slight hydrophobic sphere around it. There are a number of physical studies which show that as the hydrophobic shield is altered around the point charge, the interaction with water is affected. So my interpretation, again, is one of sort of a minimum perturbation situation. You can't afford to have something accumulated which is going to salt out all the proteins and precipitate the cytoplasm such as an ammonium sulphate type situation. Certainly, you can't have a situation where a compound would salt in everything and cause all the enzymes to be denatured. So, presumably, you're looking for the middle ground and one can only assume that these compounds operate in this way. There's quite decent evidence that sulphonium compounds act in this way as well.

R. W. Breidenbach: Is there anything more than serendipidy involved in the particular osmoregulating substances that organisms use to solve the problem? It seems that these sorts of questions should be explored if at all experimentally feasible.

G. Wyn Jones: Yes, very much so. I tend to feel that with the glycinebetaine present there are membrane interactions involved, however it is not clear whether this applies to proline. There is the concept of physiotypes which Albert and Popov have advocated and these compounds seem to delineate physiotypes. Precisely what the molecular basis of that adaptation is, I think, we should ask an ecologist.

R. H. Lawrence: Could I ask Dr. Wyn Jones a question? You made a comment about the importance of organic solutes and the importance of ion fluxes. I read the title and I see "Genetic Engineering of Osmoregulation: Impact on Plant Productivity." I come to the meeting and I hear Mary-Dell Chilton talking about agrobacterium as a possible gene transfer mechanism and I'm led to believe that things are going to be relatively simple. Some of that may be based on some over enthusiasm about the possibility of a proline plasmid that may be able to be transferred to plant cells and therefore have some sort of osmoregulatory action. And what I hear is that things are much more complicated than that and that really the definition of osmoregulation and an understanding of its ubiquity in plants appears to be of great significance. I wonder where the term genetic engineering got into this meeting because when you talk genetic engineering of plants, is it going to be eventually something that can be discretely pinpointed such as mutation selection work for the production of a specific organic solute? Does it seem realistic to take some of the bacterial data and imply that there will be osmoregulation imparted to plants where you can have proline overproducers or betaine overproducers? Can you quantitate and somehow select for this osmotic capacity that Peter Steponkus was talking about. And if not, I wonder where the term genetic engineering came in.

C. B. Osmond: The following is an experience based on a series of experiments done in the Waite Institute by Paleg and Aspinall and colleagues. In the early 70's these people suggested that a change in the proline content of barley genotypes was an indicator of drought tolerance of barley genotypes and the plant breeders were just delighted. They had something to select for. After about three or four years of close investigation, it was recognized that this was purely a fortuitous correlation. Gareth this morning told us how it comes about; rapidly stressed barley in the hands of plant physiologists not in the hands of the field micrometerology, produces a lot of proline when rapidly stressed. The breeders were completely disaffected; the physiologists were left wondering what they had in their hands and we got nowhere. We simply don't understand what's going on in higher plants with osmoregulation. I hope that our discussions here will help us see what some of the problems are and maybe go away and investigate some of them. As far as engineering osmoregulation, we don't know what we want to engineer at this moment.

Audience: We've been very diligently looking at the positive aspects of osmoregulation and I think would all agree that it is a benefit. Maybe we can ask a question. Are there any negative effects of osmoregulation in terms of higher plant productivity? I will restrict my question to just drought adaptation.

P. Steponkus: I think very definitely that there is a negative effect of drought adaptation. Plants have to pay a price for solute accumulation. If one takes the very simple observation of plants which have been drought adapted one will see that their productivity is going to be limited.

C. B. Osmond: We have to differentiate between survival and performance. In terms of performance osmoregulation costs a bit. Then we have to determine if we're interested in that product. Is it worth persisting a little longer at lower performance or pushing the system to the limits? Are we looking at the performance of plants in natural communities or are we looking at goals in terms of productivity? You specify the problems and we can then speculate on the way things are maybe related.

R. H. Nieman: With salt stressed plants the bigger the osmotic adjustment the plant has to make, the more the cost in energy and you can almost correlate the reduction in growth with the size of that osmotic adjustment. And anything you do to decrease the osmotic adjustment like increasing relative humidity on the shoot will decrease the osmotic adjustment by the shoot. Consequently salinity on the roots is affecting growth that much less. Or if you can just decrease the salinity in the root media, the plant has to make a smaller osmotic adjustment and growth is reduced less. So you can find cases where there is almost a linear relationship between osmotic adjustment and reduction in growth.

E. Epstein: I'd like to make a comment about the purpose of this discussion about proline and other indices of salt tolerance. It seems to me there is a lot of putting carts before horses here, because we pretend to know what to look for; for example, proline. It seems that what we first need to do is make the selections and find the genotypes that are tolerant, find the ones that are particularly insensitive and then evaluate how the plants respond. Mind you, I'm talking about a single species, not widely different plants. And then ask these plants, genotypes within a species, what do you know about living with all that salt that your brother under the skin doesn't. You should find significant differences that will tell you something about the osmotic adjustment. Do not pretend beforehand that we know and then tell the breeders what to look for; they may be disappointed. I want to ask a question of Peter Steponkus: unified concept, it says here, of stress in plants, I don't know, if it is a fair question. It seems to me to be one of those questions like have you stopped beating your wife? What are the stresses you are talking about? You're talking about salt stress; you're talking about cold stress; and you are talking about dehydration. In other words, you are talking about the stresses that basically mean dehydration and then you ask, is there a common concept? Of course, there is -- dehydration. But let's assume we're interested in plant nutrition, boron deficiency, for example, that's a stress. Now how does that fit in?

P. Steponkus: I hope that I made it clear -- that the title was an invited title and I think that anybody who would actively seek that title as a presentation has some latent masochistic traits, scientifically speaking. But what I did was respond to the charge. I don't necessarily subscribe that there is a unified concept to stress resistance. I think that one can demonstrate that there are some commonalities at the cellular level in terms of stresses. There are most certainly unique repercussions in terms of salt stress. One question that is not being addressed specifically at this conference in the area of salinity is why is salinity injurous. We have seen graphs and figures where growth is impeded. We have seen graphs and figures where death has been effected. But I can effect death in many, many ways. I can impede growth in many, many ways, but specifically why. And until we can answer the questions of why growth is impeded, why death is effected, I think it somewhat wasteful to try to figure out how we can prevent death. I think there are many enemies in a salt environment, and there are going to be many strategies for neutralizing these enemies. There may be some questions of concerns of fundamental understanding of osmotic adjustment. What are we expecting of this process? If one goes to drought resistance, the maintenance of turgor is the important consequence. If a plant that has osmotically adjusted is confronted with continued drought stress, what has that osmotic adjustment bought for that plant. It has deferred cellular dehydration, but of what order of magnitude? If osmotic adjustment, even though it defers the inevitable, stays static, it hasn't bought you much time. In the area of freezing resistance osmotic adjustment at one osmolal will buy you at 20 C the same volumetric reduction that you get with 0.5 osmolal at minus 2 C. So I think the fundamental question is what do you expect from osmotic adjustment? If drought resistant plants show osmotic adjustment is this the important factor? In terms of water potential, the favorite dogma of people studying drought resistance is that water moves in response to potential gradients. Do plants grow in response to potential gradients?

The fundamental effector of growth is water. If water cannot be removed in response to a water potential gradient, the plants are not going to grow regardless of the amount of photosynthate. Perhaps osmotic adjustment is important in first moving water, and maintaining water flux to the aerial portions of that plant and the osmotic adjustment that is observed in drought hardening is able to maintain the necessary osmotic disequilibrium to provide for these water fluxes.

C. B. Osmond: I think that the relative humidity, the activity of water in the atmosphere moves all the water that a plant needs to move. Let's be a bit careful about what we are attributing to osmoregulation.

P. Steponkus: It will move it through that conduit and right into the atmosphere. Whether it's going to be diverted into that elongating region is going to depend not on the atmosphere.

C. B. Osmond: A plant does grow by fixing carbon and carbon cost water and in all of these circumstances, I think. If we're asking what is osmoregulation about and what is any concept of stress, we are asking how long can that plant assimilate carbon. So all we can expect of osmoregulation is a few more days, maybe, depending on the circumstances, weeks, of normal type of gas-water relations in tissues. Now if we consider that survival of meristems is important, then the plant has to continue to assimilate carbon and get that carbon back to the meristem where it will be used for osmoregulatory purposes. This is some sort of protective function. If we're considering growth and the productivity of the organism, then the osmoregulation of the leaf presumably allows the leaf to continue photosynthesis to gain a few more days of carbon assimilation before it reaches the point when the stomata close or the photosynthetic apparatus because of a combination of circumstances is no longer able to assimilate carbon. It appears we are dealing with marginal benefits in osmoregulation and I don't know that yet we have the appropriate economic theory or information to really understand how these things should be put together.

J. Boyer: I might point out however that in the life of a corn plant a few more days during grain fill translates into somewhere between 20 to 40 bushels to the acre for the farmer. It is significant and rather surprising, but considering the plant's life cycle as a whole we are inclined to think that only a few days more or less isn't going to make that much difference. As far as reproductive success and also the quantity of that material that ends up in the grain, a few more days, particularly during grain fill, make a big difference.

Audience: I'd like to ask a question of you, John, or any one else on the panel. When would be the most effective time to study osmoregulation? At germination, John, as you discussed about today, at grain fill, what I would call partitioning of photosynthate between what is commonly harvested and what is not harvested, or would it be during the vegetative growth of the plant prior to reproductive physiology?

J. Boyer: Bob, I think that's a good question. One of the reasons that we're so interested in the germinating system is it's a very simple system. And I think in view of the unknowns in osmoregulation we want to start with as simple of a system as we possibly can. I however think that osmoregulation could play a significant role during other times of the life cycle. My hope is that it is understood and is a reasonably simple system. We could then proceed to the later phases of the life cycle and obtain a better understanding of what are the issues. I am not sure I can give you a much better answer than that, but that's really our attraction to this system and that is to attempt to understand the early phases of this stage.

R. H. Nieman: May I make a comment on osmotic adjustment to salinity? The main effect of salinity is on growth rather than on housekeeping operations of mature cells. The maximum effect is obtained when the plants have maximum growth. A young tissue like John is working with at the seedling stage may give you the optimum chance to see a response or do something to change that response, but there are some little problems because some plants differ with respect to their response to osmotic stresses and salinity. Some plants are more affected at the seedling stage, some in the vegetative or even in the fruiting stage. I think with soybean it's pretty much the same at all stages of growth, but in general, a maximum effect of osmotic adjustment will occur when there is maximum growth of the plant.

J. Raven: Can I make a point about turgor regulation. I think it might be important to distinguish between plants with turgor regulating and autoregulating activities which have no influence on how much water is available to them and those which are greatly altered by how much water is available in the future depending, of course, on how much it rains. I'm contrasting here things like marine, or particularly esturine algae, and sea grasses, including *Halophila* which we know turgor regulates very well. In those cases, what the plant does is not going to influence how much salt there is tomorrow, but the tide will still come in and there's the same absolute amount of water there. The extensive factor of how much water there is, is not going to be altered. The plant is simply adjusting its turgor to the instantaneous water potential there. The same may be true to a more limited extent of emergence halophytes, again, because plenty of water is actually in the soil and the plant can respond to the amount of salt in the soil inasmuch as that's altering water potential. I think that many of the crop plants, especially ephemerals, that we use come from fairly arid environments, and perhaps, may have a very careful balancing job to do to decide how much water to use today in case there's none left for tomorrow. This brings in the concept of altruism, but I still think it is important that in some cases precisely what the plant is doing today does influence how much water there will be left for tomorrow, granted that it doesn't rain again.

J. Radin: I would like to second that and would like to point out that there are some very successful crop plants in dry situations, (semi-arid areas) which do not osmoregulate at all, but rather are drought avoiders. Witness the soybeans or cowpeas and this again is drought avoiding. Closing their stomates is a strategy which buys time but loses productivity. In an irregular environment, it can be very successful, on the average, say over five years or so and either osmoregulation or drought avoidance would accomplish the same thing.

R. Jefferies: I'd like to add a point to something Barry mentioned this morning. Perhaps the most extreme case is Park Nobel's work on the CAM plant (Agave) which shuts down completely under drought conditions, and in fact has a higher water potential than the soil water potential. It is more positive than the soil water potential.

Audience: Assuming that I buy the idea that vegetative growth is the place to study osmoregulation and also assuming that we're talking about annual crop plants, where would be the most effective site to study osmoregulation? Would it be root ion fluxes, root xylem exudate, stem xylem exudate, osmotic potential of leaves, stomatal conductance, photosynthetic carbon dioxide uptake? There are probably others, I just listed those few. After all we are talking about research priorities in aspects of osmoregulation.

P. Steponkus: The highest priority should be studies in the vacuole, if you want to get right down to it.

Audience: In what way? Would you study fluxes? Would you study fluxes in the vacuole? Should you study osmotic potentials in the vacuole, everything in the vacuole?

P. Steponkus: I thought that the comment was where you would start and I would say that there has to be some understanding and if you want to get down to a very primary level of understanding, one should go directly to the vacuole as the arena in which one should work and most certainly you would want to consider fluxes; you would want to consider net accumulation. You'd want to consider transformations occurring within the vacuole. That may be the point where one should start and then with that foundation then try to integrate it into some organizational scheme as you progress up the organizational ladder.

I think that it depends upon what you are studying when osmotic adjustment is studied. Why do you study it? What do you want to do with it?

E. Epstein: I'm again moved to say that we pretend to know far more than we really know. Obviously we don't know the answers, so we have to ask them. And we have to ask them in an intelligent fashion. And the intelligent way to proceed is to select plants preferably within the same species so that they aren't different in everything and select those that are very sensitive to salinity stress and very tolerant to it and then ask the questions that you are asking, namely how do you alter these plants which are virtually identical in the geonme except that they differ in soil tolerance and sensitivity, respectively. How do they differ in these various processes we are discussing? That is the way to get answers, but not to pretend ahead of time that we know what we ought to ask the vacuoles in the root and what we ought to ask the vacuoles in the shoot. We don't know enough so let us first get at contrasting genotypes in only that respect, namely salt tolerance versus sensitivity, but preferably otherwise very similar in genome and then go about asking the questions. We are far to conceited and we ought to be far more humble and first get those contrasting genotypes and then ask the questions.

Audience: I agree, now let's assume you have them. What is the question you ask first? You personally, what would the question be you personally would ask first at that point?

E. Epstein: I probably would go to the root first, because it is immediately exposed to that stress, but I would ask all those other questions in my good time.

P. Steponkus: One might respond to that and ask would it be an efficient utilization of one's time to characterize two diversely contrasting genotypes even within the same species? Surely there are going to be many characteristics which are irrelevant to the particular trait that you are looking for. And yesterday in the discussion on microorganisms the inclination was to gravitate toward the escapes, the ones that were unique. There was little attention directed to the majority of the population or the genetic pool and the question of why are you hurting; why can't you go on further was not asked. I think that again in studies on cold hardiness and freezing injury, the investigations have been largely directly from a perspective of asking why are they different. I had a personal revelation when in 1966 Sakai published an article in NATURE where he froze plant tissues to -196. The initial reacton was, damn! Look at that! But when you stop and reflect upon it you say, if they can be frozen to -50 or -60, why not -196. What arises from these experiments, is a more interesting question and that is why can one plant cell type be frozen to -20, but not -21?

Audience: Perhaps this pre-empts Bill Rains' talk but perhaps one way to get a handle at the cellular level is to establish and select for those contrasting genotypes that you describe through mutation selection work that can be accomplished at the cell culture level and this would give you several populations or many populations of variance plus a wild type that you can compare at the cellular level. What has happended from the work of Nabors and others that have been involved in mutation selection work in culture where you don't need to go to something as complex as an intact plant where you can work at the cellular level and get these contrasting genotypes that you're talking about and then ask specific questions about the biochemistry of these specific mutants and genotypes.

Audience: I personally would be interested in each of the panel addressing what they think is a key research problem that needs to be solved. Just their comments; whether it's the most important in the world, I don't care. I would be interested in how you view the field and what needs to be done.

C. B. Osmond: If we're asking a question about the significance of osmoregulation in respect to plant production whether it be for food, or fuels, or chemicals we are really asking questions about photosynthesis. And I think that it's clear that we haven't yet done the types of experiments that establish how effective osmoregulation is in prolonging the photosynthetic life of plants in defined habitats. If I had two genotypes which differed in their capacity to osmoregulate, I would grow them side by side in a defined habitat and I would ask whether or not that capacity made any significant difference to the total production of the organism. This sort of experiment brings it close to the people who are really worrying about the question of productivity, the agronomist. It makes us aware of the potential limitations to yield of the display of photosynthetic surfaces and very much more complicated issues. As a physiologist, I have a whole different range of priorities that I would study and one of the more fascinating ones is the one Professor Epstein alluded to in his introduction. What allows a halophyte to absorb salt to the concentrations that it does? We have a suggestion that turgor and concentration in the vacuole appear to control influx in cells of higher plants. We know that halophytes do osmoregulate and withstand low water potentials and maintain turgor if their water potential is 2 to 3 to 5 times the concentration of salt in the cell vacuole. It seems to me an obvious question and a series of straightforward experiments will increase the understanding of the relationship between concentration in the vacuole, turgor and influx in halophytes on the one hand and nonhalophytes on the other. These are simple physiological research questions. Some of them may be related to problems of osmoregulation. If you worry about osmoregulation in the vacuole, which is after all the biggest osmoregulatory job, although not the most complex, but certainly the biggest osmoregulatory job, those experiments have to be done. This is one way of approaching the problem. But, again, some questions.

R. Jefferies: I would like to support Barry on this and discuss it further. Many of the characters in which the agronomists might be interested are polygenic so far as we know. And this seems to raise very big difficulties compared to what was being discussed yesterday. And it seems to me that another aspect to this is not just genotypic variation, but response of these genotypes under a range of conditions, not just one habitat. One of the key issues may be how the plants are moving their carbon and nitrogen around. One of the things I found on doing research on nitrogen metabolism was that there's really very little information about many aspects of nitrogen metabolism. Certain sections of the pathways have been intensively investigated but there are large components of common pathways in nitrogen that really receive very little attention in higher plants. It seems to me that particularly in natural environments or for that matter in cultured systems nitrogen is a major input into the system and we should pay a lot more attention to nitrogen metabolism.

G. Wyn Jones: It seems to me that if you follow the logic of what Barry was discussing concerning the high electrolyte concentrations in the vacuole and if we recall what John Raven was saying about the likely ionic composition of cytoplasms then we are implying that halophytic species that are able to osmotically regulate with salts, must be able to generate very specific and large gradients across the tonoplasts. I think that one of the most interesting things is to try and characterize the mechanisms across the tonoplast and possibly the chemical composition of tonoplasts which allow one to generate these gradients, and allow plants to actually take advantage of electrolytes in their environment. A simple calculation shows that if plants use sodium chloride as electrolytes in their vacuole, this is a hell of a lot cheaper than building up sugars and this sort of thing that the drought tolerant plants do. So I think that my research priority would be in fact to get out, purify tonoplasts and characterize these in plants with different salt tolerance. I think this may give some sort of criteria which will be useful for breeding stress tolerant plants.

J. Boyer: We know that the osmoregulatory phenomenon occurs, but it's clear from what we've heard today that essentially no one knows what initiates the signal in higher plants, nor what the metabolic regulatory acts are that foster the phenomenon. I would make those my first two research priorities -- to find a signal and to find a regulatory phenomena that controls the system, or enhances the system. The reason why I would make those my first two research priorities is because I think if we know what initiates osmoregulation and what the steps are in metabolism that control it, we then stand a much better chance of enhancing it -- enhancing it if this is a desirable thing from the standpoint of agriculture. And this brings me to my second research priority. I think that the benefit of osmoregulation needs to be demonstrated. We can surmise that it has some beneficial effects and I think it can be pointed to on a short term basis, but from the agricultural standpoint, from the standpoint of the entire life cycle of the plant we don't yet know what its impact is, and finally, I'd like to know how widespread the phenomenon is. We don't yet know even in a single plant why it seems to occur in some places and not in others and we certainly don't know across a wide range of species.

R. W. Breidenbach: Everyone has taken a very wholistic approach so far, and I guess I would like to preface my comments by saying that in view of some other comments that have been made earlier in the evening, I think it's important to present the philosophical view that confusion at this point doesn't necessarily mean that we should throw up our hands and turn our backs to the problem. In my experience most problems start out seeming fairly simple because we know very little about them. They become confusingly complex and ultimately simplify into some generalizations which can be exploited for benefit of whatever consumers there are. Certainly it is important to relate this to agronomic solutions. I guess I'll have to take a little bit of the reductionist view just simply to present more of a contrasting viewpoint and say that it seems to me that if I had those two plants, I would start to look at molecular relationships that seem logically potential signal receptors, primary lesions or mechanisms. Some of those were mentioned while looking at the vacuole, its membrane and its properties. I don't think I would stop there. I mean, the rest of the membranes in the cell may also be important and it is necessary to look at those under situations where we can see how real membranes, and real plants really respond to the various kinds of environmental influences. So, in specific terms, I have a whole list that I had at the end of my talk which seem somewhat important to me. We don't know anything about the cytoskeleton in plants, and we need to know something about how that operates and whether it has any role. Do these osmoregulating compounds interact with structures of the cell? The biophysics of the interactions between these compounds with proteins and membrane structures in multimeric complexes in cell I think are very important questions that will perhaps lead us over this hump of confusion to a point where we can start to make some simple generalizations. What are the effects of these kinds of variables on the membrane molecular order? How general is the hypothesis for a constant viscosity and what are the influences of these various environmental variables on the viscosity of membranes of plant cells. That's certainly not an inclusive list but a few things that I think are important, objectives to be attained before we can understand these problems and also may shed some light on other problems that we have not considered.

P. Steponkus: I think my approach would be obvious and it would ask a very fundamental question of what is wrong with the plant. I don't think that I would go to halophytes and ask why are you so superior because the plant may be superior for many reasons that may not have very much to do with why it's sick. So my approach would be first to very carefully define the focus of this meeting -- what constitutes salinity injury? Is it an osmotic stress? Is it specific in terms of ion specificity? I don't think there's going to be a well defined question. It's going to be a combination of both, but there still has to be some further definition, focusing of the primary question of what is the problem. I think that probably I would gravitate toward a cellular level initially with the proviso that I may not be able to answer the question directly for field applications at this time, but the efforts, the time that I invest are not necessarily going to be ill invested. They're going to provide a foundation upon which we can build.

J. Radin: May I ask the panel the agronomist question here? Is the complex strategy of osmoregulation in a crop plant any more effective than spraying them with agent orange to make them defoliate or maybe with white wash to reduce the radiation load on the plant?

R. H. Nieman: I would just like to continue with the original question that started all this and I concur with Pete, I thought nobody would bring that all up. What in the world does salinity do to plants? You know that's the first question. Like he was saying earlier in the day, the lesion that has hardly been mentioned may not be in this environment. You may be surprised but there are literally hundreds of plant and agricultural scientists around the world that don't believe that salinity has an osmotic effect on plants. They think that it's strictly an ionic effect. And this is really sort of an unusual environment for me to find this unanimity of opinion, this apparent unanimity of opinion that it's basically an osmotic phenomenon. But just, I think, the question you have to look for first is what's hurting, what in the world does salinity do to plants?

G. Wyn Jones: I don't think that I would subscribe to that view. I mean, I don't think that's what Pete said. He was saying that in fact it's very difficult to disentangle the ionic and the osmotic.

R. H. Nieman: Yes, I understood that, but we've been talking a couple of days now about osmotic effects, right, and osmotic adjustment. We really haven't even gotten into the possibility of ionic effects. I think that it's good to bring that point up. This is not a settled issue. One does not really know that it's totally an osmotic phenomenon. Undoubtedly they're both ionic and osmotic effects that work here.

G. Wyn Jones: Yes, one could easily make a scenario; for instance, where, in fact, the build up of ions in the cell wall significantly changes the elastic modulus. That would lead to a decrease in growth which should be both ionic and osmotic in its effect. We haven't in fact talked about the cell wall at all and its importance as a transducer of the osmotic potential into growth and the changes in the elastic modulus with salinity, clearly something which is hardly been looked at and is an important factor. And one can go on in this vein. One can easily see a situation where the build up of ions in the cell wall could lead to perhaps an increased leakiness of the membrane and therefore the cell would have to work harder to build up its osmotic pressure within the cell in order to maintain the same turgor. There are many ways in which one can see molecular interactions between ionic effects and osmotic effects and I would concur that it is extremely difficult to disentangle them. But I certainly wouldn't wish to interpret salt damage necessarily in totally osmotic terms.

R. H. Nieman: No, that was just the point I was going to make, Gareth. First of all, there is adjustment which may be an osmotic response. It could be a passive or an active response, but then there is accumulation of salt in the cell and with glycophytes that can't tolerate very much salt, I think that it is very injurious to them and then as you were saying, Gareth, the next thing that happens is the membranes get leaky and in my experience you can see this very easily by comparing things like a salt sensitive or a sodium sensitive plant like carrot with one that is more tolerant like beet. The beet can maintain its membranes and very comfortably adjust and not expend an awful lot of energy in the presence of salt concentrations that cause the membranes of carrot to go to pieces. Look at the salt glands of some of these halophytes that can tolerate saturated sodium chloride solutions of 6 molar. You put the roots of most glycophytes in a tenth of that concentration and they go to pieces. So here is the big difference and I think the other very crucial thing that needs to be looked at is what are the energy and carbon costs to the plant in terms of making the osmotic adjustment. Is the plant having to pay a big price for this? Survival comes first but you can almost make a direct correlation, as I was saying earlier, between the degree of osmotic adjustment a plant has to make and the degree the growth is decreased.

G. Wyn Jones: But again, not in halophytes. I mean, that's a glycophytic tale essentially, isn't it? If you're talking about crop plants, I certainly wouldn't argue with you. You know much more about it than I do, but if you look at *Atriplex* that isn't true. So, aren't we in fact hoping in the long run to get into a crop plant, or a useful productive plant the capability of using cheap electrolytes as osmotica so that the price you have to pay is relatively small.

R. H. Neiman: But you're still paying a price, and halophytes, by in large, are very slow growers.

C. B. Osmond: I would like to contest that. You have to specify the circumstances. You focus effectively on the problem of what is salt doing other than interfering, or possibly making water relations different in a nonhalophytic plant. I think that one of the most significant points is Manny's. I'm quoting Manny now, if I may please. Sorry to be formal earlier on. In that graph comparing *Lycopersicon cheesmanii* and *L. esculentum*, *L. cheesmanii* is doing well with a leaf salt content which is practically the same as the leaf salt content which kills *L. esculentum* and I think if that doesn't simply say that the essential problem in salinity tolerance is getting the salt in the shoot and putting it in the right place and if that's not the simple message, then I have a lot to learn; and that's true too. The question of compartmentation, the ability to get the salt first from the soil without interfering too much with what else the root has to do, getting it into the translocation stream without too much trouble getting it to the shoot and getting it into the vacuole of shoots without too much interference is significant. The compartmentation ability is the basis of the success of halophytes in high-salt, low-water activity environments. If we knew some answers to that question, I think we'd be a long way along the way. When the plant is unable to do that then your question comes in. What is the salt doing? OK. We have to ask question of the losers as well as the winners.

G. Wyn Jones: I'd like to support very strongly what Barry said. That would certainly be my interpretation of the situation. The compartmentation issue is the crucial one and getting osmotica into the vacuole is supported by considerable data that suggests that salt tolerance in those halophytes which absorb high levels of electrolytes depends on compartmentation. The problem has been, of course, basically a technological one. We haven't had the facility to get out plasma membrane and to get out tonoplast until very recently. And it is only now that the technological capability is available to get out purified membranes. The classical plasmamembrane method using PTA staining has been extremely helpful. Now that we've got protoplast methods and vacuole isolation methods there's a real possibility of doing decent biochemistry on the membranes of a variety of plants and this has really only occurred in the last few years. It seems to me this is a crucial technical development which should in time enable us to answer some of the questions.

R. W. Breidenbach: The point of intersection of the membrane at the molecular level is something that we need to look at very critically. Because you're talking about the ionic effects of monovalent cations, for example, causing some change in the permeability of the membrane and, in fact, in model membranes there is certainly some information that would suggest that as you increase the concentration of monovalent cations the area occupied by the phospholipids change and that kind of a change causes a change that results in a phase separation. At a phase separation there is a maximum compressibility of the membrane and that could in fact lead to leakiness. The model systems say it happens, but no one has really tested whether it happens with real membranes. I think that it's similar to the temperature transitions as expressed in the molecular order changes in membranes. If one takes the fatty acid composition of membranes and does a simple theoretical examination of that data, one could say that there would be no transition at all in the range of temperatures where apparently the molecular probe says something very important happens. Apparently in the very complicated plant membrane there is something different, there are properties that really need to be looked at at the molecular level. Exactly how these membranes respond to the ionic strain, the presence of other kinds of osmotic regulating compounds and a variety of other factors that we're concerned with is an important area of research.

SECTION V

APPLICATIONS FOR PLANT IMPROVEMENT

SELECTION OF SALT-TOLERANT PLANTS USING TISSUE CULTURE

D. W. Rains, T. P. Croughan, and S. J. Stavarek

Plant Growth Laboratory/Dept. of Agronomy & Range Science
University of California, Davis
Davis, California 95616

INTRODUCTION

Plants in nature have evolved a number of adaptive mechanisms to cope with the presence of salt in their environment. The most significant of these mechanisms entails actual tolerance by the plant of high levels of salt within the tissue. This "true tolerance" has been found to provide the greatest tolerance of salt, and is characteristic of most plants inhabiting extremely saline environments. These plants accumulate large amounts of ions from the soil, and maintain growth despite internal concentrations considerably higher than levels that are lethal to plants which utilize other strategies for coping with salinity (Levitt, 1972).

Plants able to grow and complete their life cycles in the presence of high levels of salt (300+ mM) are termed halophytes (Greek, "salt-plant"), and those lacking such tolerance are termed glycophytes ("sweet-plant") (Flowers et al., 1977). Halophytes not only tolerate salt but grow better in its presence than in its absence. Halophytes in solution culture attain optimal growth at salt levels ranging from 20 to 500 mM, depending on the species (Black, 1960; Flowers et al., 1977; Webb, 1966). The higher optimum of 500 mM is equivalent to the concentration of NaCl in seawater, where NaCl accounts for about 80% of the dissolved salts. Although many halophytes grow in the absence of salt, growth is reduced under this condition, is improved with the addition of salt, and is eventually inhibited by very high salt levels (Black, 1960). In contrast, the growth of glycophytes is generally optimal in the absence of salt, being inhibited by the presence of any significant amount of salt.

Halophytes and glycophytes when exposed to saline environments must contend with three basic problems: 1) maintaining favorable water relations; 2) coping with potentially toxic ions; and 3) obtaining the required nutrient ions despite the predominance of other ions in the external media.

Osmoregulation by Plants

To maintain favorable water relations despite high levels of osmotically active ions within the soil solution, the water potential within the plant must be maintained at a more negative value than that of the external environment. To accomplish this, the plant may produce its own organic osmotica (e.g., organic and amino acids) or accumulate ions from the external medium, thereby acquiring an internal concentration of osmotically active solutes sufficient to maintain water flow into the plant (Rains, 1972, 1979). The production or absorption of sufficient osmotica is metabolically expensive, potentially limiting the plant by consuming significant quantities of carbon that could otherwise be used for growth (Rains, 1978).

A representative sample of compounds which serve as organic osmotica in various biological organisms are presented in Table I.

Hiatt and Hendricks (1967) found that the organic acid content of excised barley roots increased in the presence of sodium. Osmond (1967) proposed that oxalate production was in response to excess cation accumulation in the vacuole. He determined that oxalate production was sufficient to account for 75% of the necessary charge balance. Williams (1960) suggested that there is a strong correlation between organic acid levels and the concentration of salt in the medium. Rains (1972) supported that and suggested that alterations in organic acid levels due to salinity in the environment may prove to be an important mechanism for coping with salt.

Numerous workers have reported correlations between altered nitrogen metabolism and increased salt tolerance when plants are exposed to salt. Rafaeli-Eshkol (1968) found that the harmful effects of salinity can be obviated by exogenous betaine. Storey and Jones (1975) suggested that betaine is not a waste product or ammonia detoxification compound but, instead, a molecule with a distinct physiological function in osmotic adjustment. Halophytes have been found to contain high levels of betaine, and exogenous betaine may enhance salt tolerance.

Similar results have been reported for proline (Thompson et al., 1966; Barnett and Naylor, 1966; Stewart et al., 1966; Singh et al., 1972). Stewart and Lee (1974) systematically examined the role of proline accumulation in halophytes. They found proline to be the most common amino acid in these plants, accounting for as

Table I. Microorganisms and higher plants in which the concentration of an organic compound (or compounds) increases in response to increasing external osmotic pressure, thereby possibly significantly contributing to osmotic adjustment.*

Organism	Compound
Bacteria	
(both nonhalophiles and halophilic forms)	amino acids (glutamate, GABA, proline)
Protozoa	
Miamiensis avidus (marine)	amino acids (glycine, alanine, proline)
Tetrahymena pyriformis (freshwater)	amino acids (glycine, alanine, glutamate)
Fungi	
Asteromyces cruciatus (marine)	polyhydric alcohols (mannitol, arabitol, glycerol)
Chaetomium globosum (terrestrial)	polyhydric alcohols (mannitol, arabitol, glycerol)
Dendryphiella salina (marine)	polyhydric alcohols (mannitol, arabitol)
Saccharomyces rouxii (osmophillic)	arabitol
Lichens	
Lichina pygmeae (marine)	mannosidomanitol
Microalgae	
Chlorella pyrenoidosa (freshwater)	sucrose
Cyclotella spp. (euryhaline)	proline
other diatoms:	amino acids
Dunaliella spp. (marine and halophilic)	glycerol
Monochrysis lutheri (marine)	cyclohexanetetrol
Ochromonas malhamensis (freshwater)	isofloridoside
Platymonas spp. (brackish/marine)	mannitol
Scenedesmus obliquus (freshwater)	carbohydrate (sucrose + rafinose, glucose, fructose)
Macroalgae	
Ascophyllum nodosum and Fucus spp. (marine)	mannitol
Iridophycus flacidum and Porphyra perforata (marine)	floridoside and isofloridoside
Angiosperms	
Atriplex spongiosa, Spartina townsendii, Suaeda monoica, (halophytes)	betaine
Chloris gayana, Hordeum vulgare (glycophytes)	betaine and proline
Various halophytes	
including Aster tripolium, Mesembryanthemum nodiflorum, Salicornia fruticosa	proline
Triglochin maritima	proline

*Adapted from Flowers et al. (1977)

much as 30% of the total amino acid pool. This, they found, was only true for halophytes, suggesting that proline accumulation is an adaptive response to salinity. Similar suggestions have been forwarded by Stewart and Boggess (1978).

Recently Hanson and co-workers (Hanson et al., 1979) have examined the role of proline in osmoregulation in plants. The exposure of plants to environmental stress such as drought does result in an increase in proline levels in the tissue of barley plants. When barley varieties with different tolerances to drought were compared however, it was not apparent that proline levels were positively correlated with the drought tolerance of these varieties. The unique role of proline as an osmoregulating compound is discussed in considerable detail in this volume (Csonka, Jefferies, Wyn Jones).

The alternative to producing organic osmotica is to accumulate a sufficient concentration of ions from the external medium. This poses the second problem in that such high ionic concentrations may interfere with normal biochemical activities within the cell (Poljakoff-Mayber, 1975).

The third problem faced by a plant in a saline environment is meeting the plant's nutritional requirement for certain ions (e.g., K^+) that are present at much lower concentrations in the external environment than other chemically similar but nutritionally unnecessary ions (e.g., Na^+). Halophytes, as mentioned, can absorb and tolerate high levels of ions. Furthermore, and again in contrast to the glycophytes, they selectively concentrate considerable quantities of K^+ as well as Na^+, despite the much lower concentration of K^+ in the environment (Rains and Epstein, 1967).

Plants exposed to salinity tend to absorb salt from their environment, varying from small amounts for the salt excluders to considerable quantities for the halophytes. In general, as the NaCl level in the plant's environment increases, both the Na^+ and Cl^- levels within the plant increase. Concurrently, the levels of Ca^{++} and K^+ tend to decrease (Waisel, 1972). It is of interest that at all concentrations of NaCl, halophytes not only tend to accumulate more Na^+ and Cl^- than do glycophytes, but also manifest little if any interference by Na^+ on the uptake of K^+. Furthermore, in the absence of salt, halophytes tend to accumulate higher levels of NO_3^- and SO_4^-, which may function to substitute for Cl^- in its absence (Flowers et al., 1977).

It is apparent from the previous discussion that many of the basic mechanisms associated with salt tolerance operate at the cellular level. Advances in understanding of these mechanisms will be enhanced considerably by studies of plant cells and their processes.

APPLICATIONS OF PLANT TISSUE CULTURE

Cell-culture techniques have proved useful in a broad spectrum of plant research areas. These applications range from basic physiological studies on enzymes and biochemical pathways at the cellular level, to large-scale vegetative multiplication of desirable genotypes at the commercial level (Scowcroft, 1977). The major areas in which the cell-culture approach has been used effectively are: plant biochemistry, breeding, plant development, disease control, and plant propagation.

Perhaps the greatest potential of cell-culture will be realized in its applications to genetics and plant breeding. Two aspects of plant breeding in which cell-culture is particularly applicable are the selection of mutants in cell cultures and the acquisition of haploid plants (Bottino, 1975). Haploid plants acquired through cell culture can be treated with colchicine to produce homozygous diploids for use as pure genetic lines (Jensen, 1974).

The utilization of haploid cells in cell selection programs is generally believed to enhance the mutant acquisition rate. The reasoning is that most if not all of the genes in such cells would be expressed, allowing selection for recessive mutations that might otherwise be masked by dominant alleles (Bottino, 1975). The issue is confused, however, by the unexpectedly high mutant acquisition rates obtained with diploid cells of fungi and animals (Scowcroft, 1977). The effects of ploidy on mutant acquisition rates in cell cultures of higher plants remains to be systematically studied, as is the effect of ploidy on the capacity to regenerate.

Regeneration of cell cultures that have been subcultured for more than a few months is currently a problem with many regeneratable crops, since their capacity to regenerate diminishes rapidly with time in culture. This problem is not apparent for all crop plants. Recently alfalfa, a difficult plant to regenerate from tissue after more than a few months in culture, has been regenerated from plant tissue cultured for over 3-1/2 years in our laboratory.

In examining the potential application of cell culture to plant breeding, an area showing considerable potential for contributing to the genetic variability available to plant breeders is the selection of mutant individuals from cultures of plant cells. This use of cell-culture is based on the application to cultured higher plant cells of techniques of mutant isolation developed in microbial and fungal systems (Bottino, 1975). Since the totipotency of individual cells has been firmly established (Nickell, 1977), each plant cell within a culture can be considered an individual organism. It is possible then, with appropriate techniques, to recover a mutation in any one of these large number of cultured plant cells. The general selective technique used is to grow cells that have or have not been treated with a mutagen on a medium containing a toxic substance at a concentration that normally inhibits the growth of all cells. Any cell that can proliferate under these conditions is considered a presumptive mutant. The proliferating cells are continually challenged on the media, and survivors are eventually regenerated into plants where subsequent genetic analysis can be done. The important point here is that mutations are selected at the cellular level (Bottino, 1975).

Although several experiments illustrate the tremendous potential of cell-culture as a plant breeding technique, this approach has certain limitations and problems. For example, only certain types of characters can be selected, and at present there appears no way to directly select for increased yields with cell-culture methods. Nonetheless, if it is established that the alteration or increase in quantity of a particular enzyme (a biochemical marker) would boost yields, cell culture could be used to select for mutants at this enzymatic level. Other problems include a difficulty in distinguishing between truly mutant cells and cells that have not changed genetically but have epigenetically adapted to the selection pressure applied. The final proof of a genetic basis for a particular cellular trait requires regeneration of whole plants from the selected cell lines, and subsequent physiological and genetic analysis at the whole-plant level.

Carlson (1973) used this approach in selecting mutant tobacco cells that were less susceptible to methionine sulfoximine, an analog of methionine related to the toxin produced by Pseudomonas tabaci. Plants regenerated from those resistant cells showed single factor semidominant inheritance of this character. Gengenbach et al. (1977) used exposure to the toxin produced by Helminthosporium maydis race T to select maize cells resistant to the toxin. Using Texas male-sterile (cms-T) maize, which is susceptible to this disease, they obtained resistant callus cultures which regenerated into resistant plants.

CELL CULTURE AND SALT TOLERANCE

The development of salt tolerant crop plants may ameliorate the growing problem of salt in the agricultural environment and the techniques of cell culture appear applicable to developing these salt tolerant crops.

Cell culture has several advantages as a technique for studying salt tolerance. It allows the processes or markers involved in salt tolerance to be characterized at the cellular level. The relative lack of differentiation in the cultured cells eliminates complications arising from the morphological variability and highly differentiated state of the various tissues of whole plants. The culturing of plant cells on rigidly defined culture media also permits a relatively uniform and precise treatment of these tissues with salt.

It is now well established that cell culture techniques can be used to select salt tolerant lines from nontolerant agricultural plants. Nabors et al. (1975) obtained tobacco cells that showed superior tolerance to 0.16% NaCl (w/v) and, subsequently, to 0.52% NaCl. Dix and Street (1975) selected lines of tobacco and pepper cells that grew in liquid medium containing up to 2% NaCl. Both studies measured growth as an increase in cell number or change in packed cell volume. Dix and Street observed that the packed cell volume tended to decrease after exposure to salt despite an increase in cell number. This was attributed to a decrease in cell size. The results of these studies demonstrate that selection for salt tolerance at the cellular level is feasible.

A number of plant species have been selected for salt tolerance through procedures of cell selection. Some examples are presented in Table II.

Table II. Examples of salt-tolerant cell lines selected through tissue culture.

Species	Callus or Suspension	NaCl Level Tolerated (w/v)	Reference
Nicotiana tabacum	Suspension	0.52%	Nabors et al.
Nicotiana sylvestris	Callus Suspension	1.0% 2.0%	Dix and Street
Capsicum annuum	Callus Suspension	1.0%	Dix and Street
Medicago sativa	Callus	1.0%	Croughan et al.
Oryza sativa	Callus	1.5%	(See Figure 1, Page of this volume)

In our laboratory we have obtained data from a comparative study in which a salt-selected line of alfalfa callus cells grew better in the presence of salt than did the unselected cell line from which the variants had been selected (Croughan et al., 1978). Exposing cultured alfalfa cells to an agar-solidified nutrient medium containing a typically lethal concentration of NaCl permitted identification of a cell line with an elevated resistance to salt toxicity. The tissue to be screened was increased in quantity through growth as a suspension culture. Inoculating salt-containing media in petri dishes with a slurry of these suspension-grown cells allowed this extremely large number of individuals to be screened for salt tolerance. The screening consisted of visual identification of the cells that maintained both growth and a healthy appearance despite exposure to high salinity. Consistent subculture to saline nutrient medium permitted identification of the most tolerant cell lines, ultimately resulting in isolation of the few salt-tolerant types present in the very large original population.

The salt-selected line of cells grew better than unselected cells at a high level of salt (1.0% NaCl), indicating that the selection isolated variant cells with increased capacity for growth in the presence of high levels of NaCl. This selected line additionally displayed other characteristics suggesting that the tolerance was the consequence of a shift toward a true halophytic nature. Besides tolerating high levels of salinity, halophytes actually require some salt, as evidenced by poor growth in its absence and growth stimulation when it is added (Black, 1960; Brownell, 1965). This same pattern is displayed by the salt-selected line, which grows optimally at 0.5% (85 mM) NaCl and poorly in the absence of salt.

A similar pattern was observed with rice cells selected for tolerance to salt. Two populations of rice cells, one selected in the presence of 1.5% salt and the other unselected, were grown for 42 days on media containing various levels of NaCl. The salt-selected line required the presence of 0.5% NaCl for optimal growth and sucessfully grew at 1.5% NaCl, a concentration that was lethal to the unselected line (Figure 1).

Another characteristic of halophytes is their capacity to accumulate certain ions to high internal concentrations (Flowers et al., 1977). Comparing the ionic content of the two alfalfa cell lines indicates that in the presence of salt the salt-selected line accumulated more Cl^- and K^+ than the unselected line (Figures 2 and 3). Particularly noteworthy is this capacity to maintain higher levels of K^+ in the presence of high levels of Na^+, since this trait is strongly correlated with salt tolerance in both halophilic bacteria and halophytic plants (Rains, 1972, 1979; Rains and Epstein, 1967; Brown, 1964). An elevated level of NO_3^- within the salt-selected line at the low salt levels is also consistent with

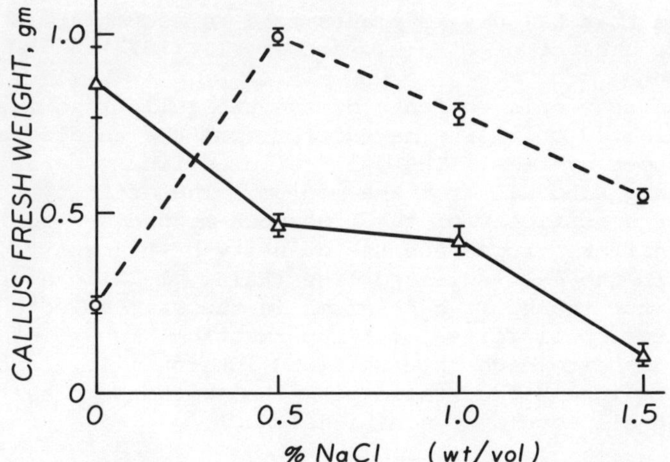

Figure 1. Rice callus fresh weight as a function of salt concentration in the medium, comparing the nonselected cell line (Δ—Δ) to the salt-selected line (O--O).

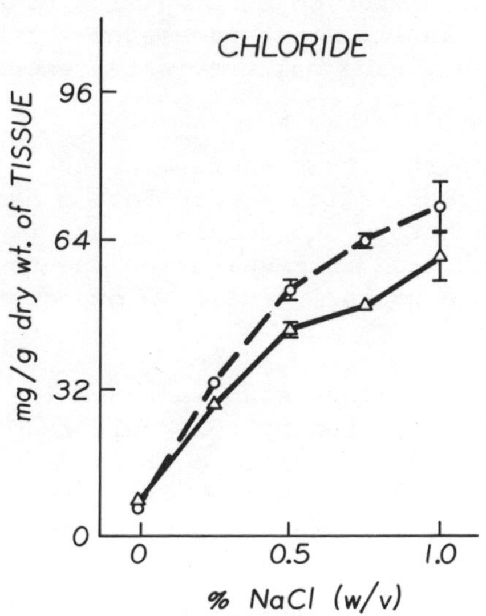

Figure 2. Chloride concentration in alfalfa callus as a function of salt concentration, comparing the nonselected cell line (Δ—Δ) to the salt-selected cell line (O--O).

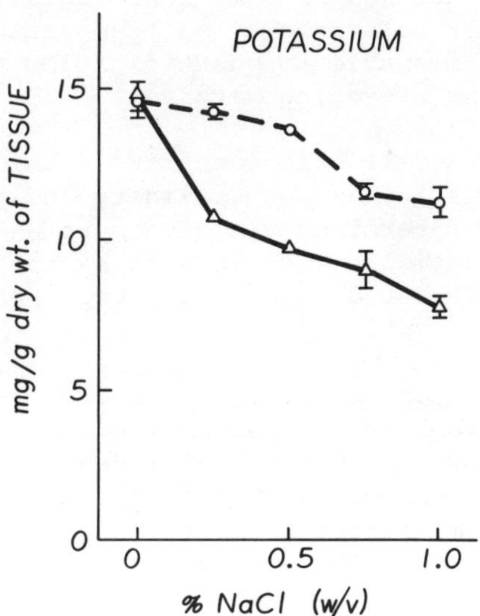

Figure 3. Potassium concentration in alfalfa callus as a function of salt concentration, comparing the nonselected cell line (Δ—Δ) to the salt-selected cell line (O--O).

observations that halophytic plants tend to accumulate NO_3^- when grown in the absence of Cl^- (Flowers et al., 1977).

Data on the ionic contents of the two cell lines suggest that the salt-selected cell line may differ from the unselected line in ionic transport systems. The salt-selected line generally accumulated more NO_3^- and Cl^- from the medium. The affinity for the substrate or specificity of the transport mechanisms for these anions may differ, or perhaps the quantity of transport mechanisms is greater in the salt-selected line (Rains, 1972). As to the Na and K ions, the transport mechanisms in the salt-selected line may show a greater specificity for K^+, permitting higher internal levels of this ion despite high external (and internal) levels of Na^+. This resulted in a higher ratio of K^+/Na^+ within the salt-selected cell line at all levels of additional NaCl.

DETERMINING THE ENERGY COST OF SALT TOLERANCE

Cell culture has the potential to expand observations on the effects of salt beyond simple quantitication of growth. Cell culture is also useful in identifying and quantifying physiological responses to salt -- such as energy cost, compartmentation, and production of osmotica. These, then, may be correlated with either salt tolerance or salt sensitivity. An important parameter that is amenable to quantification through tissue culture is the energy cost of coping with salt.

Cell culture offers a unique opportunity to determine the relation between salinity and plant respiration. Respiration can be determined via oxygen uptake both before and following the treatment of cells with salt. An increase in respiration due to salt would then be directly measurable as an increase in oxygen consumption.

One result of an increase in respiration is an increase in carbohydrate consumption. Glucose consumption by cell cultures provides an additional measurement of the metabolic cost of salt tolerance. As stated earlier, evidence leads us to believe that prime target areas for investigation of this nature are solute production, ion absorption, and compartmentation.

Measurement of glucose consumption is an indirect measure of respiration, allowing calculation of the energy costs for cells in saline conditions (Hunt and Loomis, 1976). All carbon skeletons and energy required for growth are derived from the glucose in the medium, and all the glucose input to the system can be accounted for by measuring glucose consumption and accumulation of dry matter.

Penning de Vries (1976) defines growth as dry weight increase, and quantification of the effect of salinity on this growth as a first step in understanding energy requirements. Also, changes in growth and growth rate from non-saline to saline conditions may not be immediately obvious as an increase in cell dry weight or CO_2 evolved. For example, a population of cells may divert carbon skeletons to the production of organic osmotica, thus increasing cell dry weight to maintain turgor, with no corresponding increase in population size. Likewise with the absorption of inorganic osmotica.

Further information could be gained, therefore, by characterizing the relative amounts of basic biomass constitutents (cellulosic materials, proteins, amino acids, lipids, organic acids, sugars, and minerals). From these, it is possible to describe basal cellular composition under non-saline conditions, and compare this with the cellular composition under saline conditions. In both cases it would be possible to calculate, and compare, integrated "production values" of the biomass produced under each conditions. Penning de Vries defines these production values as "the factor by which the weight of the organic substrate is multiplied in order to achieve the weight of a particular end product formed via a particular (least cost) conversion pathway."

Alterations in the energetics of cells exposed to salt can therefore be characterized both in terms of changes in relative amounts of basic biomass constituents, and in variation in production values. A third parameter to measure is the carbon dioxide production factor; defined as the weight of glucose needed to give the weight of CO_2 produced. Alterations in this factor would indicate an increased cellular effort towards what Penning de Vries terms "tool maintenance", the respiratory cost of maintaining synthetic systems (i.e., mRNA, enzymes, etc.).

Comparison of CO_2 production factors and productivity values as well as relative amounts of basic biomass constituents of control and salt-treated cells should elucidate the energy cost of tolerance of salt. Further comparison with cells cultured from salt-tolerant plants could also be made to further investigate the energy cost involved in tolerance.

Three cellular processes most likely to be involved in maintaining favorable water relations in saline media are: 1) compartmentation; 2) ion absorption; and 3) production of osmotica. These processes are part of the osmoregulation responses of plants exposed to stress. Determination of alterations in these processes will contribute to a better understanding of the energy costs and physiology of salt tolerance.

GENETIC ENGINEERING AND TISSUE CULTURE

The application of cell culture techniques in genetic engineering holds considerable potential for future plant improvement. Protoplasts from several species of plants have been fused following enzymatic removal of the cell wall and treatment with polyethylene glycol (PEG). The hybrid cells are plated onto media which favors the growth of the hybrid over that of either parental line. These hybrid cells are then regenerated into hybrid plants. This procedure could allow genetic combination of species which cannot be genetically combined through sexual crossing.

Genetic transformation studies with plants have involved treatments with both homologous (wild-type) and heterologous (foreign) DNA (Scowcroft, 1977). The correction of lesions in mutant plants following treatment with wild-type DNA has been accomplished, and while still equivocal, it appears that foreign DNA may be taken up, expressed, and possibly integrated into the genome of plant cells.

The application of such molecular gene manipulations as the use of insertion sequences and transposons offers the potential for extensively contributing to the development of new plant genotypes. This technology allows the transfer and integration of discrete units of genetic material, which may ultimately make possible the addition of desirable characteristics virtually at will.

The increasing problem of salt in the agricultural environment, particularly with irrigated agriculture, calls for the development of crop plants with improved tolerance to salinity. Cell culture offers a dual opportunity for both determining the physiological basis of this tolerance, while also posing the potential for the direct production of salt tolerant plants through the regeneration of salt tolerant cells.

REFERENCES

Barnett, N. M. and Naylor, A., 1966, Amino acid and protein metabolism in bermuda grass during water stress, _Plant Physiol._, 41:1222.

Black, R. F., 1960, Effects of NaCl on the ion uptake and growth of _Atriplex vesicaria_ Howard, Aust. J. Biol., Sci., 13: 249.

Bottino, P. J., 1975, Potential of genetic manipulation of plant-cell culture for plant breeding, _Radiat. Bot._, 15:1.

Brown, A. D., 1964, Aspects of bacterial response to the ionic environment, _Bacteriol. Rev._, 28:296.

Brownell, P. F., 1965, Sodium as an essential micronutrient element for a higher plant (*Atriplex vesicaria*), *Plant Physiol*., 40:460.

Carlson, P. S., 1973, Methionine sulfoximine-resistant mutants of tobacco, *Science*, 180:1366.

Croughan, T. P., Stavarek, S. J., and Rains, D. W., 1978, Selection of a NaCl tolerant line of cultured alfalfa cells, *Crop Sci*., 18:959.

Csonka, L. N., 1980, The role of L-proline in response to osmotic stress in *Salmonella typhimurium*: selection of mutants with increased osmotolerance as strains which over-produce L-proline, (In this volume).

Dix, P. J. and Street, H. E., 1975, Sodium chloride-resistant cultured cell lines from *Nicotiana sylvestris* and *Capsicum annuum*, *Plant Sci. Lett*., 5:231.

Flowers, T. J., Troke, P. F., and Yeo, A. R., 1977, The mechanism of salt tolerance in halophytes, *Ann. Rev. Plant Physiol*., 28:89.

Gengenbach, B. G., Green, C. E., and Donovan, C. M., 1977, Inheritance of selected pathotoxin resistance in maize plants regenerated from cell cultures, *Proc. Nat. Acad. Sci*., 74:5113.

Hanson, A. D., Nelsen, C. E., Pedersen, A. R., and Everson, H. E., 1979, Capacity for proline accumulation during water stress in barley and its implications for breeding for drought resistance, *Crop. Sci*., 19:489.

Hiatt, A. J., and Hendricks, B., 1967, The role of CO_2 fixation in accumulation of ions by barley roots, *Z. Pflanzenphysiol*., 56:220.

Hunt, W. F. and Loomis, R. S., 1976, Carbohydrate-limited growth kinetics of tobacco (*Nicotiana rustica* L.) callus, *Plant Physiol*., 57:802.

Jensen, C. J., 1974, Chromosome doubling techniques in haploids, *in*: Haploids in Higher Plants; Advances and Potential, K. J. Kasha, ed., Univ. of Guelph.

Levitt, J., 1972, Responses of plants to environmental stresses, Academic Press, New York.

Nabors, M. W., Daniels, A., Nadolny, L., and Brown, C., 1975, Sodium chloride tolerant lines of tobacco cells, *Plant Sci. Lett*., 4:155.

Nickell, L. G., 1977, Crop improvement in sugarcane: Studies using *in vitro* methods, *Crop Sci*., 17:717.

Osmond, C. B., 1967, Acid metabolism in *Atriplex*: I. Regulation of oxalate synthesis by apparent excess cation absorption in leaf tissue, Aust. *J. Biol. Sci*., 20:575.

Penning de Vries, F. W. T., 1972. Respiration and growth, *in*: Crop Processes in Controlled Environments, A. R. Rees, K. E. Cockshull, D. W. Hand, and R. J. Hurd, eds., Academic Press, London.

Poljakoff-Mayber, A., 1975, Morphological and anatomical changes in plants as a response to salinity stress, *in*: Plants in

Saline Environments, A. Poljakoff-Mayber and J. Gale, eds., Ecol. Series #15, Springer-Verlag, New York.

Rafaeli-Eshkol, D., 1968, Studies on halotolerance in a molerately halophilic bacterium: effect of growth conditions on salt resistance of respiratory system, Biochem. J., 109:679.

Rains, D. W., 1972, Salt transport by plants in relation to salinity, Ann. Rev. Plant Physiol., 23:367.

Rains, D. W., 1978, Adaptations of biological systems to enhance use of saline waters in arid environments, in: Microbial Conversion Systems for Food and Fodder Production and Waste Management, Kuwait Inst. for Scientific Research and Kuwait University.

Rains, D. W., 1979, Salt tolerance of plants: strategies of biological systems, in: The Biosaline Concept; An Approach to the Utilization of Underexploited Resources, A. Hollaender, ed., Plenum Press, New York.

Rains, D. W. and Epstein, E., 1967, Preferential absorption of potassium by leaf tissue of the mangrove, Avicennia marina: an aspect of halophytic competence in coping with salt, Aust. J. Biol. Sci., 20:847.

Scowcroft, W. R., 1977, Somatic cell genetics and plant improvement, in: Advances in Agronomy, Volume 29, N. C. Brady, ed., Academic Press, New York.

Singh, T. N., Aspinal, D., and Paleg, L. G., 1972, Proline accumulation and varietal adaptability to drought in barley: a potential metabolic measure of drought resistance, Nature Biol., 236:188.

Stewart, C. R., Morris, C. J., and Thompson, J. F., 1966, Changes in amino acid content of excised leaves during incubation. II. Role of sugar in the accumulation of proline in wilted leaves, Plant Physiol., 41:1585.

Stewart, C. R. and Lee, J. A., 1974, The role of proline accumulation in halophytes, Planta (Berl.), 120:279.

Stewart, C. R. and Boggess, S. F., 1978, Metabolism of (5-^3H) proline by barley leaves and its use in measuring the effects of water stress on proline oxidation, Plant Physiol. 61:654.

Storey, R. and Jones, R. G. W., 1975, Betaine and choline levels in plants and their relationship to sodium chloride stress, Plant Sci. Lett., 4:161.

Thompson, J. F., Stewart, C. R., and Morris, C. J., 1966, Changes in amino acid content of excised leaves during incubation. I. The effect of water content of leaves and atmospheric oxygen level, Plant Physiol., 41:1578.

Waisel, Y., 1972, Biology of Halophytes, Academic Press, New York.

Webb, K. L., 1966, NaCl effects on growth and transpiration in Salicornia bigelovii, a salt-marsh halophyte, Plant Soil, 24:261.

Williams, M. C., 1960, Effect of sodium and potassium salts on growth and oxalate content of Halogeton, Plant Physiol., 35:500.

BREEDING SALT-TOLERANT CROP PLANTS

Jack D. Norlyn

Department of Land, Air and Water Resources
University of California, Davis
Davis, California 95616

To date there has been a moderate amount of work done on the development of salt tolerant crop plants. Fuchs (1955) speculated that breeding for salt tolerance should be quite similar to breeding for drought tolerance. In the 1960's, Dewey (1962) developed a program for breeding salt tolerant crested wheatgrass, Epstein and Jefferies (1964) wrote on the genetic control of ion absorption and salt tolerance, Hunt (1965) demonstrated heritability of salt tolerance in intermediate wheatgrass, Ramage (personal communication) began his studies of developing salt tolerant barley in Arizona, and there were others, but not many. By now, there are programs under way for the improvement of the salt tolerance of several crops including alfalfa, barley, corn, grapes, and even walnuts. In the United States Department of Agriculture/Current Research Information System (USDA/CRIS) data base, there are about 30 active projects and 17 recently completed ones that include the topic of improving the salt tolerance of crops in some form of breeding program.

This recent increase of activity was stimulated by the realization that salinity is a very significant problem affecting fully one-third of all the irrigated land on earth, according to Maas and Hoffman (1977) of the USDA Salinity Laboratory at Riverside, California. It has a substantial economic impact on California agriculture, as was pointed out in a recent report (Time, March 5, 1979), in which the annual loss of crop production because of the salt buildup in the soils of the San Joaquin Valley of California was estimated at $32,000,000. In the Imperial Valley, where some crops have gone out of production because of increased salinity, a further annual reduction in yield of $7,595,000 is expected in the next 20 years due to the increase of salt levels in the irrigation water they receive from the Colorado River, (Moore, Snyder, and Sun, 1974).

Figure 1 represents the pattern of salt affected soils in California. The darkened areas in the outline map of the state represent the locations of these soils. In the eastern portion of the state along some of the coastal areas are naturally occurring salt sinks and tidal marshes. The larger area in the southeastern portion of the state represents the Imperial and Coachella Valleys. The extensive area in the central portion of the state is mainly the San Joaquin Valley, with some affected soils north in the Sacramento Valley (north of the city of Sacramento).

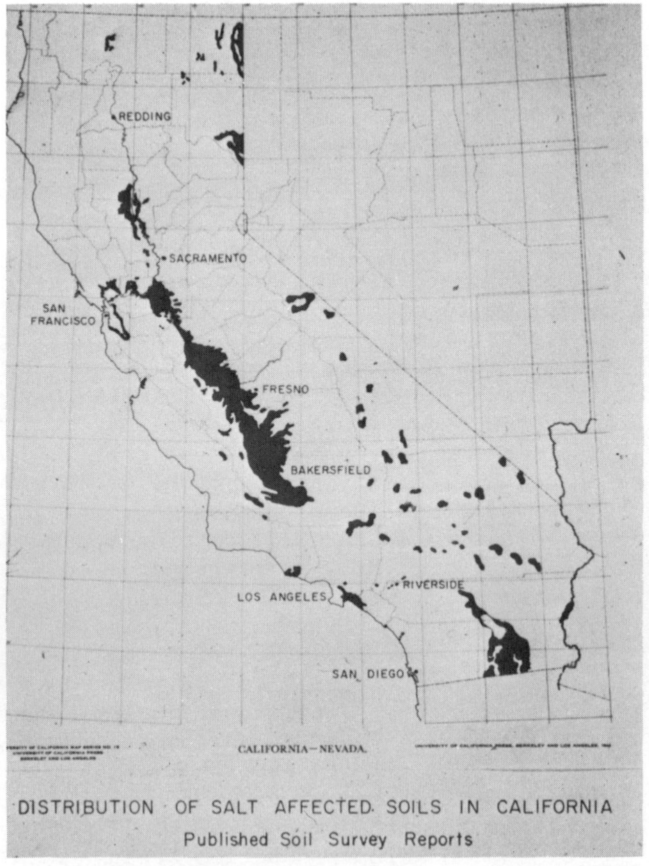

Fig. 1. Salt affected areas in California. The data represented here were compiled by G. L. Huntington, Department of Land, Air and Water Resources, University of California, Davis.

Fig. 2. Sorghum field near Stratford, California, in the San Joaquin Valley. The salinity in this area of the field prevents the growth of any plants.

The two main factors that contribute to this problem are the arid to semi-arid climate and the salt load in the water used for irrigation. According to Jim Walsh (personal communication), of the California Department of Water Resources, about 10,000,000 tons of salt are delivered in the water used in California each year, and 90% of this water is used for irrigation. In addition, the salinity of the water is expected to increase in several of the irrigation districts, particularly those using water from the Colorado River, according to Andersen and Kleinman (1978).

The impact on individual farms may be severe. Figure 2 is a picture of a sorghum field near the town of Stratford in the San Joaquin Valley. The salt buildup on the seedbeds has prevented the establishment of the crop in this area of the field. Many fields in California are spotted with saline areas like this, and many more are less severely affected, showing stunting of plants and yield depression. The total area of salt affected soils in California was estimated by Weir (1960, 1963) at 1,795,578 hectares (4,436,873 acres). Most of this area is in the agricultural regions and is variously affected with salt from a level that is just detectable by crop response to the point where nothing will grow. Also, the total area varies from year to year as some fields are improved by reclamation while others are abandoned after periods of improper management.

Reclamation procedures have been worked out by the USDA Salinity Laboratory at Riverside, California and other institutions (Scofield, 1940; Richards, 1965; and Van Aart, 1974). They include such procedures as leveling the land, providing surface and subsurface drainage (buried drain tile), applying soil amendments, and using improved irrigation practices to provide leaching. However, these procedures are not universally effective, they are expensive and they require continuous management control. As a result, complete reclamation is not to be expected; some degree of salinity will have to be endured in many areas. This realization logically leads to the idea of providing more salt tolerant crops. This may well improve yields on saline soils not fully reclaimed, and provide farmers with another management option to be used in conjunction with reclamation programs.

The original objective of the project I discuss here was to see whether the salt tolerance of an individual crop could be improved by selection and breeding. The first crop we chose to work with was barley, which is fairly salt tolerant. Also, it is easy to work with and is one of the better known crop plants in terms of genetics and physiology.

The first step in the growth of a crop is germination and emergence. By use of a system developed by Myhill and Konzak (1967) it was possible to compare many cultivars and lines of barley in the laboratory. Figure 3 presents a comparison of several lines of barley at 400 mM NaCl in the germination solution (the concentration of NaCl in seawater is approximately 500 mM). It is evident that there are differences among the lines in terms of their performance in this test. In addition, the line marked F_2, the progeny from a cross between the line marked "A" ('Arivat') and the line marked "CMS" ('California Mariout' selection), shows a mixed response. Seeds of line "A" are unable to germinate at this salt concentration while nearly all the seeds of line "CMS" germinate and emerge. The F_2 progeny appear to be segregating for this ability, some seeds not germinating, others only germinating, and still others germinating and establishing seedlings. This suggests that the ability to germinate and emerge under extreme salt pressure is under some form of genetic control, which we are now studying to see if the pattern of the control can be clearly understood.

In a similar test several hundred seeds of four cultivars were subjected to increasing levels of NaCl in the solution. Emergence was observed at the end of 21 days. The results are shown in Figure 4. 'Arivat' was the most sensitive cultivar of barley and showed the poorest performance in this test whereas 'California Mariout' performed best. The other two entries, 'Numar' and 'Briggs' were developed by Schaller and Prato (1968a,b) from two separate backcrossing programs at the University of California, Davis starting with a cross between 'Arivat' and 'California Mariout'. 'Briggs'

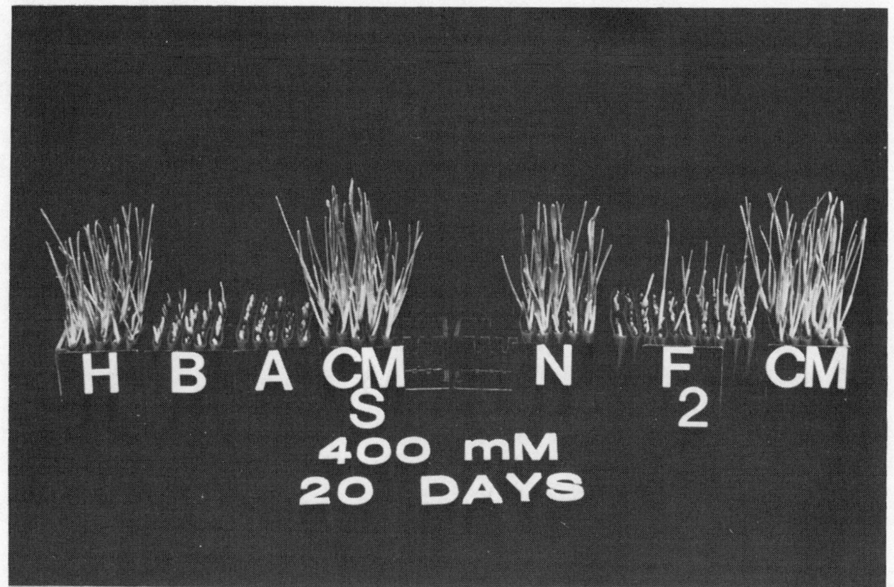

Fig. 3. Barley seedlings after 20 days in a 400 mM NaCl solution. Test line "H" was later released by the University of California as 'U. C. Signal,' "B" is the cultivar 'Briggs,' "A" is 'Arivat,' "N" is 'Numar,' "F_2" is the progeny of a cross between 'Arivat' and "CMS," and "CM" is the cultivar 'California Mariout.'

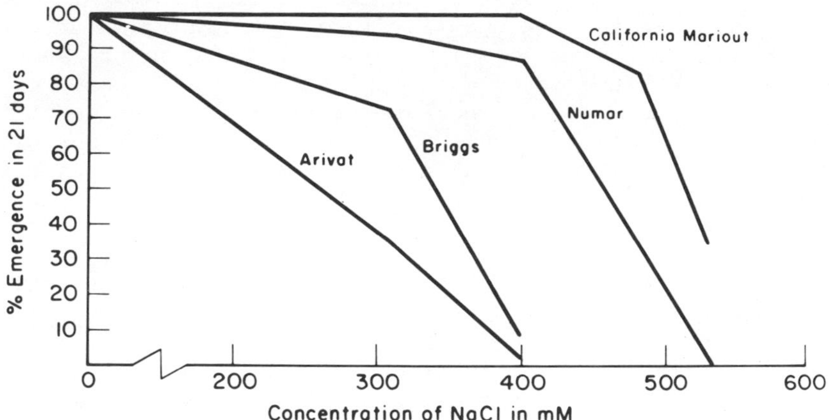

Fig. 4. Percent emergence of four cultivars of barley when subjected to different levels of NaCl in the germination solution for 21 days.

has 'California Mariout' two times in its pedigree and performs better than 'Arivat.' 'Numar' has 'California Mariout' four times in its pedigree and shows a better performance than either 'Arivat' or 'Briggs' in terms of this test. The evidence suggests that the more frequently 'California Mariout' occurs in the pedigree, the better the progeny is able to germinate and emerge under salt pressure. Some form of genetic control of germination and emergence is suggested by these data.

In further experiments it was found, in line with earlier evidence (Pearson, Ayers, and Eberhard, 1966; Akbar and Yabuno, 1974) that not only are there differences in salt tolerance among the various lines of barley, but that individual lines may also differ in their reaction to salt during the various stages of their life cycle; i.e., that tolerance at any one stage may not necessarily indicate tolerance at another stage. Therefore, if a particular line of barley is to be claimed to be salt tolerant, it must demonstrate this capability throughout its life cycle under salt stress.

In order to determine if the ability to yield under salt stress was under genetic control, 'Arivat' and 'California Mariout' were crossed, producing F_1 seed. This was grown out to produce F_2 seed. To provide adequate seed for testing in various ways, 88 of these F_2 seeds were grown out producing F_3 seed lots that were numbered separately so that the performance of the individual F_2 seed could be characterized by its progeny performance.

These individual F_3 seed lots were represented in a test in which the seeds were germinated and the seedlings established with a normal nutrient solution. As soon as all the lines were established, a stepwise salinization of the solution was begun, ending with 200 mM NaCl in the solution at the end of two weeks. The solution was maintained at that level for the remainder of the life cycle. Some of the same seed lots were subjected to the same type of test in the F_4 generation, and the performance in terms of seed yield in grams per plant was compared by regression between the two generations, as shown in Figure 5. The coefficient of correlation was 0.58, which was significant at the 5% level. This suggests that the ability to yield under salt stress is heritable, and that the genetic control appears to be fairly complex.

Several other comparisons were made among the individual seed lots. The comparison between the yield under salt stress in one generation and emergence under salt stress in the next was observed to have a coefficient of correlation of 0.12, which was not significant. This indicates that these two characters are inherited separately. Another comparison was between the rate of chloride absorption by six-day-old roots and the ability of the F_3 individuals to yield under salt stress. In this comparison the coefficient of correlation was 0.15 and not significant.

Another question is whether lines of barley respond similarly to different salt mixtures. The emergence values, shown in Table I, were observed in a Myhill and Konzak (1967) type test where one line of barley "CMS" was selected from the 'California Mariout' cultivar using the Rila Marine Mix salt (Rila Products, Teaneck, NJ) and the other was the unselected 'California Mariout.' When both lines are subjected to various levels of sodium chloride in the solution, the 'California Mariout' appears to be able to show better emergence. However, when the salt used is Rila Marine Mix, then the line "CMS" appears to do a little better. This suggests that the type of salt that the plants are exposed to may well make a difference in their response. Therefore, the type of salinity (the specific salts) that the plants are likely to encounter in the field should be the type used in the selection and breeding program.

To begin the program of selection, it was necessary to start with germination and emergence, as this is the necessary first step in crop production. To survey a large pool of barley germplasm, it was decided to screen the World Collection, which contains about 22,000 entries. The system devised to cope with the numbers involved is shown in Figure 6. It consists of 700-liter tanks, with plastic diffusion grates from light fixtures supporting cheesecloth on which batches of 35 seeds of each line are placed with a vacuum seed counter. The grates are covered with another layer of cheesecloth after seeding. Rila Marine Mix is dissolved in the solution to bring the concentration to 90% seawater. The solution is aerated, and water is added to bring the level of the solution up to wet the seeds. The cultivar 'California Mariout' is used as a check line in each tank; any line that germinates and emerges before 'California Mariout' is marked for selection. When the check line is well emerged, the salt solution is diluted gradually until it is near the concentration of salts in soil, at which time the seedlings are transplanted to soil filled pots and grown out for seed.

To insure that the selected plants are able to complete their life cycle under salt pressure, a second screening procedure is required. The seed recovered from the selected plants in the germination and emergence test is germinated at a reduced level of salts, such as 50% seawater, and the individual seedlings that survive are transplanted to the large tanks which contain the same level of salts, as well as nutrients, so that the plants can be tested throughout their life cycle. Figure 7 shows the survival of one plant, wheat in this case, in the test system.

Ralph Kingsbury, in our laboratory, has used this two step, sequential screening system on 5,000 lines of wheat and ended up with 42 lines that survived both steps and produced viable seed. Twelve of these lines appear quite good, and a few have been used by Dr. C. O. Qualset in his breeding program to improve the salt tolerance of wheat.

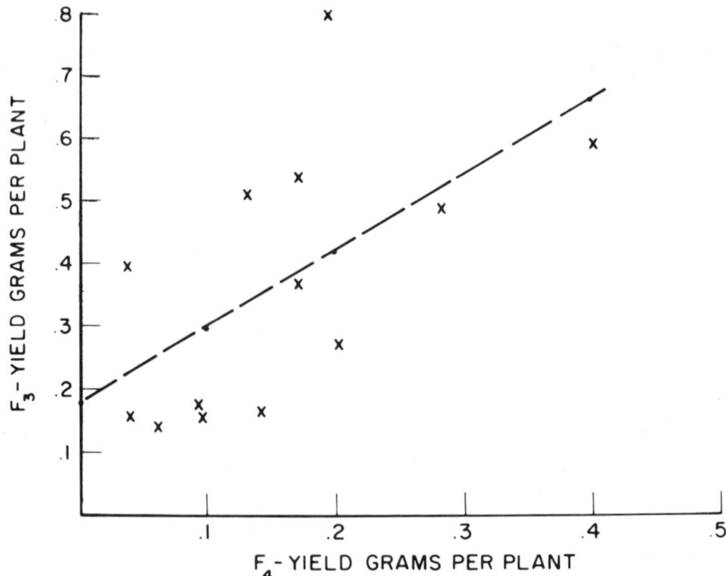

Fig. 5. Regression of yield for thirteen F_3 lines of barley on the yield for the same lines in the F_4 generation when subjected to 22 mM NaCl, after the seedlings were established in a normal nutrient solution.

Table I. Percent emergence using different salts and lines of barley EC: specific electric conductance, mmhos/cm.

a. Sodium Chloride

Cultivar	Control	0.40 M	0.48 M	0.53 M
'California Mariout'	100	95	76	30
"CMS"	100	99	49	0

b. Rila Marine Mix

Cultivar	Control	EC=30	EC=40	EC=50
'California Mariout'	100	94	92	0
"CMS"	100	100	100	2

Fig. 6. The screening system used in the greenhouse for subjecting large numbers of lines of seeds to a high level of salts during germination.

Our selection work was done with a salt mixture similar to seawater to explore the possibility of seawater culture of crops. Having had some success in selecting salt tolerant strains by the methods just described, we decided to try a field experiment to determine whether some yield was possible using seawater and dilutions of it on dune sand. The University of California maintains a marine laboratory near Bodega Bay, California, north of San Francisco. The Director of the laboratory, Dr. Cadet Hand, provided us with a suitable site. An aerial photograph of Bodega Head is shown in Figure 8. The Bodega Marine Laboratory is marked #1, #2 is the old seawater intake point, and #3 is the new seawater intake point in Horseshoe Cove. The location marked #4 is the student housing area near which sufficient space was available for our field experiment.

Figure 9 is a view of the test site looking east from the top of a dune. The cereal plots are visible in the foreground, with small plastic greenhouses for the tomato trials just behind. Tomatoes are unable to set fruit in that climate outside, but with plastic greenhouses and modest climate controls, fruiting is possible.

Fig. 7. A solitary wheat plant that has survived and set viable seed when subjected to 50% seawater salinity from germination to maturity.

Figure 10 shows three rows of barley irrigated 24 times with pure seawater during the 1977 crop year. That year was the second of two drought years, so very little rain fell on the crop during the season. However, after each rain, the seawater plots were re-irrigated to reduce dilution by rainwater, but subsequent studies have shown that the small amount of rain did have a beneficial effect during seedling establishment. Barley yields were low with seawater irrigation; the best line yielded about 1500 kg/ha. However, the fact that any yield at all was obtained was a surprise. For a more detailed discussion of the initial (1976) experiment, see Epstein and Norlyn (1977).

In addition to the program with cereals, Dale Rush, also of our laboratory, has been working with tomatoes, thus giving the

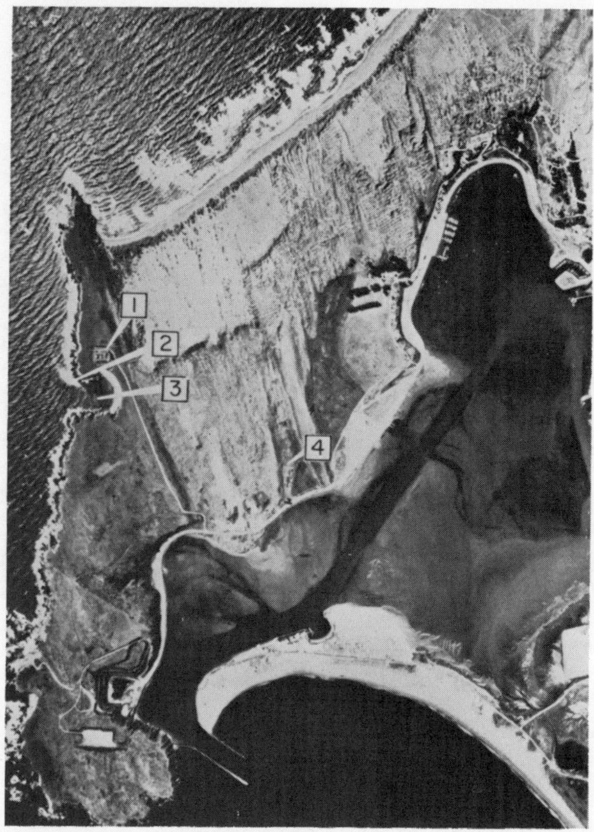

Fig. 8. Aerial photograph of Bodega Head, with the Bodega Marine Laboratory, 1, sites of seawater intake, 2, and 3, and the location of the field research site in a dune area, 4.

program a broader botanical base. Mr. Rush tested several commercial cultivars for salt tolerance. They were very similar in response, and not highly tolerant. Dr. Charles Rick, Department of Vegetable Crops at the University of California, Davis, has collected many seeds of tomatoes and related species from many areas of the world. He suggested that the material he collected from various plants in the Galapagos Islands might have the desired character (Rick, 1972). Seed from a wild tomato *Lycopersicon cheesmanii*, collected only 3 meters from high tide, produced plants that could survive with their roots in full strength seawater, or flower and fruit when subjected to 50% seawater in the nutrient solution.

The salt tolerance of this species is quite high, but it cannot be used directly, bearing tomateos that are commercially useless (see the upper right hand portion of Figure 11). The fruit is about 6 to 10 mm in diameter when fully mature, yellow to orange

Fig. 9. Field plot site in dune area near Bodega Head.

Fig. 10. Three plots of different barley lines subjected to undiluted seawater irrigation during the 1977 crop season.

Fig. 11. Fruit at the upper right are from the wild Galapagos tomato. The two large fruit at the bottom are from the domestic cultivar 'Walters.' The three fruit at the upper left are from offspring fo a cross between the other two lines, selected for high salt tolerance.

in color, tough skinned, and bitter -- hardly a saleable product. Fortunately, this species could be crossed with a large fruited, hand picked cultivar, 'Walters' (see the two large fruits just below the hand in Figure 11). The offspring, shown in the upper left of Figure 11, are the second backcross generation which has been subjected to screening for salt tolerance at three different times during the breeding program. Some of the plants derived from this cross have been able to produce promising fruit when irrigated with as much as 70% seawater. These tomatoes are the size of cherry tomatoes, red in color, and highly flavorful.

The survival results among the lines tested at Bodega are shown in Table II. Some of the Galapagos tomato plants are able to

Table II. Plant survival, in percent of total planted.

% Seawater	EC	Galapagos	Walters	F_2
0	2	100	100	100
30	16	100	100	100
50	26	50	33	83
70	34	17	0	67

survive when irrigated with 70% seawater, whereas the cultivars 'Walters' cannot. However, a high percentage of the selected F_2 progeny of a cross between the two tomatoes durvived. The data presented in this table for the F_2 tomatoes are somewhat twisted in that all the F_2 plants used in the field test were cloned from one plant that was able to survive under a very high level of salt stress. Without cloning of selected plants there would be no possibility of a reasonable comparison among treatments for F_2 individuals.

The reduction in yield when these tomato lines are subjected to increasing levels of salinity in the irrigation water is presented in Table III.

The Galapagos tomato survives but does not produce fruit during the summer in California because the day length is not right. This wild plant will fruit in the spring and fall when the day length is similar to that in the Galapagos Islands. The 'Walters' tomato did not survive at the highest level of salinity. The selected F_2 progeny produced fruit when irrigated with 70% seawater, though the yield was much reduced compared with that from the unsalinized controls (Table III).

The results of the analysis for total dissolved solids is presented in Table IV. The significance of total dissolved solids (TDS) is that these components make for flavor in the tomato.

Now that several barley, wheat and tomato lines can give some production using seawater or dilutions of it, where would these types of plants be useful if continuation of the experiments should prove encouraging? Meigs (1966) described the nearly 30,000 kilometers of the world's coastal deserts that are generally located in underdeveloped areas of the world. Therefore, if some form of crop production could be started in a few of these areas where nothing is produced at this time, it might make a contribution to those local economies. However, a good deal of additional experimental

Table III. Yield reduction, percentage of control -- fresh wt. basis.

% Seawater	EC	Galapagos	Walters	F_2
0	<2	---	100	100
30	16	---	17	52
50	26	---	8.7	43
70	34	---	0	15

Table IV. Percent total dissolved solids.

% Seawater	EC	Galapagos	Walters	F_2
0	<2	---	6.0	7.4
30	16	---	8.2	10.5
50	26	---	11.0	11.9
70	34	---	---	15.4

work needs to be done before farming the coastal deserts is anything more than an interesting speculation.

Because California is afflicted with salinity in a substantial portion of its agricultural soils we have started a program at the University of California, Davis, for the development of more salt tolerant crops for the inland situations in the San Joaquin, Coachella, and Imperial Valleys. This program includes breeders from the Departments of Agronomy & Range Science, and Vegetable Crops, in addition to our own group which will assist with selection programs and physiological criteria.

If this program is successful, and we are confident that it will be, based on the preliminary data that have been developed, it may provide a format for use by other countries of the world that suffer from soil salinity problems. The International Rice Research Institute (1974) reported that in many of the developing nations of the world, the use of more salt tolerant crops may be the only way to improve yields in salt affected areas in the near future, as the money required for the needed reclamation programs is not now available, and probably will not be available for many years.

The evidence that has been presented on barley, wheat, and tomatoes suggests that there is a significant amount of variation for the character of salt tolerance within the collection of seeds that have been studied. If this variation proves insufficient with any particular crop it is clear from the presentations at this conference that there are additional methodologies which can be brought to bear on the task of developing salt tolerant crops.

ACKNOWLEDGMENTS

Work from this laboratory referred to in this paper was supported by the National Science Foundation and Office of Sea Grant, NOAA, U. S. Department of Commerce. I thank E. Epstein for comments.

REFERENCES

Akbar, M. and Yabuno, T., 1974, Breeding for saline-resistant varieties of rice. II. Comparative performance of some rice varieties to salinity during early development stages, Japan J. Breed., 24:176.

Andersen, J. C. and Kleinman, A. P., 1978, Salinity management options for the Colorado River, Water Resources Planning Series Report P 78-003, Utah Water Research Laboratory, Utah State University, Logan, Utah.

Dewey, D. W., 1962, Breeding crested wheatgrass for salt tolerance, Crop Sci., 2:403.

Epstein, E. and Norlyn, J. D., 1977, Seawater-based crop production: a feasibility study, Science, 197:249.

Epstein, E. and Jefferies, R. L., 1964, The genetic basis of selective ion transport in plants, Ann. Rev. Plant Physiol., 15:169.

Fuchs, W. H., 1955, Züchtung auf Trocken und Salzresistenz, in: Report, Fourteenth International Horticultural Congress, Vol. I, H. Veenman and Zonan, Wageningen, Netherlands.

Hunt, O. J., 1965, Salt tolerance in intermediate wheatgrass, Crop Sci., 5:407.

International Rice Research Institute, 1974, Annual Report for 1973, Los Banos, Philippines.

Maas, E. V. and Hoffman, G. J., 1977, Crop salt tolerance: evaluation of existing data, in: "Managing Saline Water for Irrigation," H. E. Dregne, ed., Texas Tech University Press, Lubbock, Texas.

Meigs, P., 1966, Geography of Coastal Deserts, Arid Zone Research XXVIII, UNESCO, Paris.

Moore, C. V., Synder, J. H., and Sun, P., 1974, Effects of Colorado River water quality and supply on irrigated agriculture, Water Resources Res., 10(2):137.

Myhill, R. R. and Konzak, C. F., 1967, A new technique for culturing and measuring barley seedlings, Crop Sci., 7:275.

Pearson, G. A., Ayers, A. D., and Eberhard, D. L., 1966, Relative salt tolerance of rice during germination and early seedling development, Soil Sci., 102:151.

Richards, L. A., ed., 1954, Diagnosis and Improvement of Saline and Alkali Soils, Agriculture Handbook No. 60, United States Department of Agriculture, Washington, D.C.

Rick, C. M., 1972, Potential genetic resources in tomato species: clues from observations in native habitats, in: "Genes, Enzymes, and Populations," A. M. Srb, ed., Plenum, New York, pp. 255-269.

Schaller, C. W. and Prato, J. D., 1968a, Registration of Briggs barley, Crop Sci., 8:776.

Schaller, C. W. and Prato, J. D., 1968b, Registration of Numar barley, Crop Sci., 8:776.

Schofield, C. S., 1940, Salt balance in irrigated areas, J. Agric. Res., 61:17.

Time, 1979, Briny burden: peril for the San Joaquin, Time, March 5, 1979, 113(10):90.

Van Aart, R., 1974, Drainage and land reclamation in the lower Mesopotamian plain, Nat. Resources, 10(2):11.

Weir, W. W., 1960, Extent of saline and sodic soils in California, Soil Survey No. 14, University of California, Department of Soils and Plant Nutrition, Berkeley, California.

Weir, W. W., 1963, Extent of saline and sodic soils in eastern Fresno County soil survey area, Supplement to Soil Survey No. 14, University of California, Department of Soils and Plant Nutrition, Berkeley, California.

GENETIC METHODS TO BREED SALT TOLERANCE IN PLANTS

R. T. Ramage

Science & Education Administration, U.S.D.A.
Department of Plant Sciences
University of Arizona
Tucson, Arizona 85719

Increasing demands for plant products for food, chemicals and energy will require that many of the arid and semi-arid regions of the world be used for crop production. Soil salinity is a perennial problem in arid and semi-arid areas. Crops grown in these areas are quite often irrigated and irrigation frequently compounds difficulties with soil salinity. Also, in many irrigated areas, substantial amounts of brackish water are available to augment irrigation supplies. Soil salinity and the potential use of brackish water for irrigation in the arid and semi-arid regions of the world have created a great need for salt tolerant crops.

Consequently, much effort has been expended in recent years searching for crops and varieties that are salt tolerant. Much of this work has been done in countries that have extensive arid and semi-arid areas that are irrigated, such as the U.S.S.R., India, Pakistan, Israel and Egypt. There exists an enormous volume of literature pertaining to plant responses to salinity. The U. S. Department of Agriculture (USDA, 1978) has published an indexed bibliography that lists 2,357 references concerning plant responses to salinity. Many additional references may be found in the various lists of titles and abstracts such as Field Crop Abstracts and Plant Breeding Abstracts.

According to this volume of literature, over 1,500 species have been used in studying plant responses to salinity. Over 50 crop species have been evaluated for varieties that exhibit salt tolerance. These include cereal crops, fiber crops, forage grasses and legumes, both small- and tree-fruits, oil-seed crops, pulses, sugar crops, vegetable crops and trees (for reforestation and for streetside plantings where salt is used for deicing).

PLANT RESPONSE TO SALINITY

Different salts have been used in determining plant responses to salinity. In addition to NaCl, $CaCl_2$, KCl, K_2SO_4, $MgCl_2$, $MgSO_4$, Na_2CO_3, $NaHCO_3$ and Na_2SO_4 have been used, either singly, in combination with each other or in combinations with NaCl. Plant responses have also been studied by growing plants in saline and alkaline soils and by irrigating with water containing various salt solutions and concentrations. A number of studies report using seawater as irrigation water. Plants usually show greater sensitivity to single salts than to salt mixtures, probably because nutritional inbalances and specific ion toxicities are more likely to occur if one salt predominates under saline conditions.

Plant responses to salinity have been reported from plants germinated and grown in many different media, containers and treated fields. Seed have been germinated in nutrient solutions, petri dishes, growth pouches, blotter sandwiches in plastic boxes, and pots and trays in growth chambers and germination cabinets as well as in naturally and artifically salinized soils. Plants have been grown to various stages of maturity in nutrient solutions, growth pouches and pots in greenhouses and growth chambers and in saline and alkaline fields.

Growth stages where responses to salinity have been measured include imbibition, germination, emergence, seedling, tillering and mature plant stages. There is disagreement in the literature concerning the relationship of salt tolerance to the developmental stage of a plant. Some growth stages in some plants appear to be more tolerant to salinity than other growth stages. A number of experiments have shown that salinity may reduce the rate of germination but have little, if any, effect on total germination. Also, some plants have been reported to be tolerant at the seedling stage and susceptible at later growth stages while others are susceptible as seedlings but resistant at later stages. A number of authors insist that salt tolerance at germination gives the best measure of general salt tolerance on the assumption that, if the plants are to be grown in saline soils or with brackish water, a lack of salt tolerance for germination makes any potential tolerance of mature plants of little consequence.

A large number of measurements, or indices, of salt tolerance have been used. These include rate of imbibition measured by weight gain or by seed size increase, percent germination, rate of germination, seedling survival, seedling growth, fresh weight of seedlings, root and shoot growth, plant height, tillering capacity, leaf area, spike or panicle weight, number of seed per spike or panicle, and, actual field yields. Also, salt tolerance has been estimated by uptake of potassium, tetrazolium staining and electric potential.

Salinity appears to affect rate of germination more than total germination and rate of development or growth more than total growth

BREEDING FOR SALT TOLERANT PLANTS

Some authors have reported that hybrids are more tolerant than their parents while others report that tolerance is intermediate between the parents, is similar to the more tolerant parent, or that tolerance is similar to the less tolerant parent. Wild relatives of some crops are reported to be more tolerant to salinity than cultivated varieties while wild relatives of other crops appear to be more susceptible than cultivated varieties. Apparently, in the domestication and breeding of certain crops, such as carrot and celery, salt tolerance has been inadvertently bred into the cultivated varieties.

Salt tolerance has been correlated with both drought tolerance and winter survival. Usually, salt tolerant varieties have been found to be more drought and winter tolerant than salt susceptible varieties.

Very little actual breeding for salt tolerance has been done. Few deliberate efforts have been made to test potential parents for tolerance or to select tolerant plants in the segregating progeny of crosses. The development of the existing salt tolerant varieties has been mainly a matter of chance as most of them have been selected from among varieties that had been bred for other reasons. A few salt tolerant varieties have been intentionally selected from local land races. The literature records several attempts to isolate induced mutations for salt tolerance. Tissue and cell culture techniques have been suggested as means of selecting salt tolerant varieties (cf. article in this volume).

The purpose of plant breeding is to increase or stabilize yields of desirable plant parts or products. Yield and traits that tend to stabilize yield are characters that are under genetic controls. A fundamental tenet of genetics is that every characteristic of every living organism is the product of an interaction of a specific genetic constitution with a specific environment. In practicing plant breeding, we should keep in mind that the phenotype of a plant is the result of all of the genes of the plant interacting with each other and with the environment. We should appreciate that a successful variety is one that has a genotype that is balanced to accommodate the particular genetic characters of the variety. A successful variety is one that has a genotype that is balanced to accommodate the particular plant height, maturity, leaf width, chlorophyll content, salt tolerance, etc., of that variety.

An optimum balance exists between the action of a particular gene and its background genotype. Each character or combination of characters has a background genotype that is most favorable for the expression of that character or combination of characters. Our premise is that there is a most favorable background genotype for every character or combination of characters and that success in plant breeding depends upon selecting the background genotype that is most favorable for the expression of the particular combination of characters that a variety must possess (Ramage, 1977).

Any breeding program consists of two distinct phases: (1) generation of variation and (2) exploitation of the variation. In a conventional pedigree or backcross program, relatively little of the total effort goes into generation of variation -- most of it goes into exploitation, including yield tests. We believe that greater, and particularly more rapid, progress can be made if more of the total breeding effort is spent in generating variability and thereby reducing the amount of effort necessary for successful exploitation.

Two of the more serious impediments of a plant breeding program are (1) the amount of time that is required between cycles of effective selection and (2) the difficulty in maintaining sufficient genetic diversity, to allow for selection of rare gene combinations in a background genotype that is most favorable for their expression.

RECURRENT SELECTION FOR SALT TOLERANT PLANTS

Salt tolerance is an extremely complex trait and no single criterion of measurement will be adequate. High levels of salt repress seed germination and plant growth by both non-specific, osmotic effects and by the toxic effects of specific ions. Also, salt affects growth stages differently. The nature of salt tolerance for imbibition is probably very different than for germination, for seedling survival or for plant growth. In breeding for such an intricate character as salt tolerance, selecting a most favorable background genotype will probably be much more important than selecting the character itself. A salt tolerant variety will be of little value if it does not produce an acceptable yield of desirable plant parts or products.

Recurrent selection is the most feasible means of simultaneous selection for a character and its most favorable background genotype. Recurrent selection shortens the time that is required between cycles of effective selection and allows the use of the numbers of plants necessary to maintain the genetic diversity that is needed for the selection of rare gene combinations. Recurrent selection is concerned primarily with generation of variation and is used to develop populations and information for exploitation by other breeding schemes.

Recurrent selection is a breeding technique designed to accumulate or concentrate genes for a particular quantitative character in a population without a significant loss of genetic variability. The procedure was proposed and developed for corn breeding (Sprague and Brumhall, 1950). As the method depends upon obtaining rather large numbers of progenies from crosses, it has generally been considered to have the greatest usefulness for breeding cross pollinated crops.

A typical recurrent selection program begins with the establishment of a base population that contains genetic variability for the character that is to be selected for. From this base population, plants that are superior for the particular character are selected and selfed seed are obtained from them. The selfed progenies of the selected plants are grown in a plant-to-row design. All possible intercrosses are made among the selfed progenies. Equal quantities of seed of all of the intercrosses are bulked and the progeny grown for one or more generations to allow genetic recombination to occur. This completes the first cycle of recurrent selection.

The increase of the bulked intercrosses serves as the base population for the next cycle of recurrent selection. Again, superior plants are selected, selfed seed from them obtained, planted in a plant-to-row design and all possible intercrosses between progenies are made. The crosses are bulked and the progeny grown for one or more generations. The seed increase serves as the base population for the next cycle of recurrent selection.

Cycles of recurrent selection are repeated for as long as there is improvement for the selected character. The population is then exploited by selecting plants that will be inbred to produce lines that will be used in hybrids, synthetic varieties, composites, etc. This basic program has been modified and adapted to breeding a number of crops such as sorghum, sugar beets, forage crops, and others.

The typical recurrent selection program assumes that genetic variation for the selection character is limited to that contained in the beginning base population. Supposedly, a cycle of recurrent selection will be reached where there is no longer significant improvement for the selected character. The system also assumes that a population is to be developed, exploited and then discarded.

Recurrent selection will be most effective in breeding for characters whose expression can be modified as a result of changing the background genotype, transgressive segregation or accumulation of minor genes. In many instances, the character expression of a gene or genes can be modified strikingly by changing the genetic background of the gene (Ramage, 1977). Transgressive segregants are individuals that show a more extreme development of a character than was present in the parental material. They are generally assumed to be due to cumulative and complementary effects of genes contributed by the original parents. The use of a kind of transgressive segregation for increasing and stabilizing disease resistance has been proposed as the accumulation of minor genes. Salt tolerance is a character whose expression can probably be modified both by changing the background genotype and by obtaining transgressive segregants or accumulating minor genes.

The use of a typical recurrent selection breeding program is not feasible in breeding a self pollinated crop because of limitations imposed by the way self pollinated crops must be crossed. Genetic male sterility can be used to greatly facilitate crossing in self pollinated crops. The capacity for crossing permitted by the use of genetic male sterility allows the use of recurrent selection in breeding self pollinated crops. We have used modifications of the basic recurrent selection technique in breeding barley for such characters as salt tolerance, ability to cross pollinate, plant height, earliness, seed size, etc. We have called such breeding schemes "male sterile facilitated recurrent selection (Ramage, 1977b).

A male sterile facilitated recurrent selection program begins with the establishment of a base population that is segregating for both male sterility and the character that is to be selected for. Relatively large numbers of both male sterile and male fertile plants that are superior for the particular character are selected. The selected plants are crossed using male sterile plants as females. The crossed seed are bulked and the F_1 generation is grown. This completes the first cycle of recurrent selection.

As we have been using it, a cycle of male sterile facilitated recurrent selection differs from a typical recurrent selection program primarily in two ways: (1) we make crosses between selected plants rather than between selfed progenies of selected plants, and (2) we usually use a selected plant in only one cross rather than make all possible intercrosses between selfed progenies of selected plants. As selection based on an individual plant is not nearly as effective as selection based on a progeny row, we select relatively large numbers of plants in order to maintain a high degree of variability in the population. By selecting large numbers of plants and making only single crosses between them, we may not be making the most rapid progress in selecting for the particular character, but we maintain a very high degree of variability for other characters. This increases the chances of selecting the character in a more favorable background genotype. The high degree of variability for other characters also makes the population much better suited for continuous exploitation.

Just as in a typical recurrent selection program, the F_2 of the bulked crosses between selected plants serves as the base population for the next cycle of selection. Cycles of recurrent selection are repeated for as long as there is improvement for the selected character. Plants may be selected from the F_2 of any cycle and inbred for use as varieties, parents of hybrids, parents for conventional breeding programs, etc.

Another modification of a typical recurrent selection program that we employ involves adding new sources of germplasm to the base population. In our schemes, we may introduce additional sources of germplasm into the population in any cycle. The new germplasm may be for the selected character or for any other desired character, such as disease resistance, local adaptation, large seed, etc. Usually, the new germplasm sources will be male fertile and are crossed onto male sterile plants selected from the population. The number of crosses made depends upon how much we want to dilute the base population. Also, we intercross populations that have been selected for different characters to start new populations. In most of our populations, simultaneous selection for several characters is practiced.

As we use it, male sterile facilitated recurrent selection is an extremely flexible system. There is no limit to the amount of variation that can be carried in a population. The populations should continue to improve with each cycle of recurrent selection. Our populations are designed to be continued indefinitely and regularly exploited. This makes variety development from the population a continuous process.

The basic features that make male sterile facilitated recurrent selection effective are: (1) the generation and maintenance of a wide genetic base from which to select, (2) the ability to make the large number of crosses necessary to obtain rare gene combinations, (3) the opportunity to apply extreme selection pressure to large populations generated by the system, and (4) the shortened time between cycles of effective selection.

Male sterile facilitated recurrent selection produces populations that may be exploited in a number of ways. The simplest way is to select male fertile plants from the F_2 of any cycle or recurrent selection and carry them on in a conventional pedigree program. For this type of exploitation, plants may be selected from any cycle of the recurrent selection program, thus making variety development a continuous process. Another way to effectively use recurrently selected populations is to select plants from different populations to be used as parents in conventional breeding programs.

Techniques for breeding salt tolerant crops are relatively simple. Recurrent selection can be used to accumulate genes for salt tolerance in background genotypes that are most favorable for their expression. Populations developed by recurrent selection can be readily exploited by conventional breeding programs. The limiting component of a recurrent selection procedure is the selection of plants that are superior for the character being selected for. Rapid, accurate and inexpensive selection of a very few plants from a very large population is essential to a successful breeding program. Appropriate selection methods need to be developed for salt tolerance. Before better selection techniques can be developed, a better understanding of how non-specific osmotic effects and toxic effects of specific ions affect the different growth stages of plants must be gained.

REFERENCES

Ramage, R. T., 1977a, Male sterile facilitated recurrent selection and happy homes, Proc. Fourth Reg. Winter Cereals Workshop (Barley), Vol. II:92.

Ramage, R. T., 1977b, Varietal improvement of wheat through male sterile facilitated recurrent selection, ASPAC Tech. Bull., No. 37.

Sprague, G. F. and Brimhall, B., 1950, Relative effectiveness of two systems of selection for oil content of the corn kernel, Agron. Jour., 42:83.

U. S. Department of Agriculture, SEA, 1978, Plant responses to salinity: an indexed bibliography, ARM-W-6.

SECTION VI

NSF - CORNELL WORKSHOP

GENETIC ENGINEERING OF HALOTOLERANCE IN MICROORGANISMS: A SUMMARY

A. A. Szalay* and R. E. MacDonald**

Boyce Thompson Institute for Plant Research* and
Department of Biochemistry, Molecular and Cell Biology**
Division of Biological Sciences
Cornell University
Ithaca, NY 14850

We present here an outline and summary of a workshop sponsored by the National Science Foundation and held at the Boyce Thompson Institute at Cornell University, Ithaca, NY, entitled "Genetic engineering as an approach to the study of salt tolerance in microorganisms." The workshop was organized by Dr. Aladar A. Szalay in cooperation with Dr. R. E. MacDonald, Cornell University, and Dr. Janos Lanyi, Ames Research Center.

Green plants are the major source for food, biomass and energy conversion on the earth. It has been emphasized in many articles that a major limitation of plant productivity results from sensitivity of higher plants to high temperature and decreased humidity. They are also sensitive to increased salt concentrations; thus it is not possible to use seawater extensively for irrigation. This means that a very large part of the earth's surface is unavailable for cultivation because of the unavailability of fresh water.

It is not surprising that some plants and animals have learned how to live in moderately saline environments and that only microorganisms have adapted to extreme conditions of salinity. A few higher plants can tolerate the salt concentration of seawater (0.5 M or slightly higher as discussed by E. Epstein) and if special genes are involved in the mechanisms which permit these plants to tolerate higher salt concentrations, it might be possible to transfer these genes to commercially important plants and thus greatly increase the diversity of crops in arid areas. However, since such a small number of plants appear to be adapted even to low salt concentrations it seems unlikely that this can be accomplished within the foreseeable future by conventional plant breeding techniques. Thus, it may be more realistic to attempt to do this by genetic engineering techniques. It seems likely that the much greater diversity of genetic material available from microorganisms would make this task more feasible in the near future.

Physiologically, microorganisms fall into three groups relative to their degree of salt tolerance; a) those which grow without sodium ions but can tolerate and grow in concentrations up to 20% NaCl or more, b) those which require sodium ions for gorwth and can tolerate sodium concentrations up to saturation, but which grow best at low ionic strengths, c) those which _require_ high salt concentrations for viability and will not grow at concentrations much below 10% NaCl.

A direct result of increased salinity is a decrease in availability of water for cellular function; for non-tolerant organisms such as plants even low salinities are strongly inhibitory. Tolerant organisms respond to the increased osmotic pressure in one of two ways: 1) they synthesize low molecular weight compounds such as polyols or amino acids which counteract the effect of the high external salt concentration, 2) they exchange sodium ions for potassium ions in a highly selective fashion. It seems likely that an understanding of these mechanisms in microorganisms, especially in photosynthetic blue-green algae and green algae, will give us important basic information for studying salt-tolerance in higher plant cells and suggests that when sufficient information is available it may be possible to "engineer" these properties into plants of commercial interest and thus permit them to grow in areas of high salinity.

At present little is known about the mechanisms of salt tolerance in higher plants or single-celled eukaryotes. In microorganisms a considerable body of information already exists at both the physiological and molecular level and suggests that they would be preferable as material on which to test the idea that salt tolerance is a transferable character. Furthermore, microorganisms grow relatively rapidly in simple media and they are much more amenable to manipulation with both molecular and genetic studies at present than are higher plants.

The possibility exists that genetic information can be transferred using well characterized techniques of microbial genetics, i.e., genetic information may be mobilized, _in vivo_ by conjugation or, _in vitro_ by DNA transformation methods. The DNA can be isolated, sheared randomly, or cut with restriction endonucleases and ligated into cloning vectors. The ligated DNA can then be used to construct genetic "banks" from halotolerant microoganisms in some other organism such as _Escherichia coli_. Presently such transformation systems are available in gram negative and gram positive bacteria,

Neurospora, yeast and some mammalian cells. No such transformation systems have been found in plant cells. Therefore it seemed to us that experiments using isolated genes which might participate in osmoregulation might be carried out more effectively in microorganisms and we initiated this workshop to assess the needs for research which would make such an undertaking realistic. This workshop was held at the Boyce Thompson Institute at Cornell University, October 22 to 24, 1979.

The workshop was divided into seven sessions: The biology of salt tolerance; The biophysical basis of salt tolerance; The role of membranes in energy transduction; The molecular biology of salt tolerance; A genetic approach to the study of membrane function; Genetic engineering for increased salt tolerance in non-salt tolerant organisms; Applications of genetic engineering for production of useful organisms.

The first session, "The biology of salt tolerance," was chaired by Dr. H. Larsen, Universitetet i Trondheim, Norway, who opened the meeting with a general overview of the nature of halophilic and halotolerant organisms and introduced the first speaker Dr. R. MacLeod, of McGill University, who discussed moderately salt tolerant organisms, in particular marine halophiles. He pointed out that most or all of these marine bacteria have an absolute requirement for sodium and use sodium ions as co-transporters for accumulation of various nutrients. This requires a mechanism for the active extrusion of sodium from the cells which is replaced by potassium ions. This feature seems to differentiate them from the merely salt tolerant organisms which do not have a requirement for sodium ions.

Dr. J. Hellebust, from the University of Toronto, discussed moderately halotolerant algae with particular reference to their method of osmotic regulation. These algae all synthesize polyols such as glycerol or mannitol in response to increased salt concentrations. He pointed out that adaptation to higher salinities also brings about some changes in the functioning of their photosystems and protein synthesizing machinery.

Dr. A. Ben-Amotz, of the Weizmann Institute, described some of the properties of the extremely halophilic algae _Dunaliella_ which they have been growing in a pilot study. _Dunaliella_ synthesizes glycerol in response to increased external salt concentration and may contain concentrations of up to 70% of the dry weight of the cells. He discussed the pathway of glycerol synthesis and pointed out that this pathway is activated by external salt concentration. Dr. M. Shilo of the Hebrew University discussed mechanisms of adaptation of organisms to high salt concentrations and suggested that we still had a great deal to learn. In particular not all high salt environments in which organisms are found are high in sodium ions.

Some environments can be found in which high salinity, high temperature and high sulfide concentrations are simultaneously present and organisms have been isolated which are well adapted to these environments. He described some studies with a strain of Beneckia which was one of several strains growing in the Mediterranean Sea. It also grew in a hypersaline lagoon separated from the Mediterranean but none of the other strains could grow there. He found that this strain readily mutated to higher salt tolerance and readily lost this character when returned to low salinity.

Dr. D. Kushner, of the University of Ottawa, discussed the moderate halophiles with respect to their mechanisms of adaptation to moderate (ca. 10%) salinity; in particular he discussed modifications of proteins and enzymes and some of the effects of these modifications on the organisms. For instance in one case the adaptation to high salt concentration produced a sharp change in temperature optimum for growth.

Dr. Larsen then presented a discussion of the obligate or extreme halophiles. These organisms have a lower salt limit of about 2.5 M NaCl and other properties which appear to set them apart from the other halophiles, i.e., their proteins are denatured by low salinities, and cell membrane and organelle aggregates of proteins dissociate as salt concentration is lowered.

Dr. D. Oesterhelt, Universität Würztburg, chaired the second session of the day entitled, "The biophysical basis of salt tolerance." The first speaker was Dr. J. Lanyi, who discussed the biophysical aspects of adaptation to high salt and in particular some of the thermodynamic properties of proteins of organisms adapted to high salinities. He emphasized that hydrophobic bonding appeared to be an important factor in protein stability under these conditions. He also pointed out that even though many of the proteins of these organisms are acidic, the requirement for 3.0 M or more sodium ions cannot be explained by charge shielding since this would be complete at 0.5 M or less.

Dr. B. Schobert, of Technische Universität, München, discussed a model for the mechanisms by which polyols and proline can act as osmoregulators. She suggested that polyols resemble structured water in their molecular conformation. Thus, under conditions of water stress they serve to replace the function of the hydroxyl groups of structured water in maintaining protein solubility. Proline on the other hand, increases the solubility of proteins in water stress conditions by associating with hydrophobic regions of the molecule, rendering it more hydrophilic and thus more water soluble.

The next session entitled, "The role of membranes in energy transduction," was chaired by Dr. E. Racker, of Cornell University.

The first speaker was Dr. L. Packer, of the University of California, who discussed the effect of surface charge on ion pumps and presented some very interesting data obtained with some new kinds of probes for examining surface charge and charge translocation across membranes. In particular, he applied this technique to a study of H^+ translocation in the purple membrane of an extreme halophile, *Halobacterium halobium*. These probes are spin labeled weak base anions and weak base cations, Tempamine and a carboxylic acid, Tempacid, and with them he can get an accurate measurement of the internal pH of vesicles or cells. Surface potentials can be measured with another kind of spin probe that his associates synthesized. These probes have a hydrophilic head and a hydrophobic tail and depending on the length of the hydrophobic tail, the tenacity of association with the membrane varies. A change in surface charge on the membrane will change the signal obtained from this probe and thus it can be quantitated.

Dr. R. E. MacDonald, of Cornell University, was the next speaker and he discussed some of the kinds of ion fluxes that are induced by illumination of membrane vesicles prepared from *Halobacterium halobium*. Bacteriorhodospin, upon illumination, ejects protons which recycle via a Na^+-H^+ antiporter and thus serve as an energy storage system, or can be utilized to accumulate amino acids or other nutrients via a co-transport mechanism. He also described evidence for a second pigment in this organism which serves as a light driven primary sodium pump. This sodium pump differs spectrally from bacteriorhodospin and does not appear to require the movement of any other ions in the ejection of sodium from the vesicles.

Dr. A. Cazzulo, of Universidad Nationale Rosario, Argentina, presented a discussion of the properties of several enzymes which occur in organisms adapted to several different salt concentrations. Malic enzyme was examined in detail in non halophiles, moderate halophiles and extreme halophiles. He demonstrated that the sensitivity to increased ionic concentration was the same in all organisms but that the concentration at which it exerted a maximum effect varied depending on growth conditions. This was reflected in a changed Km for substrate, and Vmax. He concluded that the enzyme was similar in all three kinds of organisms but differed in the degree of adaptation to increased salt concentrations.

Dr. Oesterhelt of Ann Arbor discussed the role of ferredoxin in the metabolism of the extreme halophiles. Ferredoxin replaces NADH and NADPH as the electron acceptor in the citric acid cycle in these organisms. It constitutes about 1% of the protein of halobacteria and can be readily isolated in water. It is inactive in the chloroplast system commonly used for assaying ferredoxin from other organisms. It is a two sulfur, two iron protein similar to spinach ferredoxin and in fact has an amino acid sequence much more homologous with spinach ferredoxin than any other ferredoxin isolated to date. However, it has a unique amino acid sequence on the N-terminal end which has 60% acidic amino acid residues and this, he suggested, might be the additional piece which induced halophilicity in this protein, and could certainly account for its inactivity in the chloroplast system since its surface charge would be too high to allow it to interact at low ionic strengths. He also presented a detailed study of the mechanism by which the ferredoxin participates in the decarboxylation of pyruvate and described a novel system not involving lipoic acid as in all other systems known but the iron sulfur centers of the enzyme and ferredoxin.

Dr. R. Spanswick, of Cornell University, discussed some of the physiological differences between higher plants and bacteria and algae. He pointed out that one of the most important functions plants have is to maintain turgor pressure and thus to be able to transport water through their tissues. There must be specific ion transport mechanisms in plant cell membranes to do this and in particular mechanisms to move sodium and potassium ions but these have not been described very well. It is clear however, that plants can extrude sodium both to the outside of their cell membranes and into their vacuoles. He sounded a note of caution in expecting a solution to problems of water stress in higher plants by attempting to modify them with ion regulating systems derived from simpler organisms until the physiological control mechanisms in plants are better understood.

The next session entitled, "The molecular biology of salt tolerance," was chaired by Dr. D. Kushner who outlined some of the earlier work in the area which had been done by several people who were unable to attend the workshop. In particular the pioneering studies of Dr. S. Bayley, of McMaster University, who studied the protein translation mechanism of the extreme halophiles and concluded that even at these very high salt concentrations there was little or no misreading of the genetic code. He then introduced Dr. M. Yaguchi, of the National Research Council, Ottawa, who described some of the work his group had done on the amino acid sequence of the ribosomal proteins of the extreme halophiles. They have studied one pair of proteins, L7 and L12, in some detail and compared the amino acid sequences to similar sequences in the same proteins from 13 other organisms. The first 38 amino acids in the sequence were compared, and show that a high degree of homology exists between the extreme halophiles and methanogens which are non halophilic. This suggests that the very acidic character of the amino acid sequence may not be related to the halophilicity as had been suggested by many other researchers.

Dr. P. Fitt, of the University of Ottawa, discussed the radiation sensitivity of members of the extreme halophiles and suggested on the basis of his observations that these organisms may lack or have extremely inefficient excision repair mechanisms in the dark.

The next session dealt with the topic, "A genetic approach to the study of a membrane function," and was chaired by Dr. W. Epstein from the University of Chicago. The first speaker was Dr. J. Weber, from the University of California, Berkeley. Dr. Weber discussed the unusual genetic traits of halobacteria. He found bacteriorubrin and bacteriorhodospin mutants which occur spontaneously at frequencies of 10^{-2} to 10^{-3}, which is unusually high among microorganisms. He studied the reversion frequencies of these mutants and found the same high frequency. He concluded that genetically, halobacteria are exceptionally unstable, and suggested that this instability is under some kind of genetic control.

Dr. G. Weidinger, from the Universität Würzburg, discussed the relationship between extrachromosomal DNA and phenotypic changes in extreme halophiles. He characterized the plasmid DNA using restriction-endonucleases and found that certain segments of the plasmid DNA appear to be transposed in various phenotypes. Homologous sequences to the missing segments can be recovered from chromosomal and plasmid DNA of these phenotypes suggesting that DNA transposition may be related to phenotypic changes. He also reported that cloned fragments of the DNA of these plasmids can be translated in mini-cells of *E. coli* although no functional protein was isolated.

Dr. T. Silhavy, Frederick Cancer Center, discussed studies on the maltose transport system in *E. coli* using β-galactosidase gene fusion techniques. Fused proteins can be identified based on the β-galactosidase activity, and antibody prepared against β-galactosidase allows the isolation of the hybrid proteins. He reported that he had localized one of the fused gene products as an outer membrane (receptor) protein of *E. coli* and presented evidence for the identification of the gene directing proteins out of the cell. He described mutants which prevent exportation of proteins and discussed the importance of the leader sequences in protein transport.

Dr. W. Epstein, from the University of Chicago, described two potassium transport systems in *E. coli*: a) a very complex system regulated by six genes scattered over the *E. coli* chromosome, and b) a well defined system which has high affinity for potassium, and is repressible. The second system is regulated by four genes, three which code for inner membrane proteins and one which codes for a regulatory protein. This system is ATP driven. Further, he discussed the effect of proline as an osmoregulator in this system.

The section, "Genetic engineering for increased salt tolerance in non salt tolerant organisms," was chaired by Dr. G. Fink, of Cornell University. Dr. Fink emphasized the importance of developing systems in halophilic microorganisms for transformation similar to those in E. coli and yeast. Dr. Fink suggested an effort should be made to attract young scientists trained in laboratories where modern genetic and molecular biology techniques are practiced to the field of salt tolerant microorganisms. He also emphasized the absolute necessity of developing the molecular biology of a plant cell system including a DNA transformation system for plants.

Dr. D. Oxender, from the University of Michigan, described the structure of the isoleucine-leucine-valine transport system of E. coli. After identification of the mutants deficient in this transport system, he cloned the total E. coli DNA into lambda phage. Specific sequences of this cloned DNA were shown to complement these transport deficient mutants. Genes coding for two binding proteins were identified and excised from the lambda derivatives and subcloned into plasmids. These plasmids cause a five- to ten-fold overproduction of binding protein when used in complementation experiments. The transport activity was fully restored. He described the processing of the binding proteins from high molecular weight precursors and discussed the mechanisms of processing in the membrane.

Dr. J. Messing, from the University of California, Davis, described the use of a single-stranded bacteriophage M13, as a useful cloning vehicle. The replicative form of the DNA can be used as a cloning vehicle, and the excreted phage DNA used directly for DNA sequencing.

The final session of the workshop, "Applications of genetic engineering for production of useful microorganisms," was chaired by Dr. A. San Pietro, of Indiana University. Dr. G. Brosseau and Dr. O. Zaborsky, of the National Science Foundation, discussed the usefulness of workshops of this kind in identifying newly developing areas for scientific research and in developing an information base for recommendations for future funding of scientific programs. Dr. R. Finn, of Cornell University, described the potential of halophilic microbial systems in environmental, industrial and pharmaceutical applications. He suggests that more information would be immediately useful for treatment of high salt sewage wastes and in sugar processing. Dr. Ben-Amotz presented results of the first pilot experiments in Israel in producing glycerol from Dunaliella growing in open ponds under controlled conditions of salinity and nutrient supply.

This workshop on the application of genetic engineering to the study of halotolerance in microorganisms was the first meeting of its kind. It brought together scientists from several areas with a wide range of expertise on function and structure of these organisms, on techniques available for gene modification, and on the potential for useful applications of organisms modified for salt tolerance.

A variety of questions were posed during the meetings and were discussed, some in considerable detail. For example: 1) Which halotolerant systems are ready for genetic engineering or for industrial exploitation?, 2) To what extent is the knowledge obtained from microbial systems applicable to higher plants?, 3) What are the ways that microbial systems might be used in pilot studies on the feasibility of transferring characteristics for salt tolerance to non salt tolerant species? The workshop did not provide unequivocal answers to these questions, but it did make clear that a great deal of basic work is still needed. For instance, basic genetic studies on halophilic organisms are rare although some observations have been made on apparent transfer of halotolerant characteristics in some species.

In contrast to the sophisticated studies now being carried out in other organisms such as _E. coli_ on the genetic regulation of membrane synthesis and function, studies on the membranes of halotolerant plants and algae and bacteria are meager. In extremely halophilic algae and bacteria there is more basic information available but even here it is hampered by inadequate genetic systems. Surprisingly little seems to be known about the enzymology and mechanism of osmoregulation in most algae and higher plants although it is clear that the mechanisms involved have many similarities within these two groups, and to bacteria.

It seemed apparent that more work was needed on plant membrane systems and that work already underway on microbial membranes should be intensified. Much more work is also necessary on single cell plant cultures both from a physiological and genetic standpoint before the potential of this material can be assessed for genetic manipulation. From the standpoint of general applications it is clear that single cell cultures of halotolerant organisms hold the most promise as producers of single osmotic components such as glycerol, or as sources of proteins or other biological materials and useful chemicals. Possible uses for halotolerant organisms and enzymes exist in industrial processes but for the most part these have not been exploited except in the food industry. A greater potential exists for the use of halotolerant organisms in high salt industrial waste treatment.

All participants emphasized the need for more basic research especially in the area of moderate halotolerance and adaptability.

A more detailed summary of the meeting will be published by the National Science Foundation.

PARTICIPANTS IN THE CORNELL WORKSHOP

A. Ben-Amotz
Weizmann Institute
Israel

G. Brosseau
National Science Foundation
Washington, D.C.

J. Cazzulo
Universidad Nationale Rosario
Argentina

L. Csonka
University of California
Davis, California

W. Epstein
University of Chicago
Chicago, Illinois

G. Fink
Cornell University
Ithaca, New York

P. Fitt
University of Ottawa
Ottawa, Canada

J. Hellebust
University of Toronto
Toronto, Canada

D. Kushner
University of Ottawa
Ottawa, Canada

J. Lanyi
Ames Research Center
Moffett Field, California

H. Larsen
Tech. Hogschule
Norway

R. MacDonald
Cornell University
Ithaca, New York

R. MacLeod
McGill University
Montreal, Canada

J. Messing
University of California
Davis, California

D. Oesterhelt
University of Würtzburg
Germany

D. Oxender
University of Michigan
Ann Arbor, Michigan

L. Packer
University of California
Berkeley, California

E. Racker
Cornell University
Ithaca, New York

A. San Pietro
Indiana University
Bloomington, Indiana

B. Schobert
University of Munich
Munich, Germany

M. Shilo
Hebrew University
Jerusalem, Israel

T. Silhavy
Frederick Cancer Research Center
Frederick, Maryland

R. Spanswick
Cornell University
Ithaca, New York

A. Szalay
Boyce Thompson Institute
Cornell University
Ithaca, New York

J. Weber
University of California
Berkeley, California

G. Weidinger
University of Würtzburg
Germany

M. Yaguchi
National Research Council
Ottawa, Canada

O. Zaborsky
National Science Foundation
Washington, D.C.

SECTION VII

DISCUSSION BY NATIONAL RESEARCH AGENCIES

INTRODUCTORY STATEMENT

Alexander Hollaender

Associated Universities, Inc.
Washington, D.C. 20036

To introduce the subject of the contribution of different government agencies to the theme of this symposium, I though I'd relate a short story about something which happened to be about 46 years ago. I went to the Soviet Union as a National Research Council Fellow to be their first Rockefeller Fellow ever sent there in 1933. I had three letters of introduction to prominent scientists in the Soviet Union. One was to Professor Joffé, the head of the famous physics institute in Lesnoi outside Leningrad. Prof. Joffé is considered one of the discoverers of semi-conductors. The second letter was to Professor Vavilov, a famous geneticist who had started a most important collection of wheat from all over the world which formed the basis for a most important germ plasm collection of grains. Vavilov travelled all over the world and was many times in the United States. He was a very prominent and impressive scientist. Of course, he didn't get along with Lysenko and in the late thirties, Stalin put him into a concentration camp in Siberia, where he later died. The third letter was to Professor Maximov whose name was mentioned at this symposium when the history of drought resistance was discussed. He published, in 1929, a report on the development of drought resistance in wheat. This wheat was very sucsessful in the Soviet Union and they planted it finally by 1932 all over the USSR. But in 1932, it was terribly wet, and a serious failure of the harvest resulted. The Soviet Union then put Prof. Maximov in jail, having recommended wheat which could not tolerate a rainy season. In any case, when I arrived there, I couldn't find Dr. Maximov and finally Vavilov pointed out that I should talk with Mrs. Maximov who explained to me what had happened and that he was in jail although he had a diabetic condition, he was treated very well and was writing a new book on drought resistance. At that time and even today, scientists in the

USSR are treated exceedingly well if the government is impressed with their accomplishments even if they may be incorrect. For example, prominent scientists may be given private automobiles, even in the '30's when nobody else had one. However, if the Soviet government does not agree with the views of its scientists, they may still imprison these individuals. Of course, much less so today. We use an entirely different approach in the United States. We don't spoil our scientists as all of you know, but we also don't punish them if things don't turn out the way they are anticipated. With this as an introduction, I'd like representatives from the various government agencies to speak about their procedures in supporting this very interesting developing field. Before we get started, I should mention, that this symposium was generously supported by the Department of Energy in addition to the National Science Foundation. And now I would like to call upon Dr. George E. Brosseau of the Problem Analysis, Problem-Focus Research Division who will give us some background in regard to how the National Science Foundation proceeded in supporting this symposium.

INTRODUCTORY STATEMENT

George Brosseau

Problem Analysis
National Science Foundation
Washington, D.C. 20550

The ability of comercially important plant species to grow and produce in less favorable environments could be a key factor in whether there will be sufficient food and fiber for future generations. These environments are often characterized by too much salt or too little water or unfavorable conditions of cold or heat. Within the plant organism, osmoregulation may be a central process for coping with these various stresses. Unfortunately, our current understanding of the mechanism of osmoregulation in plants is insufficient to mount effective programs to improve the ability of important plants to survive and produce significant yields in these pessimal environments.

In supporting this symposium, the Engineering and Applied Science Directorate of the NSF hopes that we will obtain a better understanding of where we are and to identify promising avenues for future research. In so doing, perhaps the participants can give some focus and impetus to this important area of research. The symposium brings together researchers of various disciplines. Perhaps it will stimulate productive dialogue where it did not exist before. Finally, the focus on problems that combine high scientific interest with substantial economic and social significance might encourage more researchers to seek answers to the questions that are raised in the course of this meeting.

NSF has had a continuing interest in the physiology of stress tolerance in both its basic and applied research activities. In addition, NSF has given a fair amount of attention to the Biosaline concept developed at the Kiawah Island Conference in 1977. An NSF supported workshop on osmoregulation in microorganisms was held at the Boyce-Thompson Institute and a followup to the Biosaline research conference will be held in La Paz, Mexico in the Fall of 1980. Aside from their intrinsic value to science, the reports from these conferences and workshops will assist the NSF in formulating its own research agenda concerning these important activities.

THE BIOSALINE CONCEPT: AN OPPORTUNITY FOR RESEARCH AND DEVELOPMENT*

Oskar Zaborsky**

Program Manager of the Alternative Biological Sources
of Materials Program
E. A. S. Directorate
National Science Foundation
Washington, D.C. 20550

IDENTIFICATION OF THE PROBLEM: A SHORTAGE OF MATERIALS

For centuries man has known that the earth is a infinite sphere and that there are physical limitations to its size and resources. However, only in the last few decades have we actually seen our planet and realized the harsh message that, indeed, our earth has only finite resources and that more effective management is essential. At the very same time, the world's population is increased to over four billion, and severe political, social and economic instabilities are prevalent in many parts of the world. Although everyone is aware of the "energy crisis" that has been with us for several years, an equally dramatic crisis exists in a shortage of essential materials -- organic raw materials for chemicals, polymers and other industrial commodities. Of course, the "energy crisis" and the "materials crisis" are entwined.

---

*An expanded account of the biosaline concept is found in The Biosaline Concept: An Approach to the Utilization of Underexploited Resources, edited by A. Hollaender, J. C. Aller, E. Epstein, A. San Pietro, and O. R. Zaborsky, Plenum Press, New York, 1979.

**The views, conclusions and recommendations expressed herein are those of the author and do not necessarily reflect the views and policies of the National Science Foundation.

THE BIOSALINE APPROACH: AN ALTERNATIVE SOLUTION

The biosaline concept envisions the harmonious interplay of biological systems with saline environments for the benefit of man. In particular, the desire is to provide alternative options to critical material needs from renewable resources in an environmentally acceptable manner. The essential elements of the concept are biological systems and a saline or marine environment. Usually, the environment of primary interest is at high solar flux and elevated temperatures. Biological systems represent animals, plants or microorganisms or their individual cellular constituents; in particular, the critical catalysts of concern -- the enzymes -- are of prime importance. Thus, the boundary conditions for biosaline research are biological organisms, or their essential constituents, living in or being able to adapt to and tolerate saline or marine environments. Arid lands are an intrinsic part of the biosaline concept because they comprise a resource that is currently underutilized and exists at the interface of many oceans.

Figure 1 presents the essential elements of the concept in diagrammatic form. The objective of the biosaline concept is to provide us with material needs -- foods, feeds, energy, fuels, chemicals, fertilizers, structural materials, fibers, and medicinal compounds in an environmentally acceptable fashion.

The biosaline alternative is predicated on the fact that we have not appreciated nor wisely used certain of our rather abundant physical and biological resources. In particular, vast oceans and seas of the world have not been utilized to any large extent and certainly not in a managed scheme. The oceans cover 71 percent of the earth's surface and contain many valuable minerals and living resources. Oceans and seas also account for about 97 percent of all water on this planet. Of the remaining three percent of water, about three quarters is stored in the polar ice caps and in glaciers. Likewise, the marginal lands at the interface of the oceans have been neglected both in scientific investigations and for deriving essential materials. The coastal deserts alone are an extensive domain covering the earth for approximately 30,000 kilometers. Also, the semi-arid lands of the world have not been fully recognized as having merit for potential applications. These lands account for 36 percent of the land availability or the equivalent of a land area of 134,600,000 square kilometers. As with these physical resources, except perhaps for oceans, the biological resources associated with these domains likewise have been considered only as detrimental or undesirable. Yet, there exists a great diversity of microorganisms, plants and animal life in these seemingly unfit environments which are rich for intellectual endeavors and which could lead to industrial applications.

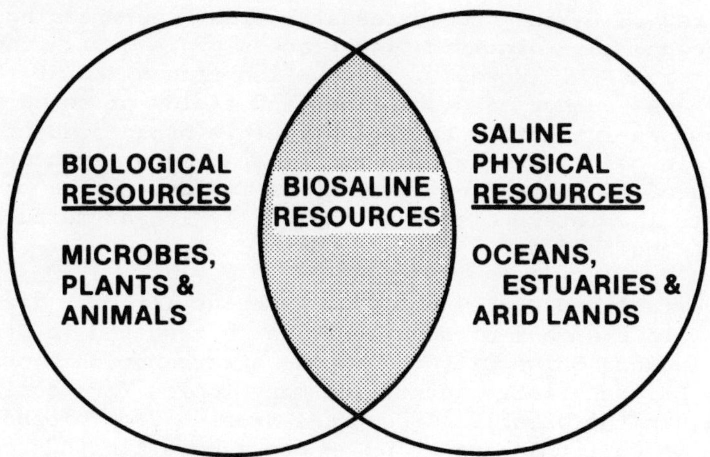

Fig. 1. The essential elements of the Biosaline concept.

In particular, microbes living in these habitats of high salinity and hot climates provide an immense reservoir of clues on how life began and how we can better utilize the remaining portions of our finite world. Even with our land-based plants, we are currently using only about 200 to 300 species for commercial purposes and only about 80 to 90 to produce crops in the U.S. which are valued at more than one million dollars. With the great diversity of biological life in the oceans, the estuarine environment and the arid regions of the world, there exists a great potential for developing new systems for our present and emerging needs. A key but often overlooked element of the biosaline concept is the use and further development of advanced technologies, such as enzyme technology and genetic engineering, to produce and process biosaline-derived resources.

RESEARCH NEEDS

In order to attain the envisioned goals, it is necessary to establish an effective information base of the relevant scientific fields and to translate basic facts into applied and developmental efforts.

With regard to the needed research areas encompassing the biosaline concept, the identification process started with the generation of the concept. However, it was also recognized at the outset that an immense number of disciplines and fields of science and engineering are intimately connected to this broad concept and that it would be best to develop any program on a limited basis. The genesis of the biosaline concept was founded on the exciting developments in biotechnology and especially in enzyme technology and genetic engineering.

In the fall and winter of 1976, numerous informal discussions were held with a number of scientists which resulted in Dr. San Pietro of Indiana University to propose a workshop on biosaline research. Dr. San Pietro identified many topics for possible discussion within the biosaline resources area -- from biophotolysis to the use of salt-tolerant crops -- but a more limited scope for the workshop had to be adopted in order to have a realistic meeting. Subsequently, a steering committee was established to select a limited number of high-priority research topics consistent with the then embryonic concept. Eventually, the following five general areas were selected:

- Land Plants and Controlled Environment Agriculture
- Photosynthetic Marine Organisms
- Salt-Tolerant Plants and Microorganisms
- Biological Waste Treatment
- Enzyme Technology

A number of relevant topics were excluded because of the time constraint of the workshop as well as the need to focus on new topics. For example, energy and climatology were not included as research topics because of a number of recent conferences, workshops and articles had been devoted to these areas. Of course, their omission from the conference did not imply that they are unimportant or that they do not form an integral part of any coherent research plan. Also, the emphasis of the research initially envisioned was to focus on nonconventional agricultural systems. The five research areas were discussed and analyzed at the International Workshop on Biosaline Research held at Kiawah Island, South Carolina in September, 1977, and a summary report was issued.

Additionally, a number of other conferences have been held. These have augmented the research needs identified initially at the Kiawah Island Workshop as well as to highlight other prime areas. An important meeting was held in Kuwait in 1977 on Microbial Conversion Systems for Food and Fodder Production and Waste Management. The meeting was organized by the Kuwait Institute for Scientific Research (KISR) and Kuwait University and was sponsored by the United Nations Educational, Scientific, and Cultural Organization (UNESCO), International Cell Research Organization (ICRO) and the International Federation of Institutes of Advanced Studies (IFIAS). The meeting focused on the regional development of the Kuwait area, as well as other countries of the Middle East, for single-cell protein, marine algae and plants as food components, arid zone and saline plant productivity, enzyme technology, and biological treatment of industrial wastes. A more recent conference, devoted in part to the biosaline concept, was the Cairo International Workshop on the Application of Science and Technology for Desert Development held in September of 1978. A prototype directory of current biosaline research projects was made available at this meeting. The identification of research areas comprising a coherent program has not stopped. In fact, this conference on Genetic Engineering of Osmoregulation: Impact on Plant Productivity for Food, Chemicals and Energy represents yet another step.

BIOSALINE RESEARCH AT THE NATIONAL SCIENCE FOUNDATION

The Foundation supports both basic and applied research in many disciplines and areas of science. Included in this diverse array of support is research on biosaline resources. During the past three decades, the Foundation has supported fundamental research on the major topic areas concerned with biosaline resources and it continues to support the disciplinary fields through several directorates. In addition, the Foundation also supports research in this area through the Division of International Programs (INT) and the newly created Engineering and Applied Sciences (EAS) Directorate.

The Middle East Section of INT through Mr. Gilbert Devey initiated the development of a possible program on biosaline research in 1976 and continues to play a major role in its further development within the Foundation. Further, this Section has stimulated organizations in biosaline-rich countries, such as Saudi Arabia and Kuwait, to undertake both programmatic and experimental activities. Other international activities on biosaline research, such as the joint U.S.-Mexican conference to be held in 1980 in Mexico, are being pursued by other INT programs.

With regard to more applied aspects of biosaline resources and to those topics that relate to on-going problem-focused research activities, the EAS Directorate has several programs that are either sponsoring experimental research or are identifying more precisely the Foundation's role. The Problem Analyses Group (through Dr. George Brosseau) is defining research needs as they pertain especially to the general area of plant stress. In fact, this conference at Brookhaven is being sponsored by the program.

In terms of producing chemicals and materials from biomass, the Alternative Biological Sources of Materials (ABSM) Program currently sponsors research on biosaline resources. The focus of this problem-focused program is to assess the feasibility of producing a wide variety of industrial chemicals and materials (intermediates, solvents, polymers, etc.) and specialty products (such as plant oils) from renewable resources (biomass). In particular, the emphasis of the ABSM Program is on bioprocessing of lignocellulose and other U.S. indigenous biomass, arid land plant products and biological nitrogen fixation. Research on using biosaline resources for food and feed applications is being pursued by the Division of Applied Research in the EAS Directorate.

The biosaline concept encompasses many diverse activities and fields of science and there should be no shortage of research opportunities for investigators. Conversely, numerous opportunities exist for funding agencies and organizations in the private sector to support essential research.

CONCLUSION

The implications of the biosaline concept are clear. With diligence and ingenuity, it can provide for a viable alternative to the present strategy of economic development and may be especially suited to those areas which have the physical and biological resources now viewed with much disdain. In fact, the oceans, arid lands, estuaries and saline soils should be viewed as a valuable resource in both physical and biological terms. The timely application of presently known science and technology, coupled with new insights and scientific facts that can be derived from indigenous biological organisms, make it an extremely fruitful endeavor for many nations. For those countries that have a biosaline environment, it provides an unequaled opportunity to develop their physical, biological and human resources for their own benefit without necessarily duplicating the patterns of the presently developed countries. At the same time, the more developed, affluent countries have an opportunity to establish for themselves and for other less-developed nations an industrial base for the future when their non-renewable fossil resources will be depleted. At the very least, the biosaline concept should serve to unify diverse elements of science toward a common objective.

THE IBR CONCEPT AND PLANT SCIENCES

Lewis G. Mayfield*

Group Leader of Integrated Basic Research
National Science Foundation
Washington, D.C. 20550

This has been an excellent conference and the papers and the participants have been stimulating. Plant research is a difficult area to pursue for several reasons. One must have enough patience and time to observe plants throughout their seasonal life cycle to have a full understanding of the plant and this experience must be integrated for various types of stress. Dr. Hollaender's story about the consequences of a plant breeding failure emphasizes the necessity of observing the productivity of a plant under various types of stress before planting on a large scale.

Some of the most difficult problems of science and engineering involve the science of interfaces. Heterogeneous catalysts is one example where precise knowledge of the surface, not the bulk properties, are of fundamental importance, yet scientific knowledge has been very difficult to obtain. Surprisingly, fundamental knowledge of enzyme catalysis is inherently easier, simply because we now have the capability to determine every atom in the catalyst. Other examples of where interfacial phenomena requires precise knowledge of the solid surface are stress corrosion, and fatigue cracking of metals. Currently tertiary oil recovery is important because of the need to increase the fraction of oil recovered from oil fields. Here again, understanding the phenomena involved is very difficult because several interfaces, rock-oil and oil-water are poorly understood. In this case it is clear that when the

---

*The views, conclusions and recommendations expressed herein are those of the author's and do not necessarily reflect the views and policies of the National Science Foundation.

droplets of oil are very small the nature of the interface is quite different than for large droplets.

Thus, the plant organism with several different interfaces, growing in soil, requiring organic and inorganic nutrients, air and sunlight is a marvelously complex system having exquisite beauty. Each step in plant growth and the ability to reproduce life is a wonder.

I lead a group called "Integrated Basic Research" (IBR) at NSF. The activities of the group cross-cut all of the basic science disciplines in NSF. We focus our attention on four major problem areas where additional fundamental understanding is likely to contribute to the long term solution. The problem areas are: (1) Advanced Measurement Investigations, (2) Biogeochemical Cycles of Carbon, Sulfur and Nitrogen, (3) Deep Mineral Resources, and (4) Population Redistribution. The proposals accepted by the program are received and reviewed by the basic research program. IBR will assist the basic research program by providing part of the funds for the proposal and in this way increases the amount of NSF support for the topic. Last year we assisted with financial support a total of 45 research efforts under the Advanced Measurement Investigations topic. Each proposal supported was by an investigator who had scientific goals to be achieved but was thwarted by existing conventional instruments which did not permit him to make the necessary measurements to achieve his scientific goals.

This experience has been an eye opener for me and I'm amazed at the number of projects, each presented in the manner of their own discipline, which impact upon the four chosen topics. It is the characteristic of basic research that it may impact a number of problem areas which makes the IBR program exciting. This leads to the belief that basic research may be usefully described in terms of its utilitarian purposes and values.

As a result of the IBR experience it is apparent that there is ongoing research outside of the plant science discipline that will have impact upon plant sciences in the forseeable future. For example, Dr. Leroy Hood (1978) at the California Institute of Technology is developing the capability for automating the sequencing of proteins utilizing very small quantities of protein. His goal is 5-10 picograms. If my crystal ball is not cloudy, it appears that this development will have a very substantial impact upon our knowledge of proteins, including the enzymes so much a key part of all living things, because the costs of acquiring sufficient high purity protein for analysis will be greatly reduced and automation will decrease significantly both the time and labor costs for the analysis. The science that results from vastly improved knowledge of the proteins and enzymes involved in plants should increase capabilities for manipulation of the plant.

These beliefs are reinforced by the paper presented by Dr. Peter Steponkus. The art of tissue culture is relatively new and combined with the cryomicroscope developed by Dr. Steponkus and the cryo-electron microscope developed by Dr. Micrea Fotino (1979) powerful new observational tools are available. It is apparent from his work that some current models of the behavior of plants under stress will have to be reexamined and that new insights requiring scientific deductions are to be gained.

Unfortunately it is much easier for the plant scientist to study in real time the above ground activity of the plant. From what I've read, it is very costly to make pertinent observations of the growing root system of plants in soil, although observations can be made at intervals during growth, but not on the same plant during its life cycle. A consequence is the less than optimum integration of knowledge of the real world plants grow in. I think we all realize this deficiency. However, from a technical view point making meaningful observations of the growing root in soil is a very difficult task. The more important question is, would it make any difference if we could observe plant roots growing in soil in a continuous fashion? What should be measured? Would our capability to improve plants increase? I suspect that if observation could be made conveniently and at relatively low cost then at least the science would be improved.

Perhaps someone can develop new techniques that will permit direct or nearly direct observations of roots in soil without the requirement of sacrificing the plant. There are several possible techniques that come to mind. Actually, the problem is one faced in one research area called non-destructive evaluation. It is also related to the effort to investigate the human body. A recent book (Preston, 1979) discusses the three major techniques of ultrasound, x-rays with computer processing, and radioisotope (nuclear) imaging. A different method which is still in research using nuclear magnetic resonance is also discussed. All of these techniques provide a graphic display of structure without the requirement of invading the body. Without prejudging the situation it may well be that variants of these techniques may find utility in plant research by providing direct observation of root growth.

Let us take an example in another area of science to see why I speculate this possibility. Oceanography has a wide variety of variety of research topics and techniques. Presently, to sample the primary biological activity in the oceans on which more complicated life forms rest, the oceanographers take a physical sample. This is expensive, time consuming and by analogue also invasive. Dr. Richard Pieper wondered why he couldn't adapt the techniques from non-invasive evaluation and medical ultrasonography to count the plankton and other small organisms in seawater. This is an example of an IBR award where (Pieper, 1978) the purpose of the grant is to conduct better sampling of interest to oceanographers -- but -- to do so requires a new measurement technique.

While no oceanographer has yet proposed extending techniques from the relatively simple task of counting the numbers and total mass of organisms into the refined task of estimating type of material, there is reason to believe that modern signal processing might permit this. There are a number of research efforts (Aller, 1977) to extend ultrasound from one of showing anatomy (structure) to one of measuring biological activity. We can speculate that if research of interest to plant scientists such as observation of root phenomena is hampered by lack of measurement ability that the IBR topic support can alter the fact.

Another topic supported by IBR and of interest to the conference participants is the biogeochemical cycles of carbon, nitrogen, and sulfur. Although our initial rationale for support of this topic came from possible climatic effects of increased atmospheric carbon dioxide, our experience is that the global cycle of nitrogen offers more opportunities for advancing the state of knowledge for materials important in growth cycles. Obviously, nitrogen fixation in terrestrial systems is well-known to the conference participants and although not all aspects of this process are completely understood, our state-of-knowledge is advanced. Less known and of considerable interest in several fields of science at present is nitrogen fixation in other environments. IBR has funded two projects which are investigating nitrogen fixation in the marine environment (Ahmed, 1979; Waterbury, 1978) and one in cypress wetlands (Brezonick, 1979). Potential benefits from this basic research may lie in areas as diverse as marine resources, salt tolerance in plant and microorganisms and genetic engineering of nitrogen fixing plants and symbionts.

Now, let me add one caveat. I could recount a number of other exciting research efforts partially supported by IBR such as one that capitalizes on the K-shell discontinuity to permit presentation of selected elements such as iodine (Mistretta, 1978). I mention this one as well as others because there is much science being developed outside the normal scope of plant sciences that may have potential benefit to plant sciences. I believe you may need eyes in your feet as well as your head if you are to advance as fast as possible. That does not make it so. We try in IBR to avoid the risk of a technique looking for an application. Rather, the professionals in the discipline must first establish that the research frontier needs the technique to advance. It is only then that IBR will provide support.

The point of making this brief list is that there is much science being developed outside the normal scope of the plant sciences having potential benefit to plant sciences. Conferences of this sort and type interaction between sciences will continue to stimulate the growth of plant sciences and we can look forward to progress in providing the food and fiber for world development.

REFERENCES

Ahmed, S. I., 1979, Comparative studies of enzymes involved in ammonia-nitrogen assimilation into amino acids in several species of marine plankton, NSF Grant OEC-7919591.

Aller, J. C., 1977, Overview of tissue characteristics research, Proceedings of Ultrasound Symp., IEEE 77CH 1265-I.S.U., pp. 226.

Brezonik, P., 1979, Nitrogen fixation by root association of trees, NSF Grant DEB-7923435.

Fotino, M., 1979, Electron imagining and cyromicroscopy in high-voltage TEM, NSF Grant PCM-7905631.

Hood, L., 1978, Development of protein microsequencing technology, NSF Grant PCM-7826484.

Mistretta, C. A., 1978, Real time x-ray image substraction techniques, NSF Grant Eng-7824555.

Pieper, R. E., 1978, Studies of ocean volume reverberation at high ocustics noise, NSF Grant OEC-7825832.

Preston, Jr., K., et al., 1979, Medical imagining techniques - A comparison, Plenum Press, New York.

Waterbury, J. B., 1978, Aerobic nitrogen fixation in the marine environment, NSF Grant OEC-7825647.

OSMOREGULATION AND BIOLOGICAL ENERGY PRODUCTION

R. Rabson*

Director of the Division of Biological Energy Research
Office of Basic Energy Science
Department of Energy
Washington, D.C.

The interests of the Department of Energy (DOE) in osmoregulation research are fairly straightforward. As you are all aware, there is much being made these days about the importance of biomass and how biomass can serve as an alternate to fossil fuels. Many of these claims are highly optimistic while others tend to minimize the contribution that biomass may make to the Nation's energy problems. Regardless of which position you take, the fact of the matter is that biomass is now used as an energy source and will be a growing energy source in the future. The magnitude of the resource is something about which we will all have to wait and see. One point which arises frequently in discussions about biomass availability concerns where these materials are to be grown and I think you can see the relationship to osmoregulation immediately. It is unlikely, in my opinion at least, that biomass crops are going to displace agricultural crops on the better agricultural lands and this means only one thing; namely, that biomass production is going to be relegated to areas not particularly well suited for agriuclture. If this assumption is correct, the implication is that biomass production will deal with an entirely different set of problems, some of which have been touched upon during this conference. This also explains exactly why DOE should be interested in this particular topic because ultimately a plant's ability to cope with drought or saline conditions through osmoregulatory mechanisms will determine whether or not plants can be grown in a

---

*The views, conclusions and recommendations expressed here are those of the author and do not necessarily reflect the views and policies of the National Science Foundation.

particular area. The other contrast to agricultural productivity will be in terms of using many different species for biomass production and probably using different techniques for growing these plants. For example, it may be that mixed cultures will be used rather than monocultures.

At the moment the overall picture in regard to biomass specifics is unclear. There is no such thing as a national biomass plan in the sense that everything is well defined. This is now in the process of definition and it would be inappropriate to exclude any possibilities at this time. The thinking goes to using indigenous species, some of which will be termed weeds and also the introduction of exotics. But there has to be much work done before such points are settled and the precise directions for biomass development are put forth. But at the same time we don't have unlimited time. We are going to be pressed for answers on all types of questions related to growing biomass of different species and in different places. Clearly, biomass production is going to be a highly regionalized affair. One aspect of the development of the biomass concept has to do with the conversion of biomass. We hear much about gasohol these days, but we must also think about other ways of converting biomass to useful fuels and other products.

In DOE there are two sizable programs that deal with questions of biomass development. The Biomass Energy Systems Program in Solar has a program in FY 80 of about $54,000,000. That program is largely a near term research and development effort which as yet has a rather small basic research component. The second program is the program that I represent. It is not a biomass program *per se* but rather a program in DOE that is part of the Office of Basic Energy Sciences. The program is just about a year old and is called Biological Energy Research (BER). The mission of the program is to develop the kinds of fundamental information in the botanical and microbiological sciences which will underpin the longer term biomass developments. The program aims at understanding how plants and microorganisms function with the idea of utilizing such information in ultimately improving production and conversion. The program has a budget in FY 80 of roughly $6,000,000. I should say also that there are a number of people in the higher echelons of DOE who are very supportative of this research area. In particular, Dr. John Deutch, Undersecretary of DOE, has been very helpful in this matter. In general, the feeling is that biological research is an area which has been neglected badly but which also has great promise and thus the prospects for an enlarged program in the future look good if the Office of Management and Budget and Congress agree.

The BER program has a component on plant adaptation, stress physiology and a variety of other topics concerning the plant and microbiological sciences. Although the national biomass effort is still incompletely defined I think it is quite clear that as far as the basic research aspect is concerned, it is not necessary to wait. Equally clear is that the number of things that we need to know related to such areas as stress physiology are numerous. Since it is the objective to develop an understanding of principles and mechanisms, it is not necessary to know precisely what species are to be used and in what manner. We do know that we shall have to be looking more closely to marginal growing areas and to more efficient use of other areas for overall production of plant materials.

I would like to make one other remark. I was particularly pleased that during the course of the meeting several people brought out the idea of energy analysis in regard to their presentations. This is to be applauded in my opinion because as a Nation it will be necessary to have the idea of conserving energy implanted in all of our thinking not only with respect to scientific considerations but in our societal activities generally. It should become common currency to look at activities in terms of energy inputs and energy yields and comparisons in terms of energy. If there is to be any resolution of the so-called energy crisis, there will have to be a close scrutiny for all aspects of our way of living to ascertain energy wasteful activities. Conservation of energy in my view is the only obvious hope for near term resolution of these problems. The longer term activities which will generate more energy with new technologies are simply too far down the road to be relied on for early results. Biomass clearly is our oldest technology. It has been superseded by cheap fossil fuels and now it is quite obvious that we must go back to biomass, but hopefully, we will consider biomass in more enlightened ways and not simply as, for example, burning wood to heat our homes. It seems to me that science can make a major contribution in showing how to grow and use biomass more efficiently and imaginatively to provide the things we need.

RESEARCH PROGRAMS OF THE U.S. SALINITY LABORATORY

Richard H. Nieman

U.S. Salinity Laboratory
United States Department of Agriculture
Riverside, California

The Bankhead-Jones Act adopted by Congress in 1935 made funds available for research on region-wide problems for agriculture. The directors of the Agricultural Experiment Stations of the 11 Western states, meeting with representatives of the USDA, identified salinity and sodicity as major problems in the West. On their recommendation, the U.S. Regional Salinity Laboratory was established in Riverside in 1937. Its mission was to acquire the knowledge and develop the agricultural skills and practices necessary for the continuation of irrigation agriculture in areas threatened by salinity and sodicity.

Very little was known at the time about the diagnosis, improvement and management of saline and sodic soils. There is still much to be learned, but it is to the credit of the original Laboratory staff that they recognized some basic problems and quickly developed tools and techniques for appraising and correcting them. The fact that irrigation agriculture continues to flourish in the West attests to the effectiveness of their work.

An important distinction was made between saline soils and sodic soils, formerly called alkali soils because of their high pH. Saline soils contain enough soluble salt to decrease crop production but they are not deficient in Ca^{2+}, so calcium amendments are of no value on them. Sodic soils, though not especially high in total salts, have a high Na^{+}/Ca^{2+} ratio and for most crops are deficient in Ca^{2+}. Consequently, they are improved by treatment with calcium amendments. Sodic soils generally must be reclaimed, by replacing sorbed Na^{+} with Ca^{2+}, before they can be used to grow crops. Seawater contains enough Ca^{2+} that it actually has been used to reclaim sodic soils.

Salinity-sodicity problems are by no means restricted to the Western U.S. In recent times, the Laboratory staff has been called upon to solve salinity-sodicity problems in many regions of the U.S. and in many foreign countries too. Aid to foreign countries has included teaching their agricultural scientists to deal with salinity and sodicity.

The U.S. Salinity Laboratory has associated with it 16 professional scientists and an equal number of support personnel. We support students from U.C. Riverside, to carry on many of the maintenance and research activities of the Laboratory. Research is divided, but not rigidly, into three main areas: 1) soil and water chemistry, 2) soil physics and engineering, and 3) plant research. The first two areas of research are concerned with 1) providing optimal soil conditions for crop production under irrigation where salinity and sodicity are hazards, 2) improving the efficiency of water use by irrigation agriculture, 3) minimizing the salt pick-up by irrigation waters. The plant research is concerned with the performance of crop plants under saline conditions and, to a lesser extent, under sodic conditions. (Sodic soils generally have to be reclaimed before they can be used for crop production but some plants can be used to aid reclamation by improving tilth.) Several different approaches have been used in the study of plant responses. I should like to mention just a few.

In the early days of the Laboratory, field tests for the salt tolerance of crops were given a high priority because growers urgently needed to know which crops would yield satisfactorily on their lands with available water. These tests have been continued to the present time; they have provided information on over 100 species regarding the reduction in yield and quality to be expected with a given soil salinity.

Some crops, the stone fruits, for example, are specifically sensitive to certain ions, so tests have been conducted to determine tolerance to specific ions, and to boron. The concentration of the latter is often high in saline soils and waters.

Work at the Laboratory has shown that plant tolerance to salinity is influenced by many climatic and edaphic factors such as: temperature, relative humidity, light intensity, soil aeration, and the concentration of certain nutrient ions, orthophosphate, for example. The method of irrigation is a factor too; sprinkler irrigation can increase salt injury by depositing salt on the foliage.

A few years ago we were able to add a plant geneticist to our staff. He is now selecting lines of forage grasses, grains, and vegetables that show promise of increased salt tolerance. This work is difficult because, as Tom Ramage stated earlier, we don't really know how to select for salt tolerance. An increase in general vigor can be mistaken for an increase in salt tolerance.

In recent times we have tried to focus more attention on the mechanisms of salt injury to plants and on the plant characteristics that protect against this injury. If we understood these, selection for increased salt tolerance would be much easier and surer.

Three salt effects that probably have a common basis and that are of particular interest to us at present are: damage to the cell membrane, increase in respiration, and decrease in the adenylate energy charge, in some cases to levels that could limit CO_2 fixation and growth. Among the crop plants that we have studied, these effects tend to be less severe in the more tolerant species, as expected.

POSTER PRESENTATIONS

Symposium on Genetic Engineering of Osmoregulation: Impact on Plant Productivity for Food, Chemicals and Energy

The Contribution of Organic Solutes to Osmotic Balance in Some Green Algae: L. Brown and J. A. Hellebust, University of Toronto, Canada.

Temperature Stress Analysis in Plant Tissue Culture: A. A. Goloff, Jr. and R. H. Lawrence, Jr., Union Carbide Corporation.

Determination of Glycine Betaine by Pyrolysis-Gas Chromatography in Cereals and Grasses: W. D. Hitz, and A. D. Hanson, MSU-DOE Plant Research Laboratory, Mighigan State University.

Transfer Cells in the Epidermis of Plant Roots -- A Common Response to Nutritional Stress?: D. Kramer, Institut fur Botanik, Darmstadt, Germany.

Interaction of Nitrogen and Water Deficits on Stomatal Behavior in Cotton: J. W. Radin, USDA, Western Cotton Research Laboratory.

ABA in Leaves of Field-Grown Soybean Under Water Stress: J. S. Samet, T. R. Sinclair and P. M. Cortes, Cornell University.

Proline Accumulation in Halophytes: S. Treichel, Technische Hochschule, Darmstadt, Germany.

THE CONTRIBUTION OF ORGANIC SOLUTES TO OSMOTIC BALANCE IN SOME GREEN ALGAE

Lewis M. Brown and Johan H. Hellebust

Atlantic Regional Laboratory
National Research Council of Canada
Halifax, Nova Scotia, Canada

ABSTRACT

Sorbitol and proline accumulate to osmotically significant concentrations in two *Stichococcus* species and *Klebsormidium marinum*. The extent of osmotic balance by these and other organic solutes depends on the isolate within this group of predominantly non-vacuolate algae. Sucrose, proline and sorbitol accumulate in the lichen alga *Hyalococcus dermatocarponis*, but concentrations on an osmotic basis are low. Sucrose and glutamic acid are significant solutes in the likely non-vacuolate *Klebsormidium flaccidum* and *Klebsormidium sterile*. A similar set of solutes accumulates in *Ulothrix fimbriata*, a freshwater algae that is highly vacuolate and less euryhaline. Two small eustigmatophycean algae and one small chlorophyte accumulate organic solutes to a moderate degree during salinity stress; mannitol and proline in the marine eustigmatophyte and proline in the marine chlorophyte. *Enteromorpha intestinalis*, a seaweed, accumulates sucrose, but at low levels. Osmotic balance in these algae may also include other solutes, compartmentation, or other factors. The role of organic solutes in osmotic balance is variable, and depends on details of physiology and ultrastructure.

TEMPERATURE STRESS ANALYSIS IN PLANT TISSUE CULTURE

A. A. Goloff, Jr. and R. H. Lawrence, Jr.

Union Carbide Corporation
Tarrytown, NY 10591

ABSTRACT

A thermogradient plate originally designed and built for temperature studies on seeds has been adapted for use with plant tissue cultures. The plate is capable of being programmed to establish near linear gradients within a range of 1 C - 40 C. In addition a polar switching device permits temperature cycling around a diagonal line of constant temperatures. Among the type of studies suitable for the apparatus are: characterization of temperature responses of specific genotypes; detection and analysis of thermosensitive responses; and selection of temperature stress tolerant variant cells/plants.

Results are presented which demonstrate the use of the thermogradient plant in characterizing the morphogenetic response of lettuce tissue culture with respect to temperature.

DETERMINATION OF GLYCINE BETAINE BY PYROLYSIS-GAS CHROMATOGRAPHY IN CEREALS AND GRASSES

William D. Hitz and Andrew D. Hanson

MSU-DOE Plant Research Laboratory
Michigan State University
East Lansing, MI 48824

ABSTRACT

Glycine betaine content of 22 species of cereals and grasses was determined by pyrolysis-gas liquid chromatography assay.

Betaines as a class were separated from aqueous plant extracts by a three resin ion exchange procedure. Ionic compounds other than betaines were bound on a mixed bed of strong anion and weak cation exchange resins. Betaines pass through the mixed resin and were bound on a strong cation exchange resin, neutral compounds were washed away and the purified betaines eluted with ammonia.

N,N,N-trimethyl betaines purified by the ion exchange procedure were determined by GLC measurement of trimethyl amine resulting from the pyrolysis of the betaine fraction at 350 C. The lower limit of the pyrolysis-GLC assay was about 2 nmole of glycine betaine.

Glycine betaine content of the 22 cereal species tested ranged from 1 to 48 μmole per g dry wt. Glycine betaine content before water stress was a fair predictor of accumulation during a stress period.

TRANSFER CELLS IN THE EPIDERMIS OF ROOTS: A STRUCTURAL DIFFERENTIATION TO OVERCOME NUTRIENT DEFICIENCY

Detlef Kramer

Institut für Botanik
Technische Hochschule
Darmstadt, West Germany

ABSTRACT

Transfer cell is a term describing cells with an increased plasmalemma surface area due to the formation of a wall labyrinth. It is presumed that this corresponds to the role of such cells in transport processes, either in secretion (e.g., glands) or absorption (e.g., xylem parenchyma cells). Although the main function of root epidermal cells is absorption of nutrients, transfer-cell characteristics have not been reported previously for this tissue.

However, we have recently found that a series of halophytes or salt tolerant plants (*Atriplex hastata*, *A. hortensis*, *Mesembryanthemum crystallinum*) develop epidermal transfer cells in the absorbing part of the root under conditions of salt stress. These species belong to the group of salt-tolerant plants with a high K^+/Na^+ selectivity. For *A. hastata*, the increased K^+/Na^+ selectivity was demonstrated by the use of x-ray microanalysis of freeze fractured root segments. Other cytological adaptations to salt stress are heavily thickened vessel walls and large amounts of cytoplasm in xylem-parenchyma cells. The latter feature can be correlated from the x-ray spectra with the functioning of these cells in the selective K^+ secretion into the vessels.

In another series, so-called iron-efficient plants were investigated after being subject to iron deficiency. These species (e.g., sunflower, *Capsicum*) are known to increase their Fe uptake efficiency by a factor of 10 to 1000 after growth in the absence of iron for several days. Also in this case, root epidermal cells have an extensive wall labyrinth.

These results show that root epidermal cells can, contrary to previous suggestions, differentiate as transfer cells in response to non-ideal nutrient conditions.

INTERACTION OF NITROGEN AND WATER DEFICITS ON STOMATAL BEHAVIOR IN COTTON

John W. Radin

United States Department of Agriculture
Science and Education Administration
Western Cotton Research Laboratory
Phoenix, AZ 85040

ABSTRACT

Stomata of normal and mildly N-deficient cotton plants (Gossypium hirsutum L.) reacted very differently to water stress. In N-deficient plants, stomata closed at water potentials up to 10 bars higher than in normal plants. Data presented here show: 1) stomata of both normal and N-deficient plants responded little to intercellular CO_2 concentrations (C_i) before the imposition of drought; 2) during drought, stomata became sensitive to C_i, closing at elevated concentrations; 3) stomata of N-deficient plants began to respond to C_i at a higher water potential than those of normal plants; 4) sensitivity to C_i was also induced by application of abscisic acid (ABA). The results strongly imply that in N-deficient plants, ABA was synthesized at higher water potentials than in normal plants.

Although the upper leaves of the plants did not visibly senesce during drought, the lower leaves rapidly yellowed and abscised. This process also occurred at higher water potentials in N-deficient plants than in normal plants. The responses to water stress of both stomatal behavior and leaf area adjustment (leaf abscission) indicate that N-deficiency promotes a "water-saving" type of drought avoidance.

ABA IN LEAVES OF FIELD-GROWN SOYBEAN UNDER WATER STRESS

J. S. Samet, T. R. Sinclair, and P. M. Cortes

Department of Agronomy
Cornell University
Ithaca, NY 14853

ABSTRACT

Abscisic acid (ABA) and leaf water potential were measured in leaves from field-grown soybean (Glycine max (L.) Merrill) during the latter part of the 1978 growing season in Varna, NY. Two cultivars were compared, 'Chippewa 64' which was fairly drought hardy and 'Wilkin' which was drought susceptible. Water was withheld from stressed plots from July 20 to August 25. Topmost fully expanded trifoliate leaves were sampled from unstressed and stressed plants of both cultivars at 0700 hr, 1600 hr, and 2200 hr on August 11 and 16. Three replications were made for each treatment. Leaf water potential measurements were made at sampling with a pressure chamber, samples were immediately placed on dry ice, subsequently freeze-dried and analyzed for ABA. Continuous in field measurements of CO_2-exchange rate (CER) were made with a differential infrared gas analyzer on comparable leaflets attached to plants of the two cultivars. At the time of sampling, soil water potential at 10 cm depth was -0.05 MPa and -0.46 MPa for unstressed and stressed plots, respectively. While plants were more stressed on August 16, on both days a diurnal cycle in leaf water potential was observed in unstressed and stressed plants in each cultivar. ABA fluctuated diurnally in leaves from stressed plants but in unstressed plants there was little diurnal change. On August 16 there was a linear correlation between ABA and leaf water potential. The relationship was different in stressed leaves of the two cultivars. 'Wilkin' did not accumulate as much ABA as 'Chippewa 64' yet the leaves from 'Wilkin' had more negative water potentials. CER in leaflets from stressed plants was reduced almost equally in both cultivars when compared to leaflets from unstressed plants.

PROLINE ACCUMULATION IN HALOPHYTES

S. Treichel

Institut für Botanik
Technische Hochschule
Darmstadt, West Germany

ABSTRACT

The phenomenon of accumulation of "compatible solutes" in plants has turned out to be an interesting subject in stress physiology. The synthesis of proline occurs in a great number of halophytes, but its regulation is unknown up to now. The influence of NaCl on pyrroline-5-carboxylate reductase, the enzyme synthesizing proline can be regarded as one possibility of regulation. The activity of p-5-c reductase in salt-treated or salt-shocked plants is considerably higher than in non-treated plants, which can explain the increased proline concentration after NaCl treatment. The different *in vitro* NaCl inhibition of p-5-c reductase from NaCl-treated and non-treated plants points to an alteration of the enzyme during salt adaptation.

Leaves of *Mesembryanthemum nodiflorum* submerged in solutions of different osmotic potential proved to be suitable plant material for the study of osmotic adaptation. After transfer into NaCl solutions these leaves lose water immediately. But corresponding with proline synthesis and the uptake of NaCl (correlation coefficient r = 0.92 - 0.98) the water content of the cells starts to increase within a short time. After 15 h (400 mM NaCl) the initial loss of water is compensated. During the next hours the water content and the turgor of the cells increase furthermore until they contain up to 60% more water than the original leaves. In sucrose-stressed leaves no restoration of initial water content occurs. This demonstrates that halophyte cells under salt stress only are able to adjust their osmotic potential to the surrounding solution by uptake of NaCl and accumulation of proline. Under sucrose stress no rapid rise in vacuole osmotic potential is possible, although proline (in the cytoplasm) is synthesized in a high degree.

Tissue suspension cultures of different halophytes were tested for the purpose of their applicability to salt-tolerance studies as a undifferentiated homogeneous plant material. First results show that isolated cells -- enclosed by a cell wall -- behave like normal cells of halophyte tissue: plasmolysis occurring after salt stress is overcome within a short time, and growth (measured as production of protein and of cell wall material) increases with advancing salt tolerance of the parent plants and with increasing salt concentrations of the growing solution. Proline synthesis is highest in the most salt-tolerant cells and increases with salt concentration. The highest proline/protein ratio occurs at 400 mM NaCl. Up to now there seems to exist no difference between whole plant cells and isolated cells of halophytes in suspension culture concerning their salt tolerance and proline synthesis.

PARTICIPANTS IN THE SYMPOSIUM

ACKERSON, R. C.
Central Res. and Dev. Dept.
E. I. du Pont de Nemours & Co.
Wilmington, DE 19898

ANDERSEN, R.
Dept. of Botany
Brigham Young University
Provo, UT 84601

ANDERSON, C. W.
Biology Dept.
Brookhaven National Lab.
Upton, NY 11973

ARMOND, P. A.
Central Research
Pfizer, Inc.
Groton, CT 06340

BEN-AMOTZ, A.
Dept. of Biochemistry
Weizmann Inst. of Science
Rehovot, Israel

BOERBOOM, M.
Pfizer Genetics, Inc.
Olivia, MN 56277

BOYER, J. S.
USDA/SEA
Botany Dept.
University of Illinois
Urbana, IL 61801

BRADLEY, D.
NASA/Goddard Inst. for Space Studies
2880 Broadway
New York, NY 10025

BREIDENBACH, R. W.
Plant Growth Laboratory/
Dept. of Agronomy & Rance Sci.
University of California
Davis, CA 95616

BROSSEAU, G.
Office of Problem Analysis for Engineering & Applied Sciences
National Science Foundation
Washington, DC 20550

BROWN, A. D.
Dept. of Biology
University of Wollongong
Wollongong, N.S.W. 2500
Australia

BROWN, L.
Atlantic Regional Lab.
National Research Council
Halifax, Nova Scotia
B3H 2Z1 Canada

CAVALIERI, A. J.
Dept. of Biology
University of South Carolina
Columbia, SC 29208

CHILTON, M. D.
Dept. of Biology
Washington University
St. Louis, MO 63130

CROUGHAN, T.
Dept. of Agronomy & Range Science
University of California
Davis, CA 95616

CROWTHER, D.
Biology Dept.
Brookhaven National Lab
Upton, NY 11973

CSONKA, L.
Plant Growth Laboratory/
Dept. of Agronomy & Range Sci.
University of California
Davis, CA 95616

DILWORTH, M. F.
National Science Foundation
1800 G Street, NW
Washington, DC 20550

DRAKE, B.
Smithsonian Institute
12441 Parklawn Drive
Rockville, MD 20851

DUNN, J. J.
Biology Dept.
Brookhaven National Lab.
Upton, NY 11973

EPSTEIN, E.
Dept. of Land, Air and Water Resources
University of California
Davis, CA 95616

FOBES, J. F.
Dept. of Horticulture
Michigan State University
East Lansing, MI 48824

FREY, N. M.
Pioneer Hi-Bred Intl., Inc.
Box 85
Johnston, IA 50131

GLURGEVICH, J. R.
Smithsonian Institute
12441 Parklawn Drive
Rockville, MD 20851

GOLOFF, A., JR.
Tarrytown Tech. Center
Union Carbide Corp.
Tarrytown, NY 10591

GROSS, D.
Pfizer Genetics, Inc.
Doniphan, NE 68832

HIND, G.
Biology Dept.
Brookhaven National Lab.
Upton, NY 11973

HITZ, W. D.
MSU/DOE Plant Research Lab.
Michigan State University
East Lansing, MI 48824

HOLLAENDER, A.
Associated Universities, Inc.
1717 Massachusetts Ave., NW
Washington, DC 20036

HUANG, A.
Dept. of Biology
University of South Carolina
Columbia, SC 29206

HULL, R. J.
Dept. of Plant & Soil Science
University of Rhode Island
Kingston, RI 02881

JEFFERIES, R.
Dept. of Botany
University of Totonto
Toronto, Ontario
M55 1A1 Canada

KADKADE, P. G.
GTE Lab., Inc.
45 Lambert Circle
Marlboro, MA 01752

KECK, R. W.
Dept. of Biology
Purdue University
Indianapolis, IN 46205

KIRBY-SMITH, J. S.
Environmental Protection Agency
401 M. Street, SW
Washington, D.C.

KRAMER, D.
Inst. fur Botanik, FB10
Technische Hochschule
D-6100 Darmstadt
West Germany

KUDIRKA, D. T.
Biology Dept.
Brookhaven National Lab.
Upton, NY 11973

LACKS, S. A.
Biology Dept.
Brookhaven National Lab.
Upton, NY 11973

LAMM, S. S.
Biology Dept.
Brookhaven National Lab.
Upton, NY 11973

LAWRENCE, R. H., JR.
Agricultural Sciences Lab.
Union Carbide Lab.
Tarrytown, NY 10951

LEDBETTER, M. C.
Biology Dept.
Brookhaven National Lab.
Upton, NY 11973

MAC DONALD, R.
Wing Hall
Cornell University
Ithaca, NY 14853

MARX, J. L.
Science
1515 Massachusetts Ave., NW
Washington, DC 20005

MAYFIELD, L.
National Science Foundation
1800 G Street, NW
Washington, D.C. 20550

MEREDITH, C. P.
Stauffer Chemical Co.
Richmond Research Center
1200 S. 47th Street
Richmond, CA 94804

MITTEN, D. H.
Monsanto Agricultural Products
800 N. Lindbergh Blvd.
St. Louis, MO 63166

MONTAGUE, M. J.
Monsanto Agricultural Products
800 N. Lindbergh Blvd.
St. Louis, MO 63166

NAUMAN, A.
Biology Dept.
Brookhaven National Lab.
Upton, NY 11973

NIEMAN, R. H.
U. S. Salinity Lab.
United States Dept. of Agric.
4500 Glenwood Drive
Riverside, CA 92501

NOGGLE, G. R.
American Society of Plant Physiologists
9650 Rockville Pike
Bethesda, MD 20014

NORLYN, J. D.
Dept. of Land, Air and Water Resources
University of California
Davis, CA 95616

OSMOND, C. B.
Res. School of Biol. Sciences
Australian National University
Canberra City, Australia

PETERSON, T.
Dept. of Plant and Soil Science
University of Rhode Island
Kingston, RI 02881

QUALSET, C. O.
Dept. of Agronomy and Range Science
University of California
Davis, CA 95616

RABSON, R.
Div. of Biological Energy Research
Office of Basic Energy Sciences, J-309
U. S. Dept. of Energy
Washington, DC 20545

RADIN, J. W.
USDA/SEA
Western Cotton Research Lab.
4135 E. Broadway Road
Phoenix, AZ 85040

RAINS, D. W.
Plant Growth Laboratory/
Dept. of Agronomy & Range Sci.
University of California
Davis, CA 95616

RAMAGE, R. T.
Plant Sciences Dept.
University of Arizona
Tucson, AZ 85721

RAVEN, J. A.
Biological Science Dept.
University of Dundee
Dundee, DD1 4HN Scotland

RICE, T. B.
Pfizer Central Research
Eastern Point Road
Groton, CT 06340

ROTH, J.
Dept. of Biology
University of Utah
Salt Lake City, UT 84112

SAMET, J. S.
Dept. of Agronomy
Cornell University
Ithaca, NY 14853

SCHALLER, C. W.
Dept. of Agronomy & Range Sci.
University of California
Davis, CA 95616

SCHAUER, A. H.
U. S. Dept. of Agriculture
1300 Wilson Blvd.
Arlington, VA 22209

SCIAKY, D.
Cold Spring Harbor Lab
Cold Spring Harbor, NY 11724

SETLOW, J. K.
Biology Dept.
Brookhaven National Lab.
Upton, NY 11973

SETLOW, R. B.
Biology Dept.
Brookhaven National Lab.
Upton, NY 11973

SHAHAK, Y.
Biology Dept.
Brookhaven National Lab.
Upton, NY 11973

SHEATH, R. G.
Dept. of Botany
University of Rhode Island
Kingston, RI 02881

SIEGELMAN, H. W.
Biology Dept.
Brookhaven National Lab.
Upton, NY 11973

SLOVACEK, R. E.
Biology Dept.
Brookhaven National Lab.
Upton, NY 11973

STAPLES, R. C.
Boyce Thompson Institute
Upton, NY 11973

STASSI, D. L.
Biology Dept.
Brookhaven National Lab.
Upton, NY 11973

STEPONKUS, P. L.
Dept. of Agronomy
Cornell University
Ithaca, NY 14853

SWANSON, E.
Dept. of Botany
University of Rhode Island
Kingston, RI 02881

SZALAY, A. A.
Boyce Thompson Institute
Cornell University
Ithaca, NY 14850

TREICHEL, S.
Inst. fur Botanik, FBIO
Technische Hochschule
D-6100 Darmstadt
West Germany

UCHIMIYA, H.
Agricultural Science Lab.
Union Carbide Corp.
Tarrytown, NY 10591

VALENTINE, R. C.
Plant Growth Laboratory/
Dept. Agronomy & Range Sci.
University of California
Davis, CA 95616

VEEN, B. W.
Centre for Agrobiological
Research
6700 AA Wageningen
The Netherlands

VINOPAL, R. T.
Dept. of Microbiology U44
University of Connecticut
Storrs, CT 06268

WAGNER, G. J.
Biology Dept.
Brookhaven National Lab.
Upton, NY 11973

WEBER, D. J.
Dept. of Botany
Brigham Young University
Provo, UT 84602

WEINRAUCH, Y. Z.
Biology Dept.
Brookhaven National Lab.
Upton, NY 11973

WYN JONES, R. G.
Dept. of Biochemistry and
Soil Science
University College of
North Wales
Bangor, Gwynedd
Wales

ZABORSKY, O. R.
Directorate for Engineering and Applied Sciences
National Science Foundation
1800 G Street, NW
Washington, DC 20550

INDEX